Applied Second Law Analysis of Heat Engine Cycles

Applied Second Law Analysis of Heat Engine Cycles offers a concise, practical approach to one of the two building blocks of classical thermodynamics and demonstrates how it can be a powerful tool in the analysis of heat engine cycles.

Including real system models with the industry-standard heat balance simulation software, the Thermoflow Suite (GTPRO/MASTER, PEACE, THERMOFLEX) and Excel VBA, the book discusses both the performance and the cost. It also features both calculated and actual examples for gas turbines, steam turbines, and simple and combined cycles from major original equipment manufacturers (OEMs). In addition, novel cycles proposed by researchers and independent technology developers will also be critically examined.

This book will be a valuable reference for practicing engineers, enabling the reader to approach the most difficult thermal design and analysis problems in a logical manner.

Applied Second Law Analysis of Heat Engine Cycles

S. Can Gülen

CRC Press
Taylor & Francis Group
Boca Raton London

CRC Press is an imprint of the
Taylor & Francis Group, an **informa** business

Designed cover image: © istock

First edition published 2023
by CRC Press
6000 Broken Sound Parkway NW, Suite 300, Boca Raton, FL 33487–2742

and by CRC Press
4 Park Square, Milton Park, Abingdon, Oxon, OX14 4RN

CRC Press is an imprint of Taylor & Francis Group, LLC

ISBN: 978-1-032-16188-4 (hbk)
ISBN: 978-1-032-16185-3 (pbk)
ISBN: 978-1-003-24741-8 (ebk)

DOI: 10.1201/9781003247418

Typeset in Times
by Apex CoVantage, LLC

Contents

SECTION II   *Obeying the Law*

Preface

Everything should be made as simple as possible, but not simpler.
—Albert Einstein

Heat engines have dominated modern life starting with James Watt's steam engine at the dawn of the industrialization around the turn of the nineteenth century. Coal-fired steam engines were the stars of the show for most of the next hundred years primarily as the motive force in the industry and in transportation via railroad locomotive and ship propulsion. At the beginning of the last century, the internal combustion engine, the piston-cylinder variant, entered the scene and put its mark on the entire century primarily through the automotive industry. The other variant, the gas turbine jet engine, emerged on the eve of the World War II, and after the war, it was the increasingly preferred technology for aircraft propulsion. During that time and, in fact, until the turn of the twenty-first century, coal-fired steam turbine power plants, along with the nuclear power plants (after 1950s), were the workhorses of electric power generation. At the time of writing, natural gas–fired gas and steam turbine combined cycle power plants have pushed coal-fired power plants out of the generation portfolio in the USA, the UK, and continental Europe. Nuclear power is in a limbo, mainly due to the astronomic cost and long construction period of the utility scale projects but not helped by the public fear (to a large extent, irrational) of the technology. However, small modular reactors (SMR) continue to be developed by various entities.

Interestingly, in the case of heat engines, practice preceded the theory by more than a century. (Note that the emergence of the steam engine can be traced back to the 1712 design of Thomas Newcomen.) The two building blocks of the science of classical thermodynamics, the first and second laws, emerged simultaneously from the labors of Rankine, Clausius, and Thompson in the 1850s. In the following two decades, contributions by Maxwell, Boltzmann, Clausius, and Gibbs established the foundations of the kinetic theory of gases and statistical thermodynamics.

In the early years of the third decade of the twenty-first century, nearly two centuries after Sadi Carnot's heat engine, 162 years after the publication of Rankine's book *Manual of the Steam Engine and Other Prime Movers*, the second law of thermodynamics is a mystery (almost a *black art*) not only to the lay public but to the students in mechanical and chemical engineering. Popular books by journalists as well as serious scientists are chock-full of mostly philosophical concepts, like time arrow and entropy of the universe (undefinable by definition). The thermodynamic property entropy is only used for compressor and turbine isentropic efficiency calculations. The other (derived) property, exergy, or availability (in early US textbooks) is only used by undergraduate students in Thermodynamics 101 homework assignments and promptly forgotten after graduation. Unless, of course, they study further to obtain doctorates and embark on an academic career. In that case, they churn out esoteric papers in obscure academic journals with pretentious titles containing terms such as *exergo-economic* or *thermo-economic*. Alas, in his professional career spanning a quarter century, the author has yet to see any of that stuff being taken seriously in the industry.

This is rather unfortunate because the concept of exergy, containing the thermodynamic state property entropy, is a very powerful tool in the analysis of heat engine cycles. Its beauty is its simplicity; that is, the fact that it can be applied without getting lost in the proverbial weeds and tedious property calculations. However, the insights that it provides is inversely proportional to its simplicity. Armed with a good grasp of the practical ramifications of the second law, distilled into the state property exergy, one can accomplish many front-end design or reverse engineering tasks rapidly and easily. The goal of the author in writing this book is twofold. First and foremost, he aims to provide the necessary background and worked out examples for self-study to arm the reader with requisite practical tools and methods. Second, he wants to install in the reader a thorough understanding of the second law and the thermodynamic property, entropy, both of which have been widely misunderstood, misinterpreted, and misapplied. This book can be a valuable study aid for engineering students as well as a handy reference for practicing engineers working in the industry or in research labs.

Before moving on, there is a simple fact that needs to be stated unequivocally. As is the case with everything I have written before and will write in the future, this book is dedicated to the memory of Mustafa Kemal Atatürk (1881–193∞) and his elite cadre of reformers. Without their vision, sacrifices, and groundbreaking work, there would not have been a fertile ground where my parents, teachers, mentors, family, and friends could shape me into the author of this book. Everything started with him; the rest was easy.

About the Author

Dr. Gülen (PhD 1992, Rensselaer Polytechnic Institute, Troy, New York), PE, ASME Fellow, Bechtel Fellow, has more than 25 years of mechanical engineering experience covering a wide spectrum of technology, system, and software design; development (GTPRO/MASTER, THERMOFLEX); performance and cost assessment; and analysis, primarily in the field of steam and gas turbine combined cycle (109FB-SS, IGCC 207FB, H-System) process and power plant turbomachinery and thermodynamics (in Thermoflow Inc., General Electric, and Bechtel). Dr. Gülen authored/co-authored three books, two book chapters, numerous internal/external archival papers, articles (50-plus), design practices, technical assessment reports, and US patents (25-plus) on gas/steam turbine power plant performance, cost, optimization, data reconciliation, analysis, and modelling.

1 Introduction

Thermodynamics is a funny subject. The first time you go through it, you don't understand it at all. The second time you go through it, you think you understand it, except for one or two points. The third time you go through it, you know you don't understand it, but by that time you are so used to the subject, it doesn't bother you anymore.

—Arnold Sommerfeld (1868–1951), German physicist

Consider the first three chapters of this book as the equivalent of a boot camp. The author will be the drill sergeant. The goal is first to bring the reader down to his/her bare thermodynamic essentials. Specifically, if there are any misconceptions about the *second law of thermodynamics* and the thermodynamic state property *entropy* obtained from fantastical descriptions found in popular science (!) books by self-proclaimed gurus and/or magazine or newspaper articles written by reporters with no scientific background; the quicker that they are gotten rid of, the better. The next phase is bringing the reader back up again to a level of thermodynamic consciousness that should enable him/her to apply those concepts to practical problems at hand.

This is by no means an easy task. Just like many recruits fail the basic training because they cannot handle the mental and physical challenges of being an infantryman, some readers may not be up to the, admittedly not too easy, task of grasping the arcane subtleties of the second law and entropy. The author will try his best to make it as easily digestible as possible, but to be fair, there is a low but not entirely dismissible probability that he will fail and the reader will decide that the effort spent on deciphering the fantastical stuff (this time, on the author's part) is wasted. In the event that this happens, it is the author's failure, not the reader's. Nevertheless, the rest of the book can still be used in a manner akin to following a cookbook recipe. After all, one does not need to be a certified internal combustion engine mechanic or thermodynamics expert to drive a passenger car. Consequently, in case you want to skip the *boring* chapters and save yourself the grief, here is a short tour guide to help you pick and choose so that the rest of the book makes sense (hopefully).

The star of the show – so to speak – is the second law, with several chapters devoted to it. In Chapter 2, a brief look at the other three laws of thermodynamics will be provided – primarily for the sake of completeness. From a perspective of practical applications, only the first and second laws are of importance. Since this is not an introductory textbook on thermodynamics, even the first law is going to get a cursory look in that chapter. It is the author's hope that most readers (if not all) are well versed in application of enthalpy and mass balance to the analysis of key cycle components encountered in thermal engineering. Nevertheless, a brief introduction into the first law with numerical examples is provided in this chapter to facilitate the

DOI: 10.1201/9781003247418-1

discussion in the rest of the book without having to switch back and forth between different textbooks.

In Chapter 3, the second law will be introduced using the classical, macroscopic arguments starting with the two statements of the second law of practical engineering interest (i.e., *Kelvin-Planck* and *Clausius* statements). This is followed by the derivation of Clausius theorem and Clausius inequality using the Carnot cycle construct as the framework. The chapter then continues with the introduction of the lost work concept by combining the first and second laws. Application of the second law to heat pumps and heat engines is demonstrated with numerical examples. The chapter is concluded with an in-depth look into the thermodynamic property entropy. It is highly recommended that one reads Chapter 3, especially Section 3.5 "Recap", for a proper understanding of what entropy is and what it is *not*. In and of itself, this is not a prerequisite for applying second law concepts to problems of practical engineering interest. Nevertheless, it is the author's belief that well-trained engineers, as *applied scientists*, must have a firm command of concepts and tools used in day-to-day activities, no matter how mundane they appear at a fist glance (to an *untrained* eye, one might add).

In Chapter 4, a deep dive into the second law via statistical mechanics will be provided. Admittedly, one may think of this as an academic overkill in an applied science book. However, for an in-depth understanding of the second law and its key building block, entropy, a proper understanding of the fundamental concepts by deriving them from scratch (for the lack of a better term) by looking at the system at a microscopic level is essential. While skipping that part is not harmful to understanding the rest of the book per se (as already stated earlier), it is highly recommended because, at least in the author's opinion, this will be the only coverage of the subject matter that you can find in a technical book in a digestible treatment that will make you grasp the real significance of entropy.

Chapters 1 through 4 comprise the first part of the book, Part I, titled *Laying Down the Law*. The second part of the book, Part II, titled *Complying with the Law*, exclusively deals with the application of concepts based on second law to practical problems associated with heat engines and power plants. For a complete coverage of the application of second law (exergy) analysis to a wide variety of industrial plants, including not only the heat engine power plants but also the chemical process plants, such as sulfuric acid and air liquefaction, the reader should consult the monograph by Kotas [1], which is pretty much the gold standard in this subject. The current work differs from Kotas's book with its strong emphasis on practical aspects on heat engine power plants, especially gas turbines in simple and combined cycle, steam turbines (both in thermal and combined cycle power plants), heat recovery from the exhaust gas of gas turbines (heat recovery steam generator, HRSG), and plant heat rejection systems. Example calculations are based on actual products. Concepts are introduced with actual application examples (e.g., the combustor of an advanced-class gas turbine). Second law and exergy are used as tools to analyze actual products offered by major OEMs.

For the reader who is perusing this book in a bookshop (quite unlikely these days) or through the Look Inside feature of Amazon (much more likely – as long as the pages that you are reading now are included in the preview), or using a copy borrowed from a friend or colleague, the author is obliged to explain why he or she should *prefer* this book over that by Kotas (ibid.), which is admittedly far superior to it in terms of scientific depth and breadth. Short answer: You should not. This book is *not an alternative* to that superb treatise. It is a *companion to it*, specifically aimed at students who have to solve a homework problem or understand a concept (or decipher a *skimpy* example) in a textbook or practitioners in the field (i.e., design and/or field engineers or researchers), who need quick answers to pressing issues. Consider the exergy analysis of heat exchangers. If the reader wants a deep dive into that particular subject, he or she can consult Chapter 3 in Ref. [1]. If the reader wants to know what exactly are the performance issues in the *heat recovery steam generator* (HRSG) of a *gas turbine combined cycle* (GTCC), the information therein will teach the reader the basics (the author studied that section and most parts of that book closely – *that is why he can write this book*), but in order to convert it to a few readily digestible formulae and graphics, the reader would have to spend hundreds, if not more, hours of extra studying, calculating, reading, rereading, recalculating *if and only if* he or she has already extensive experience in cited systems.

Furthermore, with all due respect, one can totally skip the discussion of compressors and turbines in Kotas (ibid.) because exergy analysis of those components is practically useless. Instead, one can do much better by studying the theory (using first law concepts) of those components as outlined in a widely recognized, time-tested reference, such as the *masterpiece* (no other word for it) by Saravanamuttoo et al. [2]. There are many other worthwhile references, such as the magnum opus by Traupel [3], but that book is really for the cognoscenti (plus it is in German, still waiting a worthy English translation) and can be an overkill. For a more practically oriented coverage and extensive list of valuable references, the recent monograph by the author [4] can be consulted as well.

The reason for the earlier assertion is explained via extensive calculations in Section 3.3. In that section, it is also illustrated that the exergy analysis of the combustion process, especially in the combustor of a gas turbine, is a rather frustrating exercise. Still, it is worth doing step-by-step *only once* in order to grasp what is really going on inside that component (as far as the real nitty-gritty details are concerned, including production of pollutants, such as CO and NO_x, and combustion instabilities, one must consult specialized references [5–6]). Once done, it is wiser to not do it again because it will not provide any new knowledge related to the practical aspects of gas turbine analysis. It is much better to utilize commercial heat and mas balance or chemical flowsheet simulation software packages. However, in order to do justice to that analysis, one can do hardly better than studying the relevant chapters in Ref. [1].

Finally, after reading the earlier explanation, one will be compelled to ask the next (obvious) question: If this is the case, why spend all that ink on the rigmarole in Chapter 4? The answer to that question was already answered earlier. Herein, the author would just like to add that the reader should consider that chapter as a *bonus*. As stated earlier, skipping it does not hamper the understanding of the rest of the book. Or put another way, for a particular reader, only the first four chapters can be of interest. He or she can do just fine without paying attention to the rest of the book.

REFERENCES

1. Kotas, T.J., *The Exergy Method of Thermal Plant Analysis*, London: Exergon Publishing Co, 2012.
2. Saravanamuttoo, H., Rogers, G.F.C., Cohen, H., and Straznicky, P.V., *Gas Turbine Theory*, 6th Edition, Harlow: Pearson College Div, 2008.
3. Traupel, W., *Thermische Turbomaschinen (Klassiker der Technik)* (Vol. 2), 2nd Edition, Berlin: Springer, 2000.
4. Gülen, S.C., *Gas Turbines for Electric Power Generation*, Cambridge: Cambridge University Press, 2019.
5. Lieuwen, T.C., and Yang, V. (Editors), *Gas Turbine Emissions*, New York: Cambridge University Press, 2013.
6. Lieuwen, T.C., and Yang, V. (Editors), *Combustion Instabilities in Gas Turbine Engines*, Reston, VA: AIAA, Inc., 2005.

Section I

Laying Down the Law

2 Laws of Thermodynamics

Die Energie der Welt bleibt konstant; die Entropie strebt einem Maximum zu.[1]
—Rudolf Clausius (1822–1888)

There are four laws of thermodynamics, numbered from zero to three and not, as any sane person would expect, from one to four. Furthermore, it is the *zeroth law* that was introduced last. Why this is the case has some historical reasons. Interested reader can google and find the answer on the internet quite easily. In any event, for practical purposes, we only need the first and second laws of thermodynamics. Since the second law is the pièce de résistance, so to speak, it has its own chapter (Chapter 3). Thus, this chapter is really devoted to the first law. However, for the sake of completeness, we will get the other two out of the way quickly.

Even before that, however, again, for the sake of completeness, it behooves one to make a mention of the (so called) *fourth law of thermodynamics.* In general, this is the name given to Onsager's reciprocal relations, which are derived from statistical mechanics under the principle of *microscopic reversibility.* It is not found in most textbooks on engineering thermodynamics and not even in many texts on statistical mechanics. There are even those who claim that there is no such law; that is, Onsager's theorem does not have the status of a law of physics. The author is not qualified to comment on this law one way or another other than making the interested reader aware of the existence of it (almost) so that he or she can do look into it further.

For the coverage of thermodynamic fundamentals in the present book, the primary reference is the textbook by Moran and Shapiro, specifically the first edition of it [1]. At the time of writing, the ninth edition of this time-tested reference is available (with the addition of two more authors). For most practical purposes, the first edition, which may be purchased for a few dollars from an online seller is perfectly adequate. The reason for selecting this book is its excellent coverage of the exergy concept (*availability* in the book), especially the inclusion of the chemical exergy. There is a wide selection of thermodynamics textbooks out there. One of them may very well be the favorite of a certain reader. Two other books that the author can recommend are by Van Wylen and Sonntag [2] and Çengel and Boles [3]. The former is a widely acknowledged classic (and the author's undergraduate textbook), and the latter is modern and more pedagogical in its coverage with a wide selection of worked-out examples. An indispensable reference, which is the perfect companion to Moran and Shapiro (ibid) is the book by Bejan et al. [4] (not surprisingly, late Dr. Moran is a co-author).

2.1 ZEROTH LAW

Using the definition of Moran and Shapiro (p. 16 in Ref. [1]), the zeroth law states that, when two bodies are in thermal equilibrium with a third body, they are in thermal equilibrium with one another. It is basically the idea behind a thermometer. In

DOI: 10.1201/9781003247418-3

7

other words, if there are two bodies out there, say, A and B, and we want to find out whether they are in thermal equilibrium or not, we do not have to bring them together to check for that. Instead, we use a third body (i.e., a thermometer). If each body is separately in thermal equilibrium with that third body, then they are in thermal equilibrium with each other.

2.2 THIRD LAW

The third law states (based on experimental observations) that the entropy of a pure crystalline substance is zero at the absolute zero of temperature; that is, 0 K or 0 R (p. 643 in Ref. [1]). Although the third law sounds quite mundane, it is not. Consider the formulation of the third law by German chemist and physicist Walther Nernst: It is impossible for any process, no matter how idealized, to reduce the entropy of a system to its absolute zero value in a finite number of operations. In fact, having absolute zero temperature would violate the second law. Specifically, if $T_L = 0$, one ends up with a Carnot engine such that $W = Q_H$ (i.e., 100% efficiency), in direct violation of the Kelvin-Planck statement of the second law.

Is it possible to grow a pure (perfect) crystal; that is, all lattice spaces occupied by identical atoms? In theory, the answer is yes. In practice, although some experiments have come close, this has been impossible to achieve. The lowest temperature achieved in a laboratory was of the order of mK (millikelvin); that is, a few thousandths of a Kelvin above absolute zero. In comparison, the coldest air temperature ever recorded on Earth was 184 K (around −89.2°C), logged at the Soviet Vostok Station in Antarctica in 1983. Outer space has a temperature of 2.7 K (about −270°C). There is an analogy to the speed of light, c = 299,792,458 m/s, here. Predicted by theory and confirmed by experiments, no matter how fast a body is moving, it can always be made to go faster, but it can never reach the speed of light. Similarly, no matter how cold a system is, it can always be made colder, but it can never reach 0 K.

We will stop here because, for most practical problems of interest in this book, we are not in need of determining an *absolute value* for entropy. (The only exception is calculation of the Gibbs free energy in Chapter 5.) We are more interested in *entropy differences*. (This will be discussed in-depth, quantitatively, in Section 3.4 later.) In any event, temperatures that we will deal with in thermal/power engineering are so far above absolute zero (0 K) that we can safely forget that we ever heard of the third law. There.

2.3 FIRST LAW

2.3.1 Pure Components

This is the law that spells out the famous *conservation of energy* principle. It is the bread and butter of thermal engineering. Herein, the best way to introduce the first law is by using perhaps the two most important equations in the entire body of thermodynamics – that is, the two *Tds* equations:

$$du = Tds - Pdv$$

(2.1)

$$dh = Tds + vdP \qquad (2.2)$$

Equation 2.1 is known as the *Gibbs equation* or the *first Tds* equation. It is the differential form of the first law of thermodynamics (1LT, henceforth) for a closed system undergoing an internally reversible process. Its starting point is the fundamental 1LT equation; that is,

$$dE = \delta Q - \delta W \qquad (2.3)$$

$$E = KE + PE + U, \qquad (2.4)$$

where *KE* and *PE* denote *kinetic* and *potential* energy of the system, respectively. In all applications of practical interest herein, those two can safely be assumed to be 0. Thus, the 1LT for a closed system is given by

$$dU = \delta Q - \delta W, \text{ or} \qquad (2.5)$$

$$du = \delta q - \delta w. \qquad (2.6)$$

For reversible processes

$$\delta q = Tds \qquad (2.7)$$

$$\delta w = Pdv \qquad (2.8)$$

so that one ends up with Equation 2.1; that is, the Gibbs equation

$$du = Tds - Pdv. \qquad (2.9)$$

For the derivation of Equation 2.7 from the second law of thermodynamics (2LT), refer to Chapter 3. The derivation of Equation 2.8 can be found in any introductory textbook (e.g., see pp. 33–34 in Moran and Shapiro [1]). The left-hand side (LHS) of Equation 2.1 is the internal energy change of the closed system, and it is an *exact* differential; that is, its integral can be expressed as the difference of the initial and final (equilibrium) states,

$$\int_{1}^{2} du = u_2 - u_1.$$

The first term on the right-hand side (RHS) of Equation 2.1 is the energy transfer via *heat*; the second term is energy transfer via *work*. Both terms are *inexact* differentials; that is, their integrals cannot be expressed as the difference of their initial and final values because, unlike thermodynamic properties, they are not *state functions* (or in other words, they are *path-dependent*).

To write the 1LT equation for an *open system*, we make use of the definition of the thermodynamic property *enthalpy* (for a rigorous derivation of enthalpy and its significance, see Section 14.2):

$$H = U + PV,$$

or using *intensive* properties,

$$h = u + Pv.$$

In differential terms, taking the derivative of both sides of the (intensive) enthalpy definition, one can write that

$$dh = du + Pdv + vdP \tag{2.9}$$

Substituting Equation 2.9 into Equation 2.1, the *second Tds* equation is arrived at; that is,

$$dh = Tds + vdP. \tag{2.10}$$

Equation 2.10 is the differential form of the conservation of energy for a *control volume* (i.e., the open system) for a *steady-state, steady-flow* (SSSF) process. This can be seen by rewriting Equation 2.10 as follows

$$0 = -dh + Tds - (-vdP) \tag{2.11}$$

The LHS of Equation 2.11 is the change in the internal energy of the control volume, which is 0 for an SSSF process. The first term on the RHS of Equation 2.11 is the net enthalpy transfer into the control volume via mass transfer across the boundaries of the control volume (CV, henceforth). The second term is the net heat transfer *into* the CV, and the third term is the net work done *on* the CV. Equation 2.11 is the primary building block of turbomachinery analysis, especially, pumps, compressors, and expanders (turbines), as well as other key cycle components; for example, heat exchangers and devices with chemical reactions (*reactors* in chemical engineering, e.g., *combustor* being the most common variant in heat engine applications).

Turbomachine components are *adiabatic* machines with, in most cases at least, one inlet (state 1) and one exit (state 2) so that, after integration of Equation 2.10, we have

$$h_{2s} - h_1 = \int_1^2 vdP = w_s, \tag{2.12}$$

where w_s denotes the *isentropic* work done by the component (e.g., turbine) or on the component (e.g., pump or compressor). (The *s* in the subscript denoting state 2 indicates that it is the end state of an isentropic process.) For a component handling an *incompressible* fluid, such as a water pump, Equation 2.12 is easy to evaluate; that is,

$$w_s = h_{2s} - h_1 = v_1 (P_2 - P_1) = \frac{P_2 - P_1}{\rho_1}. \tag{2.13}$$

For *compressible* fluids going through an isentropic process, we must make use of the definition of the isentropic process

$$P v^\gamma = P_1 v_1^\gamma = P_2 v_2^\gamma = \text{constant}. \tag{2.14}$$

Substituting Equation 2.14 into Equation 2.12, one obtains

$$w_s = h_{2s} - h_1 = (P_1 v_1)^{1/\gamma} \int_1^2 \frac{dP}{P^{1/\gamma}} \tag{2.15}$$

$$w_s = h_{2s} - h_1 = P_1 v_1 \frac{\gamma}{\gamma - 1} \left[\left(\frac{P_2}{P_1} \right)^{\frac{\gamma - 1}{\gamma}} - 1 \right] \tag{2.16}$$

Making use of the *ideal gas* equation of state, $Pv = RT$, in combination with Equation 2.14, and establishing the well-known isentropic relationship between temperature and pressure ratios,

$$\frac{T_{2s}}{T_1} = \left(\frac{P_2}{P_1} \right)^{\frac{\gamma - 1}{\gamma}},$$

Equation 2.16 becomes

$$w_s = h_{2s} - h_1 = R T_1 \frac{\gamma}{\gamma - 1} \left(\frac{T_{2s}}{T_1} - 1 \right), \tag{2.17}$$

which, for a *perfect gas*, can be written as

$$w_s = h_{2s} - h_1 = c_p (T_{2s} - T_1). \tag{2.18}$$

For a pump or compressor, the actual work can be found by using the isentropic efficiency; that is,

$$\frac{w_s}{w_{p/c}} = \frac{h_{2s} - h_1}{h_2 - h_1} = \eta_{s,p/c}, \tag{2.19}$$

which, for an expander/turbine, becomes

$$\frac{w_t}{w_s} = \frac{h_2 - h_1}{h_{2s} - h_1} = \eta_{s,t}. \tag{2.20}$$

For an ideal but imperfect gas, $c_p = f(T)$ so that ideal gas enthalpy is only a function of temperature, too; that is, $h = f(T)$. While ideal gas specific heat and enthalpy are functions of temperature only, entropy of an ideal gas is a function of temperature *and* pressure; that is, $s = f(P, T)$. Consequently, Equation 2.18 becomes

$$w_s = h_{2s} - h_1 = \int_1^{2s} c_p(T)dT \quad \text{such that} \tag{2.21}$$

$$s(P_2, T_{2s}) = s(P_1, T_1).$$

Equation 2.21 can be solved numerically using a property package such as JANAF tables. One can also use a polynomial relationship for $c_p(T)$; for example, see Table A-12 on p. A-19 in Moran and Shapiro [1] for fourth-order polynomials for a variety of gases, including air (valid from 300 K to 1,000 K). A more compact version of that relationship for air, valid between 15°C and 1,000°C, is given subsequently (c_p in kJ/kg-K, T in °C):

$$c_p = 1.442106 - \frac{0.440557}{\left(1 + \dfrac{T}{1158.216}\right)^{1.661673}} \tag{2.22}$$

As it will be demonstrated in Section 3.4, entropy change for a *perfect gas* is given by (also see p. 208 in Ref. [1])

$$\Delta s(T, P) = c_p \ln\frac{T_2}{T_1} - R \ln\frac{P_2}{P_1}. \tag{2.23}$$

For an *ideal gas*, Equations 2.23 must be integrated to find the entropy change; that is,

$$\Delta s(T, P) = \int_1^2 c_p(T)\frac{dT}{T} - R \ln\frac{P_2}{P_1}. \tag{2.24}$$

Compressor Control Volume

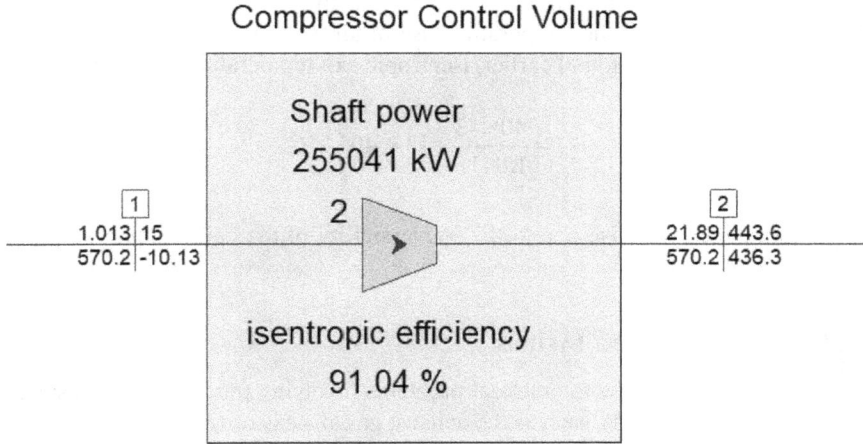

Shaft power

255041 kW

2

isentropic efficiency

91.04 %

1
1.013 15
570.2 -10.13

2
21.89 443.6
570.2 436.3

FIGURE 2.1 Compressor control volume calculation. State properties in clockwise direction from the upper-left corner are P in bar, T in °C, flow in kg/s and h in kJ/kg-K.

To put some numbers into the theory, let us look at the compressor of a gas turbine, assuming that no air is extracted for turbine hot gas path cooling. For the inputs to this calculation, we refer to the example in Section 3.2.1, where application of the second law of thermodynamics to a gas turbine will be demonstrated in ample detail. The calculation is done in Thermoflow Inc.'s THERMOFLEX flowsheet simulation software.

The implied perfect gas specific heat from Figure 2.1 is

$$h_2 - h_1 = c_p (T_2 - T_1)$$

$$436.3 - -10.13 = c_p (443.6 - 15)$$

$$c_p = 1.0416 \text{ kJ/kg-K.}$$

From Equation 2.19, the isentropic work is found as

$$w_s = (436.3 - -10.13) 0.9104 = 406.43 \text{ kJ/kg.}$$

For a compressor, the lost work is the difference between actual and isentropic work, which is calculated as follows:

$$i = (h_{2s} - h_1) \left(\frac{1}{\eta_{c,s}} - 1 \right) = 406.43 \left(\frac{1}{0.9104} - 1 \right) = 40 \text{ kJ/kg.}$$

In other words, 40 kJ/kg of work input into the system is lost and converted into heat, which manifests itself in the temperature rise of air at compressor discharge. Using the average specific heat found earlier, isentropic exit temperature is found as

$$T_{2s} = \frac{406.43}{1.0416} + 15 = 405.2°C,$$

that is, the temperature rise due to the irreversibility of the compression process is $443.6 - 405.2 = 38.4°C$.

2.3.2 MULTICOMPONENT SYSTEMS

Application of the first law to practical problems involving pure substances, such as water/steam or compounds, such as the exhaust gas of a gas turbine, in the absence of a chemical reaction is straightforward to the point of almost being trivial. This is not the case for problems involving chemical reactions or phase equilibria. (The same is true for the second law as well, of course.) As far as the analysis of heat engine cycles is concerned, analysis of the combustion reaction involving air and fuel in the combustor of a gas turbine or in the cylinders of internal combustion engine is one example.

2.3.2.1 Basic Thermodynamics

Gibbs free energy or Gibbs function, in *extensive* and *intensive* properties, is defined as

$$G = H - TS \tag{2.25a}$$

$$g = h - Ts \tag{2.25b}$$

For a process taking place involving chemical and/or phase change in a multi-component system at constant pressure and temperature, P_0 and T_0, respectively, Equation 2.25b becomes

$$g(T_0, P_0, n_i) = h(T_0, n_i) - T_0 s(T_0, P_0, n_i) \tag{2.26}$$

where n_i designates the number of moles of each component (or *species* in some references), i, in the system, which is usually referred to as a *mixture* or *solution* (depending on the context; herein, the term *mixture* is apt).

The Gibbs free energy is used to establish the equilibrium state of a system at fixed temperature and pressure; that is, T_0 and P_0. For a closed system (i.e., no material flow across the system boundaries), the fundamental *Tds* equation can be written as

$$\delta Q = dU - PdV \tag{2.27}$$

The corresponding second law equation is

$$dS = \frac{\delta Q}{T} + \delta\sigma, \tag{2.28}$$

where $\delta\sigma \geq 0$ represents the differential *irreversibility*. Combining Equations 2.27 and 2.28, we find that

$$TdS - dU - PdV \geq 0. \tag{2.29}$$

From the definition of enthalpy, $H = U + PV$, by differentiating, we obtain

$$dH = dU + PdV + VdP. \tag{2.30}$$

Similarly, differentiating Equation 2.25a, it is found that

$$dG = dH - TdS - SdT, \tag{2.31}$$

so that combining Equations 2.29, 2.30, and 2.31, the following inequality is obtained:

$$dG - VdP + SdT \leq 0. \tag{2.32}$$

Equation 2.32 tells us that, in a process taking place at constant temperature and pressure, T_0 and P_0, respectively, the only direction that the said process can proceed is the direction of *decreasing* Gibbs free energy; that is,

$$dG(P_0, T_0) \leq 0. \tag{2.33}$$

Once the system reaches an *equilibrium* state at the end of the process, Gibbs free energy is at a *minimum*, i.e., $dG(P_0, T_0) = 0$.

For applications to multicomponent systems involving phase and/or chemical equilibria, Gibbs free energy, $G(P, T, n_i)$, leads to a new thermodynamic property of prime importance, the *chemical potential*, μ_i. The chemical potential of component i in a gas mixture is defined as

$$\mu_i = \left(\frac{\partial G}{\partial n_i}\right)_{T,P,n_{j\neq i}}, \tag{2.34}$$

that is, partial derivative of G with respect to n_i while holding P, T, and number of moles of other components n_j with $j \neq i$ constant. Another term for the chemical potential is *partial molal Gibbs free energy*. Derivation of the chemical potential can be found in any thermodynamic textbook (e.g., pp. 675–676 in Moran and Shapiro

[1]). Note that the chemical potential is an *intensive* property so that, for a system comprising N components, the *extensive* property (i.e., Gibbs free energy) becomes

$$G = \sum_{i=1}^{N} n_i \mu_i.$$

(2.35)

Substituting Equation 2.35 into Equation 2.33 and rearranging terms, we find that

$$dG(P_0, T_0) = \sum_{i=1}^{N} \mu_i dn_i \le 0,$$

(2.36)

which translates the equilibrium criterion, $dG(P_0, T_0) = 0$, into the following form:

$$\sum_{i=1}^{N} \mu_i dn_i = 0,$$

(2.37)

For a single phase (e.g., gas or liquid) of a *pure substance*, $G = n\mu$; that is, the chemical potential is just the Gibbs free energy per mole, $\bar{g}$. By the same token, intensive properties on a *mass* basis, such as h, u, v, and s, can also be expressed on a *molar* basis (e.g., $\bar{h}$), which is related to h via

$$\bar{h}[kJ/kmol] = h[kJ/kg] \times MW[kg/kmol]$$

For an *ideal gas mixture*, however, using Equation 2.35 as the *template* (so to speak), we have the following relationship for the mixture *enthalpy*

$$H = \sum_{i=1}^{N} n_i \bar{h}_i(T).$$

(2.38)

(Ideal gas enthalpy is a function of temperature only.) Since the ideal gas *entropy* is a function of temperature *and* pressure, it is given by

$$S = \sum_{i=1}^{N} n_i \bar{s}_i(T, P_i),$$

(2.39)

where the partial pressure of component i is given by $P_i = y_i P$, and y_i is the mole fraction of i in the mixture (i.e., $y_i = n_i/n$). Consequently, for the Gibbs free energy, from Equation 2.25, we can write that

$$G = \sum_{i=1}^{N} n_i \left[\overline{h}_i (T) - T \overline{s}_i (T,P_i) \right], \text{ or} \tag{2.40}$$

$$\overline{g}_i (T,P_i) = \mu_i = \sum_{i=1}^{N} \overline{h}_i (T) - T \overline{s}_i (T,P_i). \tag{2.41}$$

Translated into plain language, Equation 2.41 tells us that the chemical potential of component i in an ideal gas mixture with N components ($i = 1, 2, \ldots, N$) is equal to its Gibbs free energy per mole of i, evaluated at the mixture temperature, T, and the partial pressure of i in the mixture, $P_i = y_i P$.

It is extremely important to recognize that a partial molal property, such as $\overline{g}_i$ is the value of the property of component i *as it exists in the mixture*. As verified by experiments, it is in general *not* equal to the property of i as a *pure fluid* at the temperature and pressure of the mixture. For any pure fluid property, M, as a function of pressure, P, and temperature, T, the *exact differential* can be written as

$$dM = \left(\frac{\partial M}{\partial P} \right)_T dP + \left(\frac{\partial M}{\partial T} \right)_P dT. \tag{2.42}$$

For a mixture of N components, however, the exact differential of the property, $M(P, T, n_i)$, includes an additional term and becomes

$$dM = \left(\frac{\partial M}{\partial P} \right)_{T,n} dP + \left(\frac{\partial M}{\partial T} \right)_{P,n} dT + \sum_{i=1}^{N} \left(\frac{\partial M}{\partial n_i} \right)_{P,T,n_{j \neq i}} dn_i. \tag{2.43}$$

However, as was shown earlier for the specific case of Gibbs free energy, the extensive property M of a mixture can also be expressed as a weighted sum of the partial molal property $\overline{M}_i$; that is,

$$M = \sum_{i=1}^{N} n_i \left(\frac{\partial M}{\partial n_i} \right)_{P,T,n_{j \neq i}} = \sum_{i=1}^{N} n_i \overline{M}_i. \tag{2.44}$$

Consequently, we can write the exact differential for M using Equation 2.44 as well; that is,

$$dM = \sum_{i=1}^{N} n_i d\overline{M}_i + \sum_{i=1}^{N} \overline{M}_i dn_i, \tag{2.45}$$

which, subtracting it from Equation 2.43, results in

$$0 = \left(\frac{\partial M}{\partial P}\right)_{T,n} dP + \left(\frac{\partial M}{\partial T}\right)_{P,n} dT - \sum_{i=1}^{N} n_i d\bar{M}_i. \qquad (2.46)$$

Applying Equation 2.46 to the Gibbs free energy, after some substitutions via Maxwell relationships, results in the *Gibbs-Duhem* equation:

$$0 = VdP - SdT - \sum_{i=1}^{N} n_i d\mu_i. \qquad (2.47)$$

Let us compare the Gibbs-Duhem equation, Equation 2.47, with Equation 2.37, rewritten subsequently for convenience, which provides a relationship among the properties of a system *at equilibrium* at temperature, *T*, and pressure, *P*.

$$\sum_{i=1}^{N} \mu_i dn_i = 0, \qquad (2.48)$$

For a process taking place at constant *P* and *T*, from Equation 2.47, we find that

$$\sum_{i=1}^{N} n_i d\mu_i = 0. \qquad (2.49)$$

Let us now consider a *steady-state, steady-flow* (SSSF) process. From the definition of the Gibbs free energy and Equation 2.35, we find for the enthalpy of a mixture with *N* components the following relationship:

$$H = TS + \sum_{i=1}^{N} n_i \mu_i, \qquad (2.50)$$

which, upon differentiation of both sides, becomes

$$dH = TdS + SdT + \sum_{i=1}^{N} n_i d\mu_i + \sum_{i=1}^{N} \mu_i dn_i. \qquad (2.51)$$

Combining Equation 2.47, the Gibbs-Duhem equation, and Equation 2.51, we obtain

$$dH = TdS + VdP + \sum_{i=1}^{N} \mu_i dn_i. \qquad (2.52)$$

For a reversible process (i.e., $dS = 0$) taking place at constant P and T, and making use of Equations 2.35 and 2.49, Equation 2.52 leads to the following correlation:

$$dH = \sum_{i=1}^{N} \mu_i dn_i = dG - \sum_{i=1}^{N} n_i d\mu_i = dG. \tag{2.53}$$

Equation 2.53 tells us that the minimum (or maximum, depending on the particular process) thermodynamic work that can apply to a process taking place at constant P and T with a chemical reaction is given by the change in Gibbs free energy. This is a very important finding. Equation 2.53 provides us with a powerful tool, as will be shown in Chapter 13, using important chemical processes as examples. The key condition is "with a chemical reaction". Otherwise, that is, in the absence of a chemical reaction, at constant and P and T, nothing will happen in the system. There is no change enthalpy, and one does not even have to bother about Gibbs free energy. It does not provide us with any new information.

How do we find the change in Gibbs free energy for ideal gas mixtures? Let us rewrite Equation 2.53 in a difference form using the intensive properties; that is,

$$\Delta H = \sum_{i=1}^{N} n_i \Delta \overline{g}_i, \text{ or} \tag{2.54}$$

$$\dot{W}_{rev} = \sum_{i=1}^{N} \dot{n}_i \Delta \overline{g}_i, \text{with} \tag{2.55}$$

$$\overline{g}_i (T, P_i) = \mu_i = \overline{h}_i (T) - T \overline{s}_i (T, P_i). \tag{2.56}$$

When systems with chemical reactions (i.e., with changing n_i of inlet and exit material streams) are involved, property calculations hit a snag: how to come up with a common datum for enthalpy and entropy. When we do the simple enthalpy calculation, say, for an expander or compressor (i.e., $w = \Delta h = h_1(T) - h_2(T)$), this is a non-issue. Whether the working fluid is pure (e.g., CO_2) or a mixture (e.g., air), the datum used for enthalpy cancel out. However, when a chemical reaction takes place, reactants disappear and products are formed; that is, differences cannot be calculated for all components involved. For enthalpy, this problem is resolved via the concept of *enthalpy of formation*, which involves the definition of the following:

- *Stable elements* (e.g., C, H_2, O_2, and N_2 but *not*, say, CO_2, H_2O, i.e., *compounds* that are formed from stable elements, e.g., CO_2 is formed from the reaction of C and O_2)
- *Standard reference state* (SRS), 25°C (77°F) and 1 atm (14.7 psia)
- The enthalpy of a stable element at SRS is zero
- The enthalpy of a compound at SRS is equal to its enthalpy of formation $\overline{h}_f^{\circ}$,

Consequently, for component i in a mixture, which is a compound (e.g., CO_2), at temperature, T, the enthalpy is given by

$$\bar{h}_i(T) = \bar{h}_{f,i}^\circ + \left[\bar{h}_i(T) - \bar{h}_i(T_{ref})\right] = \bar{h}_{f,i}^\circ + \Delta\bar{h}_i. \tag{2.57}$$

The first term on the right-hand side (RHS) of Equation 2.57 is the enthalpy of formation of i from its elements at the reference temperature, T_{ref}. The second term on the RHS of Equation 2.57 is the change in enthalpy of i, associated with a change of state, *at constant composition*.

The situation is somewhat tricky when it comes to entropy (i.e., *there is no entropy of formation*). The enthalpy of formation is a quantity that can be found (in principle) by measuring the heat transfer in a reaction in which the compound in question is formed from its elements. The reaction can be *endothermic* (i.e., heat is transferred into the reactor) or *exothermic* (i.e., heat is transferred from the reactor). The standard device (the *reactor*) used for such measurements is the calorimeter. (This is why the term *standard heat of formation* is used in some references.) There is, however, no corresponding device for measuring the entropy of formation. In lieu of that, the concept of *absolute entropy* is used to facilitate calculation of Gibbs free energy of formation. The concept of absolute entropy is based on the *third law* of thermodynamics, which states that the entropy of a *pure crystalline* substance is zero at the absolute zero of temperature, 0 K. Other substances have a nonzero value of entropy at 0 K. (See Section 2.2 for more on the third law.) This is the datum provided by the third law. The entropy relative to this datum is called the absolute entropy and can be determined experimentally (e.g., via specific heat measurements).

The absolute entropy per unit mole at temperature T and 1 atm pressure is designated as $\bar{s}^\circ(T)$. The absolute entropy of component i in a mixture at any pressure and temperature is found from

$$\bar{s}_i(T, P_i) = \bar{s}_{f,i}^\circ(T) + \left[\bar{s}_i(T, P_i) - \bar{s}_i(T, P_{ref})\right] = \bar{s}_{f,i}^\circ(T) + \Delta\bar{s}_i. \tag{2.58}$$

For an ideal gas, Equation 2.58 becomes

$$\bar{s}_i(T, P_i) = \bar{s}_{f,i}^\circ(T) - R\ln\frac{y_i P}{P_{ref}}. \tag{2.59}$$

Finally, using Equation 2.58 as a template and noting that Gibbs free energy is a function of temperature *and* pressure, we can write that

$$\bar{g}_i(T, P_i) = g_{f,i}^\circ + \left[\bar{g}_i(T, P_i) - \bar{g}_i(T_{ref}, P_{ref})\right] = g_{f,i}^\circ + \Delta\bar{g}_i. \tag{2.60}$$

Using Equations 2.25b and 2.60, we go through the following steps to arrive at a formula for the chemical potential of i in an ideal gas mixture:

$$\bar{g}_i(T, P_i) = \bar{h}_i(T) - T\bar{s}_i(T, P_i), \tag{2.61}$$

$$\bar{g}_i(T, P_i) = \bar{h}_i(T) - T\bar{s}_i^\circ(T) + \bar{R}T \ln \frac{y_i P}{P_{ref}}, \text{ with} \tag{2.62}$$

$$\bar{g}_i^\circ = \bar{h}_i(T) - T\bar{s}_i^\circ(T), \text{ so that} \tag{2.63}$$

$$\bar{g}_i(T, P_i) = \bar{g}_i^\circ + \bar{R}T \ln \frac{y_i P}{P_{ref}}, \text{ or} \tag{2.64}$$

$$\mu_i = \bar{g}_i^\circ + \bar{R}T \ln \frac{y_i P}{P_{ref}}. \tag{2.65}$$

For *homemade* calculations in Excel or other tools, we will need property information for key substances and compounds. Selected data is provided in Section 14.1. How to use the relationships introduced earlier will be illustrated by calculated examples in the following sections.

Chemical potential comes in handy in calculations involving chemical and phase equilibria. In heat engine calculations, the only instance where such calculations are required is the combustion process. However, such calculations are rarely done manually anymore. Furthermore, almost all combustion calculations of interest, even in heat and mass simulation software, are done under the assumption of chemical equilibrium so that combustion products include CO_2, H_2O, O_2 (if there is excess air), N_2, and Ar, where C and H in the fuel is converted completely into CO_2 and H_2O. Detailed calculations to find non-equilibrium and/or incomplete combustion products, such as CO, NO, NO_2, SO_2 (usually ignored for natural gas), and radicals, require specialty software, such as CHEMKIN. Even with such specialty software, however, such calculations require significant correction using empirical correlations obtained from combustion lab experiments on a case-by-case basis.

Thus, no use will be made of chemical potential in the practical calculation examples in this book. One reason for introduction of the chemical potential herein is, of course, for the sake of completeness. The other reason is to introduce the Gibbs function, which will come up later in the discussion of the concept of *exergy*, of which it is a very close relative (so to speak). (Exergy is the key concept that encompasses the 2LT in practical calculations.) As noted earlier, for a single phase (e.g., gas or liquid) of a *pure substance*, the chemical potential is just the Gibbs free energy per mole, $\bar{g}$, which will be used in example calculations later and in Chapter 13. The third and final reason is that Gibbs function leads to a thermodynamic equilibrium criterion for constant (P, T, N) systems closely related to the entropy maximum for constant (E, V, N) systems. For the reader interested in learning more on thermodynamics and phase equilibria of multicomponent systems (e.g., solutions), the best reference (in this author's opinion) is the book by Van Ness and Abbott [5].

2.3.2.2 Combustor Example

Application of the first law to a combustor is rather complicated because it involves a chemical reaction with multiple components. It is certainly possible (and quite straightforward) to program the chemical reaction relationships in an Excel spreadsheet (e.g., see the book by the author [4] and the references therein). Nevertheless, for most practical purposes, this is best done using a commercial software, such as THERMOFLEX, especially when one is looking at fuels other than 100% methane (CH_4). The control volume for the combustor of the example that will later be used in Section 3.2.1 to demonstrate the application of 2LT (i.e., General Electric's 7HA.01 characterized by published ISO base load data) is shown in Figure 2.2. The assumptions used in the calculation are as follows:

- 24% of compressor discharge airflow is diverted for hot gas path cooling purpose
- Fuel is 100% CH_4 (LHV = 50,046.7 kJ/kg)
- Fuel is heated to 226.7°C (440°F)
- Combustor exit (turbine inlet) temperature is 1,600°C

For this component, 1LT is expressed in an enthalpy balance as follows (using the state-point numbers in Figure 2.2):

$$0 = \dot{m}_2 h_2\left(T_2, y_{i,2}\right) + \dot{m}_5\left(LHV + \Delta h(T_5)\right) - \dot{m}_3 h_3\left(T_3, y_{i,3}\right) \qquad (2.66)$$

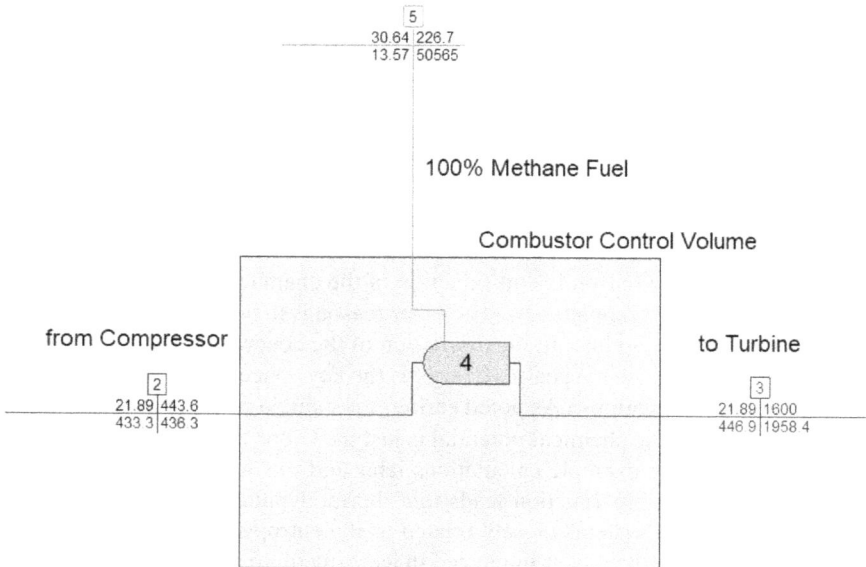

FIGURE 2.2 Combustor control volume calculation. State properties in clockwise direction from the upper-left corner are P in bar, T in °C, flow in kg/s and h in kJ/kg-K.

$$\dot{m}_3 = \dot{m}_2 + \dot{m}_5 \,(\text{Conservation of mass}) \tag{2.67}$$

In Equation 2.66,

- $y_{i,2}$ is the molar composition of air from the compressor
- $y_{i,3}$ is the molar composition of the combustion products (i.e., the hot gas)
- $\Delta h(T_s)$ is the sensible heat of fuel; that is,
  $\Delta h(T_5) = c_{p,\text{fuel}}(T_5 - T_{\text{ref}})$
- LHV is the lower heating value of fuel at the reference temperature, T_{ref}, of 25°C

Molar composition of the combustion products is determined from the reaction formula

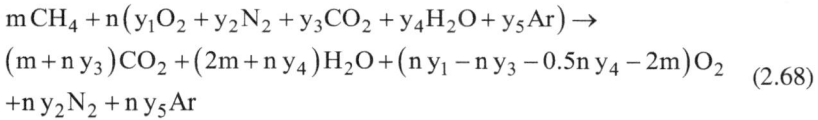

$$
\begin{aligned}
&m\,CH_4 + n\left(y_1 O_2 + y_2 N_2 + y_3 CO_2 + y_4 H_2 O + y_5 Ar\right) \rightarrow \\
&(m + n\,y_3)CO_2 + (2m + n\,y_4)H_2 O + (n\,y_1 - n\,y_3 - 0.5n\,y_4 - 2m)O_2 \\
&+ n\,y_2 N_2 + n\,y_5 Ar
\end{aligned} \tag{2.68}
$$

where n is the number of moles (per second) of air entering into reaction with m moles (per second) of fuel and y_i are mole fractions of air components. In Figure 2.2, $m = 0.846$ kmol/s and $n = 15.014$ kmol/s, or $n/m = 17.75$, which is the *air-fuel ratio* (AFR) of this control volume (on a molar – *not* mass – basis). *Stoichiometric* AFR for combustion of methane (i.e., there is no O_2 in the combustion products) is 9.524 so that this reaction is with *excess air* with an *equivalence ratio*, ϕ, of $17.75/9.524 = 1.864$ (i.e., 86.4% excess air). As we will see later in the book, this control volume encompasses the entire combustor control volume. What exactly is happening in the actual flame zone is quite different – in particular, the mixture of air and fuel going into reaction in the flame zone of a DLN combustor is *lean*.

In order to simplify the manual calculation, air is assumed to contain only nitrogen (79% by volume) and oxygen (21% by volume) such that stoichiometric combustion equation of 1 kmol of methane can be written as

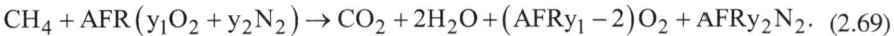

$$CH_4 + AFR\left(y_1 O_2 + y_2 N_2\right) \rightarrow CO_2 + 2H_2 O + (AFRy_1 - 2)O_2 + AFRy_2 N_2. \tag{2.69}$$

With $y_1 = 0.21$ and $AFR = 9.524$ (i.e., $y_1 \times AFR = 2.0$), we end up with

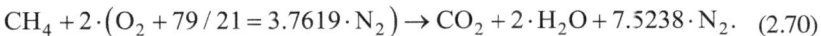

$$CH_4 + 2 \cdot \left(O_2 + 79/21 = 3.7619 \cdot N_2\right) \rightarrow CO_2 + 2 \cdot H_2 O + 7.5238 \cdot N_2. \tag{2.70}$$

Equation 2.70 tells us that 1 kmol of methane burns with $2 \times (1 + 3.7619) = 9.5238$ kmol of air to produce 1 kmol of CO_2 and 2 kmol of H_2O. The left-hand side of the equation represents the *reactants*. The right-hand side of the equation represents the *products*.

In a gas turbine combustor, methane reacts with *excess air* so that the reaction equation can be written as

$$CH_4 + 2x \cdot (O_2 + 3.7619 \cdot N_2) \rightarrow CO_2 + 2 \cdot H_2O + (2x - 2) \cdot O_2 + 7.5238x \cdot N_2 \quad (2.71)$$

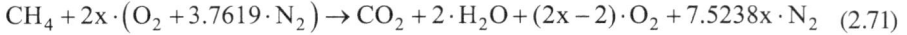

where $x > 1$ is the excess air ratio. For example, if $x = 1.25$, the combustion reaction is said to take place with 25% excess air. Note that both equations represent the *equilibrium* process of *complete combustion*. When the reaction takes place with excess air, combustion products include unreacted oxygen.

For computational purposes, a combustor can be considered as a control volume or reactor. There is no work interaction between the reactor and the surroundings. The reactor control volume can be used with two assumptions to calculate parameters of importance:

- *Isothermal* reactor (usually at a reference temperature, e.g., $T_{ref} = 25°C$)
- Adiabatic reactor

The isothermal reactor calculation is used to calculate the heating value of the fuel (e.g., methane). The adiabatic reactor calculation is used (i) to calculate the fuel flow requisite to achieve a specified combustor exit temperature with given amount of airflow or (ii) to calculate the combustor exit temperature for given air and fuel flows.

Equation 2.66 is difficult to use because streams 2 (combustion air) and 3 (combustion products) have different compositions. In other words, they are different fluids. Consequently, as discussed in detail in the preceding section, we will use Equation 2.57, written subsequently again for convenience, to calculate enthalpies:

$$\bar{h}(T) = \bar{h}_f^° + \left[\bar{h}(T) - \bar{h}(T_{ref}) \right] = \bar{h}_f^° + \Delta \bar{h} \quad (2.72)$$

In Equation 2.32, $\bar{h}$ (pronounced "h bar") denotes the specific enthalpy *per unit kmol*, which is the more convenient way (i.e., vis-à-vis *per unit mass*) to do enthalpy balance calculations involving reactions. In Equation 2.72, the enthalpy of formation is denoted by $\bar{h}_f^°$. In order to carry out manual calculations, we need enthalpy of formation and molar enthalpy data for compounds and stable elements of interest, which are provided in Section 14.1.

Next, we have to rewrite Equation 2.66 using molar flow rates instead mass flow rates. In order to be able to do that, we have to resort to Equation 2.68, which, in its simplified form, becomes

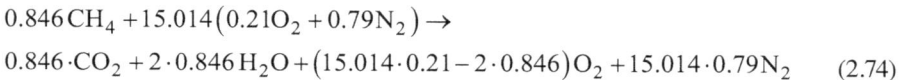

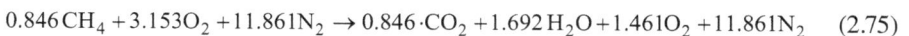

$$m\,CH_4 + n(y_1 O_2 + y_2 N_2) \rightarrow m\,CO_2 + 2mH_2O + (n\,y_1 - 2m)O_2 + n\,y_2 N_2 \quad (2.73)$$

$$0.846\,CH_4 + 15.014(0.21O_2 + 0.79N_2) \rightarrow$$
$$0.846 \cdot CO_2 + 2 \cdot 0.846\,H_2O + (15.014 \cdot 0.21 - 2 \cdot 0.846)O_2 + 15.014 \cdot 0.79N_2 \quad (2.74)$$

$$0.846\,CH_4 + 3.153O_2 + 11.861N_2 \rightarrow 0.846 \cdot CO_2 + 1.692\,H_2O + 1.461O_2 + 11.861N_2 \quad (2.75)$$

Consequently, Equation 2.66, combined with Equation 2.75, becomes

$$
0 = \begin{cases} \begin{cases} 3.153\left(\overline{h}_f^\circ + \Delta\overline{h}\right)_{O2,716.7K} +11.861\left(\overline{h}_f^\circ + \Delta\overline{h}\right)_{N2,716.7K} \\ +0.846\left(\overline{h}_f^\circ + \Delta\overline{h}\right)_{CH4,499.8K} \end{cases}_{Reactants} \\ \begin{cases} -0.846\left(\overline{h}_f^\circ + \Delta\overline{h}\right)_{CO2,1873.2K} -1.692\left(\overline{h}_f^\circ + \Delta\overline{h}\right)_{H2O,1873.2K} \\ -1.461\left(\overline{h}_f^\circ + \Delta\overline{h}\right)_{O2,1873.2K} -11.861\left(\overline{h}_f^\circ + \Delta\overline{h}\right)_{N2,1873.2K} \end{cases}_{Products} \end{cases} \tag{2.76}
$$

Noting that the enthalpy of formation for stable substances is zero, Equation 2.76 can be simplified:

$$
0 = \begin{cases} \left\{3.153\left(\Delta\overline{h}\right)_{O2,716.7K} +11.861\left(\Delta\overline{h}\right)_{N2,716.7K} +0.846\left(\overline{h}_f^\circ + \Delta\overline{h}\right)_{CH4,499.8K}\right\}_{Reactants} \\ \begin{cases} -0.846\left(\overline{h}_f^\circ + \Delta\overline{h}\right)_{CO2,1873.2K} -1.692\left(\overline{h}_f^\circ + \Delta\overline{h}\right)_{H2O,1873.2K} \\ -1.461\left(\Delta\overline{h}\right)_{O2,1873.2K} -11.861\left(\Delta\overline{h}\right)_{N2,1873.2K} \end{cases}_{Products} \end{cases} \tag{2.77}
$$

We will come back to solving Equation 2.77 using Excel later. To start, let us do an *isothermal reactor* example calculation to find the lower heating value of CH_4 using the reaction formula in Equation 2.30. The calculation is done in Excel using data tables in Section 14.1, as shown in Table 2.1.

The top part of the table (reactants) tells us the following:

- The fuel (CH_4) enters the reactor at a rate of 1 kmol/s and at 298.15 K
- The oxidant, air, enters the reactor at rates specified by the stoichiometric reaction formula, i.e., Equation 2.31, and also at 298.15 K
- Reactants carry into the reactor enthalpy at a rate of -74,850 kJ/s

TABLE 2.1

Calculation of CH_4 Lower Heating Value

Reactants	K T	kJ/kmol h_f°	kJ/kmol Δh	kmol/s $\dot{n}$	kJ/s $\dot{n}\overline{h}$
CH_4	298.15	−74.850	0	1.000	−74,850
N_2	298.15	0	0	7.524	0
O_2	298.15	0	0	2.000	0
				10.524	−74,850

Products	K T	kJ/kmol h_f°	kJ/kmol Δh	kmol/s $\dot{n}$	kJ/s $\dot{n}\overline{h}$
H_2O (g)	298.15	−241,820	0	2.000	−483,640
N_2	298.15	0	0	7.524	0
CO_2	298.15	−393,520	0	1.000	−393,520
				10.524	−877,160

The bottom part of the table (products) tells us the following:

- N_2 does not enter the reaction
- Products, H_2O and CO_2, exit the reactor at 298.15 K
- They carry enthalpy out of the reactor at a rate of −877,160 kJ/s

Thus, the net enthalpy balance of the reactor is as follows:
The negative number indicates that the reaction releases energy at the calculated rate. The amount of the energy released (i.e., the *enthalpy of combustion*) is the heating value of CH_4 at 298.15 K and 1 atm. It represents the *lower* heating value (LHV) of CH_4 because H_2O in the products in a gaseous state. If H_2O condenses to its liquid phase and releases its latent heat of condensation, the enthalpy of combustion will be higher. In that case, the bottom (products) part of Table 2.1 can be rewritten as shown in Table 2.2.

In this case, the net enthalpy balance of the reactor becomes the following:

−965,180 − (−74,850) = −890,330 kJ/s, or
−890,330/1 kmol/s of CH_4 = −890,330 kJ/ kmol of CH_4, which comes to
−890,330 kJ/kmol / 16.04 kg/kmol = −55,507 kJ/kg.

This number represents the *higher* heating value (HHV) of CH_4, and it is 55,507/50,019 = 1.11 or 11% higher than the LHV.

The second example is reproducing the THERMOFLEX calculation shown in Figure 2.2 by a manual (well, sort of) calculation. To be precise, it is the solution of Equation 2.77 using the proverbial pen and paper. Once again, the calculation is done in Excel using the data tables in Section 14.1.1, as shown in Table 2.3. In this example, we use the reaction formula in Equation 2.31 because the reaction takes place with excess air. How do we know that? From Figure 2.2, air enters the combustor (reactor) at a rate of 433.3 kg/s, which corresponds to 433.3/28.86 = 15.08 kmol/s (using the molecular weight of air as calculated by THERMOFLEX). It follows that N_2 flow rate is 0.79 × 15.08 = 11.86 kmol/s (79% of air by volume is nitrogen). However, the stoichiometric flow rate of N_2 is 7.524 kmol/s (e.g., see Table 2.1). In Table 2.3, N_2 flow rate, normalized to 1 kmol/s of fuel flow, is 14.07 kmol/s so that the reaction takes place at 14.07/7.524 = 1.87 or 87% excess air.

TABLE 2.2
Calculation of CH_4 Higher Heating Value

Products	K T	kJ/kmol $h_f°$	kJ/kmol Δh	kmol/s $\dot{n}$	kJ/s $\dot{n}\bar{h}$
H_2O	298.15	−285,830	0	2.000	−571,660
N_2	298.15	0	0	7.524	0
CO_2	298.15	−393,520	0	1.000	−393,520
				10.524	−965,180

TABLE 2.3
Calculation of CH_4 Requisite for 1,600°C Combustor Exit Temperature

Reactants	K T	kJ/kmol $h_f°$	kJ/kmol Δh	kmol/s Actual $\dot{n}$	Normalized $\dot{n}$	kJ/s $\dot{n}\bar{h}$
CH_4	499.8	−74,850	8,662	0.843	1.00	−55,785
N_2	716.7	0	12,454	11.861	14.07	147,720
O_2	716.7	0	13,061	3.153	3.74	41,179
				15.857	18.814	133,115

Products	K T	kJ/kmol $h_f°$	kJ/kmol Δh	kmol/s Actual $\dot{n}$	Normalized $\dot{n}$	kJ/s $\dot{n}\bar{h}$
H_2O	1,873.2	−241,820	65,973	1.692	2.00	−296,414
N_2	1,873.2	0	51,508	11.861	14.07	610,931
O_2	1,873.2	0	54,339	1.461	1.74	79,731
CO_2	1,873.2	−393,520	83,687	0.846	1.00	−261,133
				15.857	18.814	133,115

TABLE 2.4
Adiabatic Flame Temperature of CH_4

Reactants	K T	kJ/kmol $h_f°$	kJ/kmol Δh	kmol/s $\dot{n}$	kJ/s $\dot{n}\bar{h}$
CH_4	298.15	−74,850	0	1.000	−74,850
N_2	298.15	0	0	7.524	0
O_2	298.15	0	0	2.000	0
				10.524	−74,850

Products	K T	kJ/kmol $h_f°$	kJ/kmol Δh	kmol/s $\dot{n}$	kJ/s $\dot{n}\bar{h}$
H_2O	2,344.6	−241,820	88,891	2.000	−305,858
N_2	2,344.6	0	68,171	7.524	512,922
CO_2	2,344.6	−393,520	111,607	1.000	−281,913
				10.524	−74,850

In this case, the bottom part of the table represents the enthalpy calculation at 1,873.15 K (1,600°C) so that the Δh column contains values calculated from the property data in the tables provided in Section 14.1. In Table 2.3, there are two molar flow rate columns: actual and normalized. For the reactants, the actual N_2 and O_2 flow rates are calculated as described earlier. The flow rate for the fuel as an initial guess can be set to any value, say, 0.9 or 1.1. The values in the normalized flow rate column are calculated so that the fuel flow rate is unity. The normalized flow rate column for the products contains values calculated using Equation 2.31. Those numbers are then translated to actual flow rates using the *inverse* of the calculation for the reactants. The actual fuel flow rate can be found using the Excel What-If function to set the total enthalpy flow rate of the products equal to the total enthalpy flow rate of the reactants. The result of the calculation is 0.843 kmol/s of CH_4 or 0.843 × 16.04 = 13.52 kg/s, which is almost the same as the number shown in Figure 2.2.

The final example is calculation of the *adiabatic flame temperature* of CH_4. This calculation is similar to calculation of the enthalpy of combustion. The pertinent reaction formula is Equation 2.30. The temperature of the products is adjusted until the enthalpy rates of the products and reactants are equal to each other (i.e., the reactor is *adiabatic*). The result is shown in Table 2.4: the adiabatic flame temperature of CH_4 is 2,345 K.

NOTE

1 World's energy stays constant; [its] entropy strives for a maximum.

REFERENCES

1. Moran, M.J., and Shapiro, H.N., *Fundamentals of Engineering Thermodynamics*, New York: John Wiley & Sons, Inc., 1988.
2. Van Wylen, G.J., and Sonntag, R.E., *Fundamentals of Classical Thermodynamics*, 3rd Edition, Hoboken: Wiley, 1986.
3. Çengel, Y.A., and Boles, M., *Thermodynamics: An Engineering Approach w/ Student Resources DVD*, 5th Edition, New York: McGraw-Hill Science/Engineering/Math, 2005.
4. Bejan, A., Tsatsaronis, G., and Moran, M., *Thermal Design & Optimization*, New York: John Wiley & Sons, Inc., 1996.
5. Van Ness, H.C., and Abbott, M.M., *Classical Thermodynamics of Non-Electrolyte Solutions*, New York: Mc-Graw-Hill, Inc., 1982.

3 Second Law – Macroscopic Perspective

Since a given system can never of its own accord go over into another equally probable state but into a more probable one, it is likewise impossible to construct a system of bodies that after traversing various states returns periodically to its original state, that is a perpetual motion machine.

—Ludwig Boltzmann (1886)

The second law of thermodynamics (2LT, henceforth) is quite simple in spirit but impossible to encapsulate in a sentence. This is the reason why it is made so mysterious in popular science writings, in particular to explain the mysteries of universe with vague notions, such as the arrow of time and similar stuff, like the undefinable entropy of the universe.

There is no single equation or simple statement that expresses the 2LT, which, for most practical purposes, can be summed up by two *negative* statements (as expressed on pages 167–168 of the textbook by Moran and Shapiro, Ref. [1] in Chapter 2):

1. *Kelvin-Planck statement*: It is impossible for any system to operate in a thermodynamic cycle and deliver a net amount of work to its surroundings while receiving energy by heat transfer from a single thermal reservoir. In other words, there is no perfect heat engine.
2. *Clausius statement*: It is impossible for any system to operate in such a way that the sole result would be an energy transfer by heat from a cooler to a hotter body. In other words, there is no perfect refrigerator.

It is noteworthy that there is no mention of *time*, with or without an arrow. There is no *universe* or *entropy of the universe*. There is no mention of *entropy* or any other thermodynamic property (state function) for that matter. There are, however, several key terms:

1. A system
2. Its surroundings
3. A thermodynamic cycle
4. A thermal reservoir (or body)

The system in question interacts with its surroundings via exchange of heat and/or work. A thermal reservoir is an idealization of a closed system or body so large that no matter how much energy is transferred from or into it, its temperature does not change. The Kelvin-Planck (K-P) statement essentially tells us that one cannot design

DOI: 10.1201/9781003247418-4

a heat engine operating in a thermodynamic cycle with 100% efficiency. A natural consequence (*corollary*) of the K-P statement is the *Carnot cycle*, which defines the efficiency of a heat engine operating between a hot temperature reservoir and a cold temperature reservoir. This is the maximum possible efficiency of any conceivable, reversible heat engine operating in a thermodynamic cycle between the same two temperature reservoirs.

To be precise, there are *two* Carnot corollaries of the K-P statement:

- The thermal efficiency of an *irreversible* power cycle is always less than the thermal efficiency of a *reversible* power cycle when each operates between the same two thermal reservoirs.
- All reversible power cycles operating between the same two thermal reservoirs have the same thermal efficiency.

For our purposes, translation of these two corollaries into a single statement that would facilitate practical calculations is as follows:

There can be no real heat engine operating in a thermodynamic cycle between two temperature reservoirs and more efficient than a Carnot engine (i.e., an ideal engine operating in a Carnot cycle) between the same two reservoirs.

If the statement earlier is too much verbiage for the practically orientated reader who just needs a simple formula, here it is:

$$\eta_{Carnot} = 1 - \frac{T_L}{T_H} \tag{3.1}$$

where T_L and T_H are the temperatures of the low and high temperature reservoirs, respectively. For those who are more visually inclined, the process is illustrated in Figure 3.1.

From the diagram in Figure 3.1, one can readily observe that the efficiency of the reversible heat engine operating in a Carnot cycle is also given by

$$\eta_{Carnot} = 1 - \frac{Q_L}{Q_H} \tag{3.2}$$

because the cycle work is given by

$$W_{cycle} = Q_H - Q_L \tag{3.3}$$

Combining Equations 3.1 and 3.2, one finds that

$$\frac{Q_L}{Q_H} = \frac{T_L}{T_H} \tag{3.4}$$

System + Surroundings (Isolated)

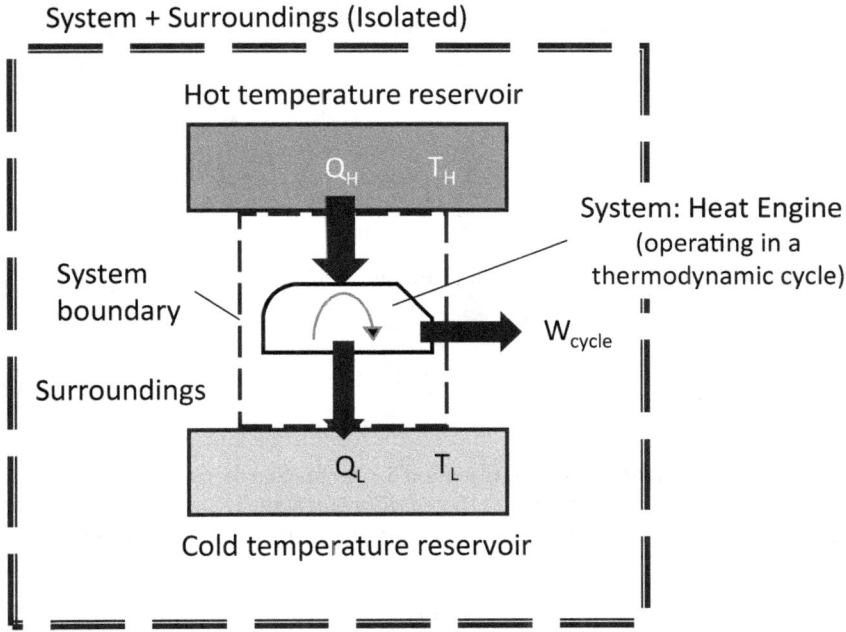

FIGURE 3.1 Heat engine operating in a thermodynamic cycle between two temperature reservoirs.

$$\frac{Q_L}{T_L} = \frac{Q_H}{T_H} \tag{3.5}$$

$$\frac{Q_L}{T_L} - \frac{Q_H}{T_H} = 0. \tag{3.6}$$

At this point, a knowledgeable reader may be feeling exasperated (rightfully so) and thinking that he wasted his money on a pile of paper parading this basic stuff. Still, the author would like to suggest to the reader to be patient for a few more paragraphs, and it will be shown how the *Carnot corollary* of the K-P *statement* of the 2LT brings home the definition of that mysterious quantity; that is, *entropy* or, to be precise, the definition of the *change* in entropy.

Equation 3.6 came from the great minds of science in the nineteenth century, such as Gibbs, who thought (roughly speaking) as follows: We have a cyclic thermodynamic process at hand – that is, we start at an *equilibrium* state, say, state 1; go through a series of *reversible* processes; exchange heat (and work) between the system and its surroundings; and end up at the same equilibrium state 1. Thus, the change in any thermodynamic property of state 1 (e.g., temperature, pressure, energy, and any *arbitrary* property that one can invent by combining *any two* properties) must be *zero*.

Now, instead of using only two heat transfers between the system and its sur-roundings (i.e., Q_H and Q_L), one can imagine a very large number, say, N, of infini-tesimally small heat transfers, δQ_i, such that one can express the equality given by Equation 3.6 as follows:

$$\sum_{i=1}^{N} \frac{\delta Q_i}{T_i} = 0 \tag{3.7}$$

In the limit of $N \rightarrow \infty$, the summation in Equation 3.7 becomes an integral; that is,

$$\int_1^1 \frac{\delta Q}{T} = \int \frac{\delta Q}{T} = 0. \tag{3.8}$$

This is the proverbial aha moment. Equation 3.8 points to a thermodynamic property, say, S, such that, when one begins at the *equilibrium* state 1, go through a reversible cyclic process, and end up at the *equilibrium* state 1 again, one can write that

$$S_1 - S_1 = \int \frac{\delta Q}{T} = 0. \tag{3.9}$$

or, equivalently,

$$dS = \frac{\delta Q}{T}. \tag{3.10}$$

This new property, S, is called *entropy*. (Note that we could have used the symbol, say, X, and called it *this*tropy, *that*tropy, or something else. At this point, since the symbol and the name have been assigned more than a century ago, this would have been a pointless gimmick.)

For any arbitrary, non-cyclic, *reversible* process that takes place between the *equi-librium* states 1 and 2, Equation 3.10 can be written as

$$\int_1^2 dS = S_2 - S_1 = \int_1^2 \frac{\delta Q}{T}$$

$$S_2 - S_1 = \int_1^2 \frac{\delta Q}{T}. \tag{3.11a}$$

Equations 3.10 and 3.11a, strictly speaking, do not define the entropy but, rather, the *change* in entropy for a reversible process. Furthermore, entropy is a thermodynamic property and, as such, a *state function*, and thus, its differential is an *exact differential*, denoted by the symbol *d*. (In contrast, *Q* is a *path function*, and thus, its differential is an *inexact differential*, denoted by the symbol δ.) It is also very important to understand that entropy, as a property (i.e., a state function), is only definable for *equilibrium* states. It should also be added that the units of *S*, an *extensive* property, in the SI unit system is J/K or kJ/K (1 kJ = 1,000 J). It can be transformed into an *intensive* property via division by the system mass, *M* (i.e., *s* = *S/M*), with the units of kJ/kg-K. (In British or US Customary system, the units of *S* and *s* are Btu/R and Btu/lb-R, respectively.) Thus, another way to write Equation 3.11a is

$$s_2 - s_1 = \int_1^2 \frac{\delta q}{T}, \qquad (3.11b)$$

where *q* = *Q/M* and has units of J/kg or kJ/kg.

What about an *adiabatic* process (i.e., δ*Q* = 0)? Does that mean that the entropy change for an adiabatic process is always zero? The answer is no. Equation 3.11, which is an *equality*, is only true for a *reversible* process. For an *irreversible* process, the entropy of the system always increases, whether the process in question is adiabatic (δ*Q* = 0) or not (δ*Q* ≠ 0); that is,

$$S_2 = S_1 + \int_1^2 \frac{\delta Q}{T} + I \qquad (3.12a)$$

where *I* > 0 quantifies the process *irreversibility* (or *entropy production*). For an adiabatic process (i.e., δ*Q* = 0), one ends up with

$$S_2 - S_1 = \Delta S = I. \qquad (3.13)$$

Since *I* = 0 for a reversible process and *I* > 0 for an irreversible process, for an adiabatic process,

$$S_2 - S_1 = \Delta S \geq 0. \qquad (3.14)$$

For the sake of completeness, in per unit mass terms, Equation 3.12a becomes

$$s_2 = s_1 + \int_1^2 \frac{\delta q}{T} + i, \qquad (3.12b)$$

where *i* = *I/M*. Note that *I* and *i* have the same units as *S* and *s*, respectively.

At this point, we should say something about the system that we are looking at. Specifically, Equation 3.12 is valid only for a *closed system*; that is, there is no mass flow across the system boundary, neither into the system nor out of it (i.e., the system mass is conserved (M = constant)). Equation 3.14, however, implies an *isolated* system, which is closed (i.e., no mass interaction with its surroundings) *and* adiabatic (i.e., no heat interaction with its surroundings). In an isolated system, system *mass*, M, and *energy*, E, are both conserved. Furthermore, system *volume*, V, must also be constant because, otherwise, even though there is no heat exchange between the system and its surroundings, there would be *work exchange* due to the moving, say, of one part of the system boundary (e.g., a piston) either *inward* (i.e., work done *on* the system) or *outward* (i.e., work done *by* the system). This type of system is referred to as an (E, V, N) system, where N is the number of microscopic particles comprising the system (e.g., gas molecules), and it is a proxy for the system mass, M. As an example, consider that, using the Avogadro's constant, $N_A = 6.02214076 \times 10^{23}$ mol^{-1}, for n mols of gas with mass of M grams, one can write that

$$M = n \times MW = n \times N_A \times m = N \times m,$$

where MW is the molecular weight of the gas in question in g/mol, and m is the weight of one molecule of gas in grams.

For a general process ($\delta Q \neq 0$) in a closed system, from Equation 3.12, one can write the following relationships for the entropy change; that is, as an *equality*

$$\Delta S = \int_1^2 \frac{\delta Q}{T} + I, \tag{3.15a}$$

or as an *inequality*

$$\Delta S \geq \int_1^2 \frac{\delta Q}{T}. \tag{3.15b}$$

Equation 3.15b is known as the generalized *Clausius inequality*. The first term on the right-hand side (RHS) of Equation 3.15a is the entropy transfer with heat transfer; it can be *positive* (heat *into* the system), *negative* (heat *out of* the system), or zero (adiabatic). The second term on the RHS of Equation 3.15a is the entropy production; it is either *zero* (*reversible* process) or *positive* (*irreversible* process). It cannot be negative. The left-hand side (LHS) of Equation 3.15 is the entropy change between states 1 and 2, which can be positive, negative, or zero. Since it is a state function, no information about the process is necessary to evaluate it, whereas the RHS of Equation 3.15 is a function of the process path.

Finally, for a *cyclic* process, since, by definition, $\Delta S = 0$, Equation 3.15a becomes

$$0 = \int \frac{\delta Q}{T} + I \tag{3.16}$$

or, equivalently,

$$-I = \int \frac{\delta Q}{T},$$

(3.17)

which can be rewritten as an inequality; that is,

$$\int \frac{\delta Q}{T} \leq 0.$$

(3.18)

Equation 3.18 is also known as the *Clausius Theorem*, which is a special case of the generalized inequality given by Equation 3.15b. The *equality* holds for a cycle comprising *reversible* processes. The *inequality* holds for a cycle comprising *irreversible* processes. For the heat engine cycle shown in Figure 3.1, Equation 3.18 can be written as

$$\int \frac{\delta Q}{T} = \frac{Q_H}{T_H} - \frac{Q_L}{T_L} \leq 0.$$

It follows that, for a reversible cycle (i.e., the Carnot cycle), as shown earlier,

$$\frac{Q_H}{T_H} - \frac{Q_L}{T_L} = 0$$

$$\frac{Q_H}{Q_L} = \frac{T_H}{T_L}$$

$$\eta_{Cyc,Rev} = 1 - \left.\frac{Q_H}{Q_L}\right|_{Rev} = 1 - \left.\frac{T_H}{T_L}\right|_{Rev}$$

For an irreversible cycle, however,

$$\frac{Q_H}{T_H} - \frac{Q_L}{T_L} < 0$$

$$\frac{Q_H}{Q_L} < \frac{T_H}{T_L}$$

Consequently, for an arbitrary, irreversible cycle operating between the same two temperature reservoirs; that is,

$$\left.\frac{T_H}{T_L}\right|_{Irr} = \left.\frac{T_H}{T_L}\right|_{Rev},$$

we conclude that

$$\left.\frac{Q_L}{Q_H}\right|_{Irr} > \left.\frac{Q_L}{Q_H}\right|_{Rev},$$

so that

$$\eta_{Cyc,Irr} = 1 - \left.\frac{Q_H}{Q_L}\right|_{Irr} < \eta_{Cyc,Rev} = 1 - \left.\frac{Q_H}{Q_L}\right|_{Rev}.$$

What we have demonstrated is the core of 2LT for thermal engineering, specifically, for analysis of heat engines. Specifically:

> *No heat engine operating in a thermodynamic cycle between two temperature reservoirs can be more efficient than a Carnot engine operating between the same two temperature reservoirs.*

Before concluding, let us spend some time on reversible and irreversible processes, which were introduced in the earlier discussion without fanfare. The following is per Moran and Shapiro (i.e., p. 169 in Ref. [1] in Chapter 2):

- A process is *reversible* if both the system and the surroundings can be returned to their respective initial states after the process has occurred.
- A process is *irreversible* if the system and its surroundings cannot be exactly restored to their respective initial states after the process has occurred.

Reversible processes are *hypothetical* processes. All *real* processes are irreversible. Reversible processes are idealizations that make the analysis of real systems manageable. The idea underlying a hypothetical reversible process this: It happens in infinitesimally small steps such that each successive step in the process is a *quasi*-equilibrium state. In other words, the process can be shown as a *solid line* on an *x-y* diagram with the *x* and *y* axes representing thermodynamic properties. (Remember that *two* properties are enough to define an equilibrium state so that a two-dimensional diagram is adequate.) This is illustrated by the *temperature-entropy (T-s)* diagram in Figure 3.2 for a generic, adiabatic *compression* process. (It is equivalent to an enthalpy-entropy, *h-s*, or *Mollier* diagram.)

The ideal process is adiabatic and reversible so that, according to Equation 3.15, $\Delta s = 0$. It is shown on the *T-s* diagram as a *solid line*. Any state between the initial and final (equilibrium) states 1 and 2s, respectively, is also an equilibrium state and

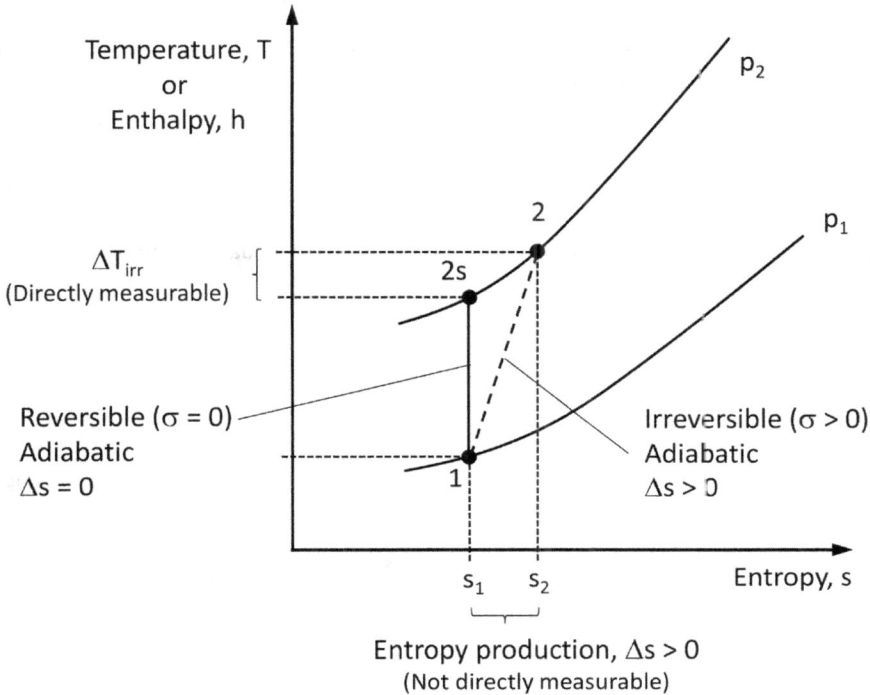

FIGURE 3.2 Generic compression process (i.e., increasing pressure) on *T-s* diagram.

can be identified on the *T-s* diagram. The actual process is also adiabatic but irreversible. While the initial and final states, 1 and 2, respectively, are equilibrium states and, thus, identified on the *T-s* diagram, the intermediate states are not equilibrium states. Therefore, the process is shown as a *broken line* because those non-equilibrium states cannot be identified on the *T-s* diagram.

The entropy generation in the irreversible process cannot be quantified by *direct measurement* (i.e., there is no entropy-meter that can do that). However, it can be *indirectly* quantified by the direct measurement of initial and final temperatures (and pressures). Those measurements can then be used to determine initial and final internal energies (closed system) or enthalpies (open system) using a suitable *equation of state*. As shown in the *T-s* diagram in Figure 3.2 (and well known by all mechanical and chemical engineers), the entropy production (irreversibility) of the compression process manifests itself as a rise in the internal energy (temperature) of the fluid (i.e., as heat). Quantification of irreversibilities or losses in the context of actual turbomachines (e.g., pumps and compressors) is done by the use of the concepts of *isentropic* (see Section 2.3.1) or *polytropic* efficiency. The reader can consult an elementary textbook of thermodynamics (e.g., Ref. [1] in Chapter 2) or a standard reference on turbomachinery (e.g., Ref. [2] in Chapter 1) for details.

3.1   COMBINING FIRST AND SECOND LAWS

Let us take another look at the Clausius inequality and start by rewriting Equation 3.10 subsequently for convenience:

$$dS = \frac{\delta Q}{T}$$

$$(3.10)$$

Recall that we took the integral of the LHS and RHS of Equation 3.10 and ended up with the Clausius inequality, Equation 3.15b:

$$\Delta S \geq \int_1^2 \frac{\delta Q}{T}$$

$$(3.15b)$$

Let us now try a different tack and, again starting from Equation 3.10, go through the following steps:

$$TdS = \delta Q$$

$$Tds = \delta q$$

$$\int_1^2 Tds = \int_1^2 \delta q = q$$

$$(3.19)$$

Equation 3.19 is an equality valid for a reversible process. Comparison with Equation 3.15b suggests that for an irreversible process, the LHS of Equation 3.19 will be greater than the RHS so that we end up with a slightly different version of Clausius inequality; that is, that for any arbitrary process between the two states, 1 and 2

$$\int_1^2 Tds \geq q_{rev}.$$

$$(3.20)$$

If asked to identify a single equation that best explains the second law of thermodynamics from a *concrete* engineering point of view (i.e., without resorting to *pseudo* philosophical explanations using the arrow of time, etc.), in this author's opinion, one should point to Equation 3.20, which tells us the same things that we already learned earlier. However, Equation 3.20 is in a form more readily available to application to the engineering problems, especially using the *Tds* equations.

As we have seen earlier, another way to write the inequality in Equation 3.20 is

$$\int_1^2 Tds = q_{rev} + i,$$

(3.21)

where the term i is as follows:

- Is zero for a *reversible* process
- Is nonzero *and* positive for an *irreversible* process
- Can *never* be negative

In addition to irreversibility, the term i is also known under several different names, such as *entropy production*, or *exergy destruction*. One other name, which is of most interest to practicing engineers, is *lost work*. To see how *work is lost*, let us start with a fundamental thermodynamic relationship (i.e., one of the two *Tds* equations)[1]:

$$dh = Tds + vdP$$

(3.22)

$$\int_1^2 dh = \int_1^2 Tds + \int_1^2 vdP$$

(3.23)

Combining Equation 3.23 with Equation 3.21 and carrying out the integration on the LHS, we end up with

$$h_2 - h_1 = \int_1^2 vdP + (q_{rev} + i)$$

(3.24)

Integration of Equation 3.22 for an isentropic process, $ds = 0$, gives us the isentropic work done by the system (or on the system); that is,

$$h_{2s} - h_1 = \int_1^2 vdP.$$

(3.25)

(Refer to Figure 3.2 for state numbering.) Thus, combining Equations 3.24 and 3.25, the result is

$$h_2 - h_1 = (h_{2s} - h_1) + (q_{rev} + i).$$

(3.26)

Reversible heat transfer
- Nonzero in general (+ or -)
- Zero when adiabatic

Rate of entropy production
(Zero when 1 → 2 is reversible)

$$h_2 - h_1 = \left(h_{2s} - h_1\right) + \left(q_{rev} + i\right)$$

Actual work between states 1 and 2

- Zero when 1 → 2 is isentropic
- Nonzero in general (+ or -)
- Greater than zero when adiabatic _and_ irreversible

Isentropic work between states 1 and 2

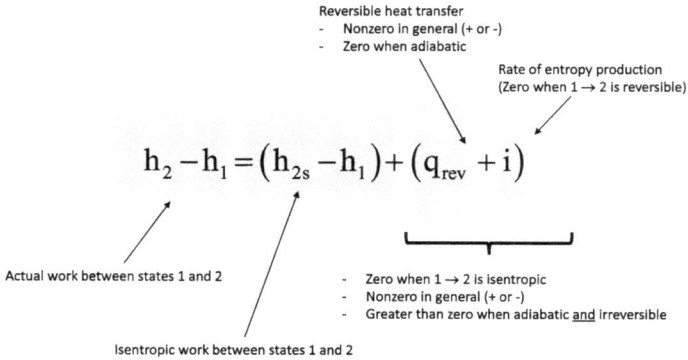

FIGURE 3.3 The second *Tds* equation for a general process between two arbitrary states 1 and 2.

The wealth of information carried inside the deceptively simple Equation 3.26 is illustrated in Figure 3.3. The two most common uses of Equation 3.26 in heat engine and turbomachinery thermodynamic analysis are for *adiabatic compression* (illustrated in Figure 3.2) and *adiabatic expansion*. In either case, setting q_{rev} to 0, Equation 3.26 becomes

$$h_2 - h_1 = \left(h_{2s} - h_1\right) + i \tag{3.27}$$

$$i = \left(h_1 - h_{2s}\right)\left(1 - \eta_{t,s}\right) \text{ for a turbine } \left(\text{expander}\right), \tag{3.28}$$

$$i = \left(h_{2s} - h_1\right)\left(\frac{1}{\eta_{c,s}} - 1\right) \text{ for a compressor,} \tag{3.29}$$

and, thus, makes the name *lost work* for *i* quite clear. In Equations 3.28 and 3.29, isentropic efficiencies for components are utilized to quantify *i* starting from the (ideal) isentropic process.

3.2 APPLICATION OF THE SECOND LAW – A PREVIEW

3.2.1 HEAT ENGINES

In the preceding section, considerable effort has been spent to demonstrate the usefulness of entropy and eliminate the unnecessary mystery and mysticism surrounding it in popular science. Now it is time to give a brief demonstration of what we learned so far. At the risk of sounding repetitious, let us restate the engineering significance of the 2LT once more. For practical engineering purposes, the gold standard is the Carnot corollary of the Kelvin-Planck statement of the 2LT, which is expressed in the definition of the Carnot cycle and its efficiency, given by Equation 3.1, and repeated subsequently for convenience; that is,

$$\eta_{Carnot} = 1 - \frac{T_L}{T_H}.$$ (3.1)

There can be no heat engine operating in a thermodynamic cycle reversibly between two temperature reservoirs and can be more efficient than a Carnot cycle operating between the same two temperature reservoirs.

How do we put this plain and simple statement and the accompanying simple formula, Equation 3.1, to good use? Let us illustrate it by a concrete example. Modern (land-based) heavy-duty industrial gas turbine used for electric power generation is an *internal combustion heat engine*. Let us consider one such engine, General Electric's 7HA.01, which was first introduced in 2012. Its key performance data at ISO base load is obtained from *Gas Turbine World's 2020 GTW Handbook*:

- 290 MWe and 42% net LHV efficiency
- Cycle pressure ratio (*PR*) of 21.6:1
- Turbine inlet temperature (*TIT*) of 1,600°C
- Exhaust temperature of 1,158°F (625.6°C)
- Exhaust flow of 1,293.7 lb/s (586.8 kg/s)
- 3,600 rpm (60 Hz)

Gas turbines operate in a *Brayton cycle*. The actual machine's operation is quite different from the predictions of the ideal cycle model, but valuable information can be derived from an ideal cycle analysis, which assumes an *air-standard* Brayton cycle, comprising the following *four* processes:

- *Isentropic* compression and expansion
- *Isobaric* (constant pressure) heat addition and heat rejection

Cycle working fluid is air modeled as *perfect gas*; that is, it obeys the *ideal gas* equation of state

$$PV = m\frac{R_{unv}}{MW}T$$

$$P = \rho\frac{R_{unv}}{MW}T$$

and its specific heat, c_p, is constant. The universal gas constant, R_{unv}, is 8.314 kJ/kg-K; air molecular weight, *MW*, can be assumed to be 29 kg/kmol. Consistent pressure units to be used in the ideal gas equation with *V* in m³, *m* in kg, and ρ in kg/m³ is kPa (1,000 Pa, 1 Pa = 1 N/m²). Temperature is, of course, always in the absolute scale (i.e., in degrees Kelvin). From any introductory textbook (e.g., Moran and Shapiro, Ref. [1] in Chapter 2), for an air-standard Brayton cycle with constant c_p, the cycle efficiency is a function of cycle pressure ratio only; that is,

$$\eta_{\text{Brayton}} = 1 - \text{PR}^{\frac{1-\gamma}{\gamma}} \tag{3.30}$$

where $\gamma = 1.4$ is the ratio of specific heats; that is, $\gamma = c_p/c_v$. Substituting $PR = 21.6{:}1$ into Equation 3.30, the ideal cycle efficiency for this gas turbine is found as 58.435%. If one could have built a Carnot engine operating between $T_H = 1{,}600°C$ and $T_L = 15°C$, its efficiency, from Equation 3.1, would be 84.617%. Thus, the *Cycle Factor, CYF*, for a heat engine operating in an (ideal) air-standard Brayton cycle with a cycle *PR* of 21.6:1, is

$$\text{CYF} = \frac{58.435}{84.617} = 0.69.$$

CYF is a quantification of the goodness of the ideal cycle in question with reference to the ultimate theoretical benchmark.

The actual cycle efficiency for 7HA.01 in our example is 42% so that the *Technology Factor, TF*, for the HA class technology with $PR = 21.6{:}1$ and $TIT = 1{,}600°C$ is

$$\text{TF} = \frac{42}{58.435} = 0.719.$$

TF is a quantification of the goodness of the actual cycle performance (as implemented in the engine hardware operating in the field) with reference to its *ideal* performance with no losses and reversible underlying processes. *CYF* and *TF* are very important factors of immediate practical use. (This will be obvious when we look at the state-of-the-art technology in heavy-duty industrial gas turbines in Chapter 10.) For the purposes of the current exercise, the manifestation of the second law is quantified by the *Cycle Factor*. In order to prove the veracity of the assertion, we have to start from the *T-s* diagrams of Carnot and Brayton cycles, which are shown in Figure 3.4.

In Figure 3.4, the Carnot cycle {1-2C-3-4C-1} comprises the following four processes:

- *Isentropic* compression expansion
- *Isothermal* heat addition and heat rejection

The deviation of the air-standard Brayton cycle {1-2-3-4-1} from the theoretical ideal, the Carnot cycle, is quantified by the two *triangular areas* on the *T-s* diagram in Figure 3.4:

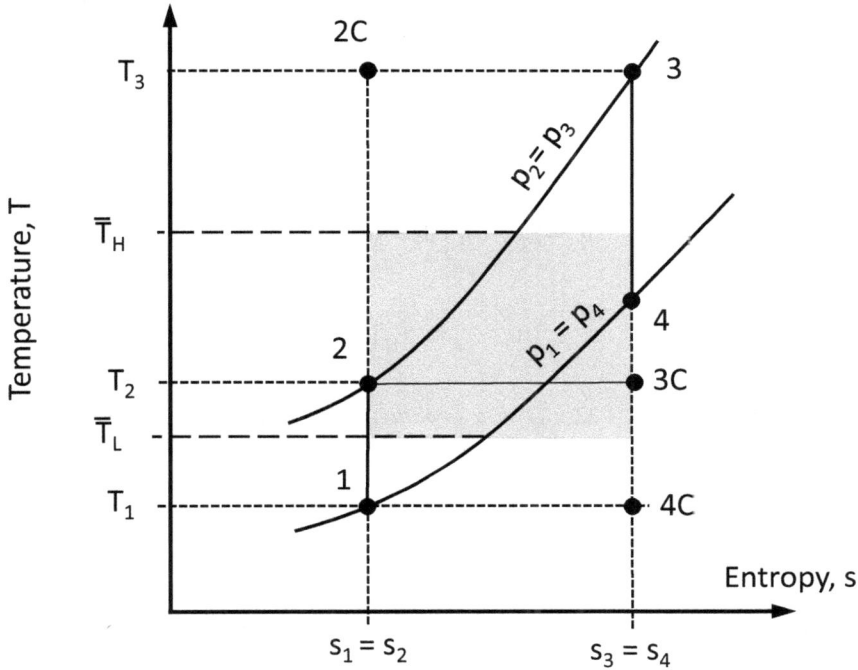

FIGURE 3.4 Air-standard Brayton cycle {1-2-3-4-1} superimposed on the Carnot cycle {1-2C-3-4C-1}.

- The *lost work* associated with non-ideal (i.e., non-isothermal) heat addition {2-2C-3-2}
- The *lost work* associated with non-ideal heat rejection {1-4-4C-1}

The first one is quantified as shown in the calculation sequence subsequently:

$$\Lambda_H = T_3 (s_3 - s_2) - c_p (T_3 - T_2)$$

$$c_p (T_3 - T_2) = \overline{T}_H (s_3 - s_2)$$

$$\Lambda_H = (T_3 - \overline{T}_H)(s_3 - s_2), \tag{3.31}$$

where we have made use of the fundamental *Tds* relationship for the *isobaric* (i.e., $dP = 0$) heat addition process between states 2 and 3, starting from Equation 3.22 introduced earlier, as follows:

$$dh = Tds + vdP \tag{3.22}$$

$$dh = Tds$$

$$\int_2^3 dh = h_3 - h_2 = \int_2^3 Tds$$

$$h_3 - h_2 = \overline{T}_H \int_2^3 ds = \overline{T}_H (s_3 - s_2)$$

$$c_p (T_3 - T_2) = \overline{T}_H (s_3 - s_2) \qquad (3.32)$$

In Equations 3.31 and 3.32, $\overline{T}_H$ is a *hypothetical* temperature, which is constant during the isobaric heat transfer process between states 2 and 3. The second loss can also be determined using a similar approach; that is,

$$\Lambda_L = \left(\overline{T}_L - T_1\right)(s_4 - s_1) \qquad (3.33)$$

In Equation 3.33, $\overline{T}_L$ is a hypothetical temperature, which is constant during the isobaric heat transfer process between states 4 and 1.

Noting that

$$\Delta s = (s_3 - s_2) = (s_4 - s_1)$$

Equations 3.31 and 3.33 can be rewritten as

$$\Lambda_H = \left(T_3 - \overline{T}_H\right)\Delta s \qquad (3.34)$$

$$\Lambda_L = \left(\overline{T}_L - T_1\right)\Delta s \qquad (3.35)$$

Equation 3.34 is the work output of a *hypothetical* Carnot cycle operating between the two temperature reservoirs at T_3 and $\overline{T}_H$ It is the *lost work* due to the deviation of the *isobaric* heat addition of the air-standard Brayton cycle from the ideal, *isothermal* heat addition process of the Carnot cycle.

Equation 3.35 is the work output of a *hypothetical* Carnot cycle operating between the two temperature reservoirs at $\overline{T}_L$ and T_1. It is the *lost work* due to the deviation of the isobaric heat rejection of the air-standard Brayton cycle from the ideal isothermal heat rejection process of the Carnot cycle.

One interesting takeaway from this exercise is that the air-standard Brayton cycle itself can be construed as a Carnot cycle itself operating between two temperature reservoirs at $\overline{T}_H$ and $\overline{T}_L$ such that its efficiency can be written as

$$\eta_{Brayton} = 1 - \frac{\overline{T}_L}{\overline{T}_H} \qquad (3.36)$$

This is not only true for the Brayton cycle but for any thermodynamic cycle, such as the old standards (e.g., Rankine, Otto, Diesel) as well as the thermodynamic cycles of the emerging technologies, such as supercritical CO_2 (closed-cycle gas turbine). This will be made clear later in the book via many examples.

Before moving on, it should be noted that $\overline{T}_H$ and $\overline{T}_L$ are the Brayton cycle *mean-effective* heat addition and heat rejection temperatures, respectively. Using basic thermodynamic relationships, one can show that $\overline{T}_H$ and $\overline{T}_L$ are logarithmic means of the initial and final temperatures of their respective heat transfer processes. For an ideal gas with constant specific heat, this is indeed a true statement. For water/steam (i.e., a real fluid), the exact formula is, $\overline{T} = \Delta h/\Delta s$, which was rigorously derived earlier. Thus, in general, $\overline{T}$ being equal to the logarithmic mean of temperatures would be conceptually true and numerically close but not exact.

When $\overline{T}_H$ and $\overline{T}_L$ in Equation 3.36 are replaced by logarithmic averages of T_2 and T_3 and T_4 and T_1 defining them, respectively, with a little algebra, it can be shown that Equation 3.30 is obtained. In other words, the air-standard Brayton cycle efficiency is a function of the cycle pressure ratio only. This is a result of the perfect gas assumption with constant c_p. In reality, c_p is a function of temperature and composition, which, when properly accounted for, will lead to the conclusion that the cycle thermal efficiency is a function of cycle pressure ratio and cycle maximum temperature. These are left as exercises for the readers. For help and/or further study, the reader is pointed to the monograph by the author (Ref. [4] in Chapter 1) or the excellent book on gas turbines by Saravanamuttoo et al. (Ref. [2] in Chapter 1).

There are two aspects of the picture painted earlier using the *T-s* diagrams of the Carnot and Brayton cycles, which are manifestations of the 2LT. First, let us recognize that the Brayton cycle is identified by two parameters: cycle pressure ratio, *PR*, and the cycle maximum temperature, T_3. The first thing that the 2LT tells us that the theoretical limit for the efficiency of a heat engine is given by the Carnot efficiency, which is based on T_3 and T_1. The Brayton cycle, even in its ideal form with four reversible processes, falls short of this theoretical limit due to the fact that its heat transfer parts are non-isothermal. However, the 2LT enables us to reformulate the ideal Brayton cycle to a *proxy* which is equivalent to another Carnot cycle but with isothermal heat transfer processes at temperatures different from the two limiting temperatures (i.e., T_3 and T_1). This is the second role of the 2LT.

The following can be shown as *PR* is increased:

- $\overline{T}_H$ gets closer and closer to T_3
- $\overline{T}_L$ gets closer and closer to T_1
- Brayton cycle efficiency gets closer and closer to the Carnot cycle efficiency

But this is small consolation for the design engineer because the work output of the cycle gets closer and closer to zero. To do the actual algebra (rather straightforward) is left as an exercise for the reader. But let us at least to do this much. Why can't we come up with another ideal cycle (e.g., {1-2-3C-4C-1})? This is a cycle with the same isentropic compression as in the Brayton cycle but with true isothermal heat addition {2 → 3C}, isentropic expansion {3C → 4C}, and isothermal heat rejection just like in Carnot cycle.

$$\eta_{Brayton} = 1 - \frac{T_1}{T_2} \tag{3.37}$$

But recalling the isentropic compression relationship between inlet and exit temperatures and pressures; that is,

$$\frac{T_2}{T_1} = \left(\frac{P_2}{P_1}\right)^{\frac{\gamma-1}{\gamma}} = PR^{\frac{\gamma-1}{\gamma}}, \tag{3.38}$$

and substituting it into Equation 3.37, we end up with Equation 3.30. Voila! The two Carnot cycles (i.e., the shaded one bounded by $\overline{T}_H$ and $\overline{T}_L$ and {1-2-3C-4C-1}) are equivalent to each other. In terms of cycle efficiency, that is. In terms of cycle-specific work, the ratio of the second cycle (let us call it cycle B) to the first (let us call it cycle A) is given by

$$\frac{w_B}{w_A} = \frac{(T_2 - T_1)\Delta s}{(\overline{T}_H - \overline{T}_L)\Delta s} = \frac{(T_2 - T_1)}{(\overline{T}_H - \overline{T}_L)}, \tag{3.39}$$

which, for PR = 20:1 and T_3 = 1,500°C, works out to 0.595.

In Section 3.1, we demonstrated how the 2LT sneaks into compression and expansion processes. Let us look at how it does that for a heat transfer process (e.g., Brayton cycle heat addition {2 → 3}). In that case, Equation 3.24 becomes

$$h_3 - h_2 = \int_2^3 vdP + (q_{rev} + i) \tag{3.40}$$

For isobaric heat transfer, the first term on the RHS of Equation 3.40 is zero. If the process is reversible, $i = 0$, and we end up with Equation 3.32, and the 2LT shows itself via Equation 3.20 or 3.21. Now, what if the process is still reversible but not isobaric? In that case, Equation 3.40 can be rewritten as

$$q_{rev} = \int_2^3 Tds = c_p (T_3 - T_2) - \int_2^3 vdP, \text{ or} \tag{3.41}$$

$$\overline{T}_H' \Delta s = c_p \left(T_3 - T_2 \right) - \int_2^3 v dP, \text{ or} \tag{3.42}$$

$$\overline{T}_H' = \frac{c_p \left(T_3 - T_2 \right) - \int_2^3 v dP}{\Delta s} = \frac{c_p \left(T_3 - T_2 \right) - \int_2^3 v dP}{c_p \ln \dfrac{T_3}{T_2} - R \ln \dfrac{P_3}{P_2}}. \tag{3.43}$$

Let us compare $\overline{T}_H'$ with the original $\overline{T}_H$ (given by $(T_3 - T_2)/ln\,(T_3/T_2)$) for the reversible isobaric heat transfer process. Noting that the specific volume ratio, $VR = v_3/v_2$, which can be evaluated from the ideal gas relationship as $VR = T_3 P_2/T_2 P_3$, or $T_3/T_2 = PR \times VR$, we have

$$\overline{T}_H' = \frac{\overline{T}_H - \dfrac{\int_2^3 v dP}{c_p \ln \dfrac{T_3}{T_2}}}{1 - \dfrac{R}{c_p} \dfrac{\ln \dfrac{P_3}{P_2}}{\ln \dfrac{T_3}{T_2}}} = \frac{\overline{T}_H - \dfrac{\int_2^3 v dP}{c_p \ln \left(1 - \varepsilon \right) \left[1 + \dfrac{\ln VR}{\ln \left(1 - \varepsilon \right)} \right]}}{1 - \dfrac{R}{c_p} \left[\dfrac{1}{1 + \dfrac{\ln VR}{\ln \left(1 - \varepsilon \right)}} \right]}. \tag{3.44}$$

where ε represents the pressure loss between state-points 2 and 3, which is usually small (e.g., 5% or 0.05) so that $ln\,(1 - \varepsilon) \sim -\varepsilon$, and noting that $R/c_{p} = (\gamma - 1)/\gamma$, Equation 3.44 becomes

$$\overline{T}_H' = \frac{\overline{T}_H - \dfrac{(\gamma - 1) \int_2^3 T \dfrac{dP}{P}}{\gamma \left(\ln VR - \varepsilon \right)}}{1 + \dfrac{\varepsilon (\gamma - 1)}{\gamma \left(\ln VR - \varepsilon \right)}}. \tag{3.45}$$

With $\gamma = 1.4$ for air, $\varepsilon = 0.05$, and $(ln\,VR - \varepsilon) \sim 1$ (it is left to the reader as an exercise to verify it for $PR = 20 \times (1 - \varepsilon)$ and $T_3 = 1,500°C$), Equation 3.45 simplifies to

$$\overline{T}_H' = \overline{T}_H - \frac{R}{c_p} \int_2^3 T \frac{dP}{P}. \tag{3.46}$$

The second term on the RHS of Equation 3.46 represents the irreversibility associated with the pressure loss during heat transfer, which actually slightly increases $\overline{T}_H$ and reduces the cycle efficiency. Evaluation of that term requires knowledge of $T(P, V)$ during the process. An approximation can be made that it takes place at an average value of $\overline{T}_H$ (which is permissible because the term originates from the change of entropy, Δs); that is,

$$\overline{T}_H' = \overline{T}_H \left(1 - \frac{R}{c_p} \ln \frac{P_3}{P_2} \right) = \overline{T}_H \left(1 - \frac{\Delta s_P}{c_p} \right), \tag{3.47}$$

where Δs_p represents the entropy increase due to the drop in pressure during heat addition. Since $P_3 < P_2$, it is a negative number and indicates an increase in $\overline{T}_H$ For the example herein, it works out to about -0.015 so that is $\overline{T}_H'$ about 1.5% higher than $\overline{T}_H$ (about 17 K). Furthermore, if we assume that cycle heat rejection is isobaric $\{4' \to 1\}$, reduction in cycle efficiency can be shown to be one percentage point. This is one quantification of the *lost work* due to pressure loss during heat addition. (Note that the isentropic expansion takes place across a $PR = 20 \times (1 - \varepsilon)$ so that $T_{4'}$ is slightly higher than T_4.) For a different perspective, the reader should consult pp. 123–125 in Kotas (Ref. [1] in Chapter 1).

3.2.2 HEAT PUMPS

Let us now look at heat pumps (or their counterpart, refrigerators) but using a different approach than we used for heat engines (e.g., gas turbines). The reason for that is that heat pumps or refrigerators have not been standardized to the degree that gas turbines, diesel engines, or even steam turbines have been in the literature (e.g., in academic as well as commercial or trade publications). The following analysis is going to be based on a *reverse Brayton cycle* gas heat pump (or refrigerator), whose *T-s* diagram is presented in Figure 3.5.

As shown in Figure 3.5, there are two key temperatures, T_1, at the inlet of the compressor, and T_3, at the inlet of the turbine. In a practical application, the hot region (i.e., the region that is heated), which is the counterpart of the hot temperature reservoir for a heat engine, is at a temperature $T_H < T_3$. Similarly, the cold region (i.e., the region that is cooled), which is the counterpart of the low temperature reservoir for a heat engine, is at a temperature $T_L > T_1$. If the cycle is intended as a *heat pump*, the objective is heat transfer *to* the hot region, Q_H. If the cycle is intended as a *refrigerator*, the objective is heat transfer *from* the cold region, Q_L. The cost of achieving the said objective is the work input to the cycle (i.e., $W = Q_H - Q_L$). The performance metric is the *coefficient of performance* (COP), which, for Carnot cycle devices, is defined as

$$COP = \frac{Q_H}{Q_H - Q_L} = \frac{T_H}{T_H - T_L} = \frac{1}{1 - \frac{T_L}{T_H}} (\text{heat pump}),$$

$$\tag{3.48a}$$

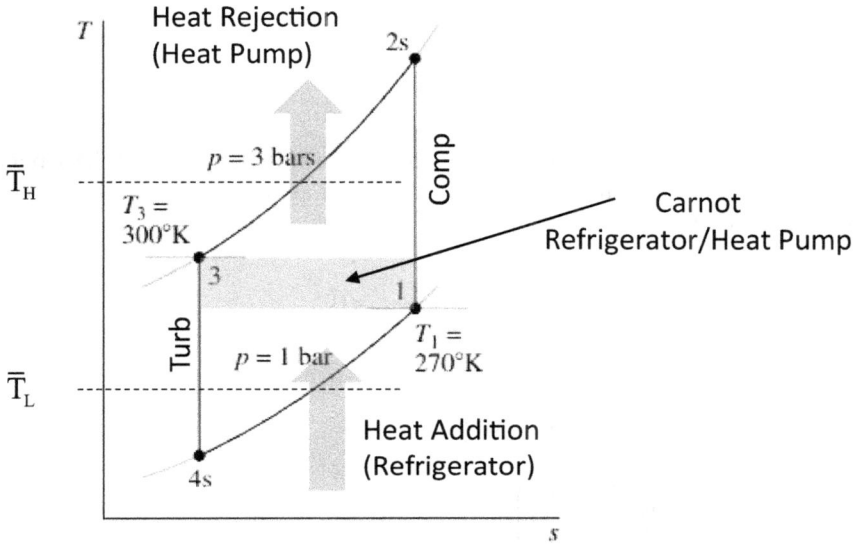

FIGURE 3.5 Reverse Brayton gas compression heat pump (refrigerator) cycle.

$$COP = \frac{Q_L}{Q_H - Q_L} = \frac{T_L}{T_H - T_L} = \frac{1}{\dfrac{T_H}{T_L} - 1}(\text{refrigerator}). \tag{3.48b}$$

For the ideal reverse Brayton cycle devices, in analogy to the heat engines discussed earlier,

$$COP = \frac{1}{1 - \dfrac{\overline{T}_L}{\overline{T}_H}}(\text{heat pump}), \tag{3.49a}$$

$$COP = \frac{1}{\dfrac{\overline{T}_H}{\overline{T}_L} - 1}(\text{refrigerator}), \tag{3.49b}$$

with

$$\overline{T}_H = \frac{T_3 - T_{2s}}{\ln\left(\dfrac{T_3}{T_{2s}}\right)}, \tag{3.50}$$

$$\overline{T}_L = \frac{T_1 - T_{4s}}{\ln\left(\dfrac{T_1}{T_{4s}}\right)}. \tag{3.51}$$

The rest of the discussion assumes that the device in Figure 3.5 is a heat pump. (The reader can do the calculations for a refrigerator as an exercise.) For starters, the Carnot cycle heat pump COP is

$$COP = \frac{1}{1 - \dfrac{270}{300}} = 10.$$

Furthermore, with $k = 1 - 1/\gamma = 0.2857$ for $\gamma = 1.4$,

$$T_{2s} = T_1\left(\frac{P_2}{P_1}\right)^{0.2857} = 270 \cdot 3^{0.2857} = 369.6K,$$

$$T_{4s} = T_3\left(\frac{P_2}{P_1}\right)^{-0.2857} = 300 \cdot 3^{-0.2857} = 219.2K,$$

so that

$$\overline{T}_H = \frac{300 - 369.6}{\ln\left(\dfrac{300}{369.6}\right)} = 333.6K,$$

$$\overline{T}_L = \frac{270 - 219.2}{\ln\left(\dfrac{270}{219.2}\right)} = 243.7K.$$

Thus, COP of the ideal, reverse (air-standard) Brayton cycle heat pump is found as

$$COP = \frac{1}{1 - \dfrac{243.7}{333.6}} = 3.71$$

for a Carnot factor of $CF = 3.71/10 = 0.371$.

Let us repeat the exercise, for an ideal cycle with nitrogen (N_2) as the working fluid. The cycle is modeled in THERMOFLEX with REFPROP property package, and the state-point data is summarized in Table 3.1.

TABLE 3.1

State-Point Data for the Ideal Cycle in Figure 3.5

	P, bara	T, K	h, kJ/kg	s, kJ/kg-K
1	1	270.0	−29.17	−0.0988
2s	3	369.7	74.28	−0.0988
3	3	300.0	1.92	−0.3157
4s	1	219.0	−82.00	−0.3157

TABLE 3.2

State-Point Data for the Cycle in Figure 3.5 with Pressure Losses and Non-ideal Components

	P, bara	T, K	h, kJ/kg	s, kJ/kg-K
1	1.0	270.0	−29.17	−0.0988
2	3.121	387.6	92.94	−0.0612
3	3.06	300.0	1.92	−0.3215
4	1.02	226.0	−74.74	−0.2889

Consequently,

$$\overline{T}_H = \frac{h_3 - h_{2s}}{s_3 - s_{2s}} = \frac{1.92 - 74.28}{-0.3157 - -0.0988} = 333.6K,$$

$$\overline{T}_L = \frac{h_1 - h_{4s}}{s_1 - s_{4s}} = \frac{-29.17 - -82}{-0.0988 - -0.3157} = 243.6K,$$

$$COP = \frac{1}{1 - \dfrac{243.6}{333.6}} = 3.70.$$

In conclusion, using a *real gas* equation of state, REFPROP package in this case, does not change the answer that can be obtained from the simple relationships derived from the *perfect gas* assumption.

Now let us look at an *actual* design with components with 90% polytropic efficiency and pressure losses in the high- and low-pressure legs of the cycle (2% each). State-point data from the THERMOFLEX model is summarized in Table 3.2.

From the data in Table 3.2, one can calculate that $\overline{T}_H = 349.7K, \overline{T}_L = 239.7K, COP = 3.18$ so that the technology factor, $TF = 3.18/3.70 = 0.86$.

3.3 MULTICOMPONENT SYSTEMS

3.3.1 COMBUSTION

In Section 2.3.2, we have calculated the enthalpy of combustion (lower and higher heating values, depending on the state of H_2O in the products) of CH_4. Let us now calculate the reversible work associated with this process. To do this, however, we have to jump the gun and introduce the concept of *exergy* (see Chapter 5). The *physical* exergy quantifies the theoretically possible *maximum* work that can be obtained by bringing the state of the fluid in question from P and T to equilibrium with the surroundings at P_0 and T_0 (the *dead state*) in a *reversible* process. For processes involving pure substances or mixtures with no reaction, the reversible work could be found from the exergy balance for the control volume in question with zero irreversibilities. The same principle applies when there is a reaction is involved with the additional computational work to evaluate the *chemical* exergy.

In this section, we will evaluate two quantities:

- Exergy of a fuel (CH_4 to be specific)
- Maximum reversible work associated with the combustion of methane (CH_4)

It should be pointed out that there are alternative models for evaluation of chemical exergy. The model used herein is taken from the textbook by Moran and Shapiro (Section 13.6 in Ref. [1] in Chapter 2). For an alternative coverage, the reader can consult section 3.4 in Bejan et al. [3]. The most difficult part in calculating the chemical exergy of a fuel is precise specification of exergy reference environments. In the book cited earlier, the authors discuss two alternative standard exergy reference environments, which they refer to as model I and model II (ibid.). (See Section 5.2 for more on this subject.)

The author has to admit that spending a lot of ink on the chemical exergy of fuel has little value for practical engineering applications. One can just say that it is 4 to 5% higher than the fuel LHV and be done with it. Nevertheless, it is an important enough parameter, and its derivation provides a good opportunity to revisit the fundamental thermodynamic considerations involved in the derivation of the important 2LT concept of exergy. Herein, we will only look at the maximum reversible work and entropy generation associated with the combustion of methane.

In order to find the maximum reversible work for the combustion of methane, we reimagine the combustor as a simple reactor, where methane reacts with oxygen to form the products CO_2 and water, $H_2O(l)$. The reaction is simply

$$CH_4 + 2O_2 \rightarrow CO_2 + 2 \cdot H_2O.$$

$$(3.52)$$

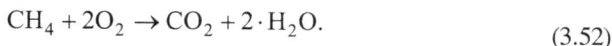

Applying the first and second law equations to the reactor and combining the two, we find the following equation for the reversible work:

$$\overline{w}_{rev} = \left\{ \overline{h}_{CH4} + 2\overline{h}_{O2} - \overline{h}_{CO2} - 2\overline{h}_{H2O(l)} \right\} - T_0 \left\{ \overline{s}_{CH4} + 2\overline{s}_{O2} - \overline{s}_{CO2} - 2\overline{s}_{H2O(l)} \right\},$$

$$(3.53)$$

which, noting that the reactants and products are at T_0 and the enthalpy of formation for O_2 is zero, simplifies to

$$\bar{w}_{rev} = \left\{ \bar{h}^{\circ}_{CH4} - \bar{h}^{\circ}_{CO2} - 2\bar{h}^{\circ}_{H2O(l)} \right\} - T_0 \left\{ \bar{s}_{CH4} + 2\bar{s}_{O2} - \bar{s}_{CO2} - 2\bar{s}_{H2O(l)} \right\}. \quad (3.54)$$

To avoid the tedium associated with calculation of the entropy terms on the RHS of Equation 3.54 (refer to Equation 2.59 for calculating entropy of an ideal gas mixture), we assume that each reactant enters the reactor separately at T_0 and P_0 and each product leaves the reactor separately at T_0 and P_0 as well. This simplifying assumption eliminates the pressure-dependent terms of entropy for each component. Thus, Equation 3.54 becomes

$$\bar{w}_{rev} = \left\{ \bar{h}^{\circ}_{CH4} - \bar{h}^{\circ}_{CO2} - 2\bar{h}^{\circ}_{H2O(l)} \right\} - T_0 \left\{ \bar{s}^{\circ}_{CH4} + 2\bar{s}^{\circ}_{O2} - \bar{s}^{\circ}_{CO2} - 2\bar{s}^{\circ}_{H2O(l)} \right\}. \quad (3.55)$$

The first term in curly brackets on the RHS of Equation 3.55 is the higher heating value (HHV) of methane so that the equation can be written as

$$\bar{w}_{rev} = HHV - T_0 \left\{ \bar{s}^{\circ}_{CH4} + 2\bar{s}^{\circ}_{O2} - \bar{s}^{\circ}_{CO2} - 2\bar{s}^{\circ}_{H2O(l)} \right\}. \quad (3.56)$$

Furthermore, recalling that the Gibbs free energy (GFE) is defined as $\bar{g} = \bar{h} - T\bar{s}$ (see Section 2.3.2.1), Equation 3.53 can also be written as

$$\bar{w}_{rev} = \bar{g}_{CH4} + 2\bar{g}_{O2} - \bar{g}_{CO2} - 2\bar{g}_{H2O(l)} = -\Delta G. \quad (3.57)$$

Using the expansion of GFE given by Equation 2.60 and since $T_0 = T_{ref} = 25°C = 298.15$ K and $P_0 = P_{ref} = 1$ atm, it can be easily shown that Equation 3.57 reduces to (noting that GFE of formation for O_2 is 0)

$$\bar{w}_{rev} = \bar{g}^{\circ}_{f} - \bar{g}^{\circ}_{CO2} - 2\bar{g}^{\circ}_{H2O(l)}. \quad (3.58)$$

Using the values listed in Table 14.7 in Chapter 14, maximum reversible work associated with combustion is calculated as

$\bar{w}_{rev} = -50,790 - (-394,380 + 2 \times -237,180) = -50,790 + 868,740 = 817,950$ kJ/kmol, or

$\bar{w}_{rev} = 817,950/16.04 = 50,994$ kJ/kg.

Entropy generation during the combustion process can be calculated as follows:

$$\dot{S}_{gen} = \sum_{R} \dot{n}_i \bar{s}_i - \sum_{P} \dot{n}_j \bar{s}_j, \quad (3.59)$$

so that the combustion irreversibility (adiabatic process with no work interaction between the system, i.e., the combustor/reactor, and its surroundings) is given by

$$\dot{I} = T_0 \dot{S}_{gen} = T_0 \left(\sum_R \dot{n}_i \bar{s}_i - \sum_P \dot{n}_j \bar{s}_j \right).$$

(3.60)

For the combustion example in Section 2.3.2, which was based on the gas turbine combustor shown in Figure 2.2, combustor irreversibility can be calculated using the data provided in Section 14.1.1. The Excel calculation block is displayed in Table 3.3. Ideal gas entropy values are calculated from the tabulated data using interpolation.

From Equation 3.45, entropy generation is found as

$$\dot{S}_{gen} = 4,055 - 3,479 = 576 \text{ kJ / s, so that}$$

$$\dot{I} = T_0 \dot{S}_{gen} = 298.15 \, 576 = 171,722 \text{ kJ / s } (kW).$$

Fuel energy input to the combustor (LHV) is calculated as

$$\dot{Q}_{fuel} = \dot{m}_{fuel} LHV = 13.52 \text{ kg / s } 50,019 \text{ kJ / kg } = 676,197 \text{ kJ / s } (kW),$$

so that 171,722/676,197 = 0.254 or 25.4%; that is, one-fourth of the fuel input to the gas turbine is lost due to combustion irreversibility. There is simply nothing that one can do to alleviate this situation.

In the calculations, we have ignored the contribution of pressure to entropy for simplicity. In order to illustrate this, we expand Equation 3.45 as follows (making use of Equation 2.59)

TABLE 3.3
Combustion Entropy Generation Calculation

Reactants	K T	kJ/kmol-K $\bar{s}$	kmol/s $\dot{n}$	kJ/K-s $\dot{n}\bar{s}$
CH_4	499.8	199.67	0.843	168
N_2	716.7	217.45	11.861	2,579
O_2	716.7	232.10	3.153	732
Total			15.857	3,479

Products	K T	kJ/kmol-K $\bar{s}$	kmol/s $\dot{n}$	kJ/s $\dot{n}\bar{s}$
H_2O	1,873.2	261.67	1.686	441
N_2	1,873.2	249.99	11.861	2,965
O_2	1,873.2	266.58	1.467	391
CO_2	1,873.2	305.90	0.843	258
Total			15.857	4,055

$$\dot{S}_{gen} = \sum_R \dot{n}_i \left(\overline{s}_i^\circ - R_{unv} \ln y_i P_R \right) - \sum_P \dot{n}_j \left(\overline{s}_j^\circ - R_{unv} \ln y_j P_P \right).$$ (3.47)

In Equation 3.47, $\overline{s}_{i,j}^\circ$ is the temperature-dependent part of the ideal gas entropy for components i and j, R_{unv} = 8.314 kJ/kmol-K is the universal gas constant, $y_{i,j}$ is the mole fraction of components i and j in their respective mixtures, and P represents the total pressure in atm (of reactants and products). Note that Equation 3.47 is written with the implicit assumption of that the reference pressure is 1 atm (1.01325 bar). Otherwise, the correct version of the pressure-dependent part is $R_{unv} \ln \left(y_{i,j} \dfrac{P_{R,P}}{P_{ref}} \right)$.

For the example earlier, contribution of the pressure-dependent terms to the entropy balance is calculated using the Excel block shown in Table 3.4. Mole fractions of components are calculated from molar flow rates (e.g., $y_i = \dfrac{\dot{n}_i}{\dot{n}_{tot}}$). Pressure of reactants and products are assumed to be the same (i.e., 21.89 bar), as shown in Figure 2.2.

From the numbers in Table 3.4, contribution of the pressure-dependent terms to entropy generation in the combustor is calculated as follows:

$$(-297) - (-315) = 18 \text{ kJ} / \text{k} - \text{s},$$

and translates to a contribution of 298.15 × 18 = 5,396 kJ/s (kW) to the combustor irreversibility. This corresponds to an increase of 5,396/172,341 = 0.0313 or 3.1%. While not insignificant per se, this contribution does not represent a meaningful

TABLE 3.4

Contribution Pressure-Dependent Terms to the Entropy of Reactants and Products

Reactants	K T	– y_i	kJ/K-s $\dot{n}_i R_{unv} \ln y_i P_R$
CH_4	499.8	0.053	0.97
N_2	716.7	0.749	275.75
O_2	716.7	0.198	38.11
Total		1.000	315
Products	K T	– y_j	kJ/K-s $\dot{n}_j R_{unv} \ln y_j P_P$
H_2O	1,873.2	0.106	11.69
N_2	1,873.2	0.749	275.75
O_2	1,873.2	0.092	8.32
CO_2	1,873.2	0.053	0.97
Total		1.000	297

change in the big picture; that is, that nearly one-fourth of fuel energy input is irretrievably lost during the combustion process.

Another small detail that must be mentioned is that we ignored the contribution of fuel pressure, which is 30.64 bar at the combustor inlet. (Typically, fuel pressure has to be higher than the pressure inside the combustor by 40 to 60% – or even higher – so that it can be injected through the fuel nozzles.) This increases the contribution of the pressure-dependent terms to the combustor irreversibility 6,101 kW or 3.5%.

In closure, it is quite clear that slogging through the tedious calculations (even in the easiest possible manner with many simplifying assumptions as illustrated earlier) to do a painstakingly detailed exergy analysis of a gas turbine is not a particularly fruitful endeavor. Everything else in the system (i.e., non-isentropic compression and expansion, mixing of coolant and hot gas streams in the turbine, etc.) pale in comparison to the exergy destruction (irreversibility) in the combustor, about which nothing can be done.

3.3.2 TURBINE STAGE

At the end of the last section, we made a strong assertion. Let us prove it. Continuing with the combustion example, let us look at the first turbine stage downstream of the combustor. The turbine stage model in THERMOFLEX is shown in Figure 3.6.

A detailed investigation of a turbine stage is an extremely tedious undertaking. One has to follow not only the main hot gas flow through the stator (comprising the nozzle vanes) and rotor (comprising the buckets or blades) but also the secondary flows through the components, rotor wheels (discs), turbine rings, and a multitude of nooks and crannies to cool the entire hot gas path. For an in-depth analysis of the turbine stage, the reader is referred to the monograph by Sultanian [1]. For cooled

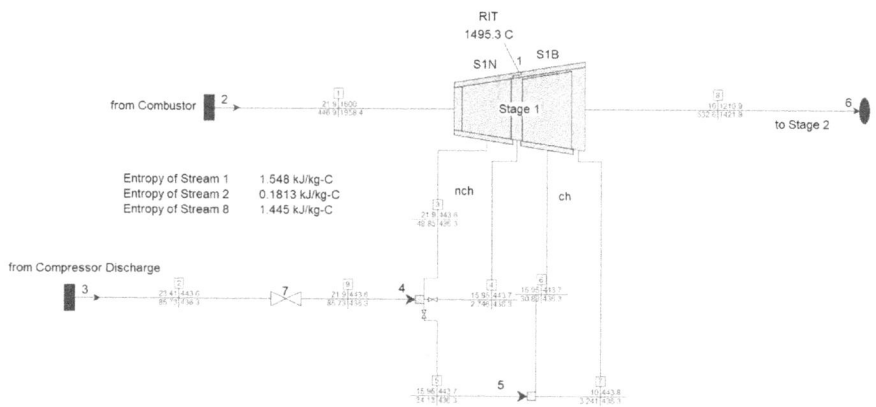

FIGURE 3.6 Cooled turbine stage (S1N: Stage 1 nozzle vanes, S1B: Stage 1 rotor blades). Nonchargeable and chargeable cooling flows are denoted by nch and ch, respectively. State properties in clockwise direction from the upper-left corner are P in bar, T in °C, flow in kg/s, and h in kJ/kg-K.

turbine stage analysis, a series of papers by the Cambridge University researchers are recommended [2–3]. For our purposes herein, analysis of the problem is much simpler. For starters, recognize that there are two intertwined mechanisms at work in a cooled turbine stage:

• Uncooled expansion with aerodynamic losses
• Cooled expansion with heat transfer and mixing of cooling air with hot gas flow

Uncooled expansion losses are quantified by the *isentropic efficiency* of the expansion process across the turbine stage; that is, from the inlet of the stator to the exit of the rotor. A detailed analysis of loss/entropy generation mechanisms inside the stage via heat transfer (between the flow streams and the components) and mixing of flow streams is extremely complicated. However, just by looking at the inlet (hot gas from the combustor) and exit (hot gas diluted by stage cooling airflows) states can give us a good idea about the overall magnitude of the losses. In order to do this investigation, the stage model is run with uncooled stage efficiencies ranging from 80 to 100%. The results are plotted in Figure 3.7.

As shown in Figure 3.7, when the uncooled stage efficiency is 100%, all losses can be attributed to the heat transfer and mixing losses associated with the coolant flow. This comes to about 16 MW or 10% of the stage shaft work output. Modern gas turbine technology is probably somewhere between 90 and 95% uncooled (isentropic) efficiency, which corresponds to about 18–19 MW of irreversibility (11–13% of stage shaft work output). The key takeaway is that the entire range of losses in Figure 3.7 corresponds to 2.3 to 3.4% of fuel LHV (or exergy) input to the combustor, which pales next to the combustor irreversibility (about 25% as calculated in the preceding

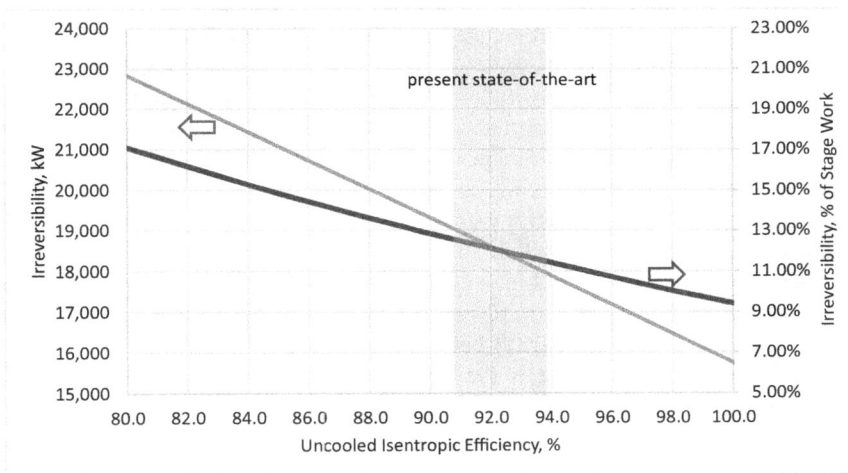

FIGURE 3.7   Cooled turbine stage irreversibility.

section). Considering that any improvements in stage performance would be unlikely to be more than one-tenth of that range, it becomes obvious that the exergy analysis of a gas turbine to identify the proverbial sweet spots of improvement is not a particularly fruitful exercise. Just do one's best and be done.

3.4 ENTROPY

Let us take a closer look at the mysterious quantity called *entropy* that we used in our sample calculations earlier. Specifically, let us look at the data in Table 3.2 and consider the heat transfer from the working fluid from state-point 2 to state-point 3. In particular,

- At state-point 2, entropy, $s_2 = -0.0612$ kJ/kg-K
- At state-point 3, entropy, $s_3 = -0.3215$ kJ/kg-K, so that
- Entropy in this process *decreases* (i.e., $s_3 < s_2$)

So much for the oft-made assertion that entropy always increases or, at best, stays the same. Depending on the process one is looking at, entropy can certainly *decrease*.

What about another commonly used explanation that entropy is "a measure of disorder"? How can one say with any scientific certainty that the gas at state-point 3 is at a state of *less* disorder than the gas at state-point 2 (or *vice versa*)? What is the measure of *order* in a system comprising billions and billions of gas molecules bouncing around at a microscopic level in either state? We must discard this one as well as a misconception.

The least convincing of all misconceptions is the one about the connection between entropy and the so-called arrow of time. As Stephen Hawking stated in his popular book *A Brief History of Time*, "[t]he increase of disorder or entropy is what distinguishes the past from the future, giving a direction to time." In the heat transfer process between state-points 2 and 3, entropy decreases. There are no ifs, ands, or buts about it. What should one conclude then? Does time flow *backward* in that process, say, unlike the process between state-points 4 and 1, where entropy *increases* and time flows *forward*? Clearly, there is something not quite right here. Since the author does not carry the intellectual weight to refute a statement made by a giant of science, all he can say is that Professor Hawking must be talking about another entropy or a highly specific case, which, of course, he was.

The Bekenstein-Hawking entropy or black hole entropy is the amount of entropy that must be assigned to a black hole in order for it to reconcile the laws of thermodynamics with the existence of black hole event horizons. It is a part of the area of study known as *black hole thermodynamics*. It must be stated that there are criticisms of the analogy (or similarity) between the properties of black holes and the laws of thermodynamics. In fact, in the beginning, even Hawking – and others – had stressed that the laws of black holes only looked like thermodynamics on paper; they did not actually relate to thermodynamic concepts like temperature or entropy.[2] This changed after two landmark findings by Bekenstein (1972) and Hawking (1974), with the latter stating that radiation emitted by the black holes (a discovery in and of itself) would have exactly the same temperature in the thermodynamic analogy.

Finally, what on earth does it really mean (if anything at all) that the value of entropy at state-point 2 has a *negative* value (i.e., −0.0612 kJ/kg-K)? Is there an entropy meter that can measure it and tell us this value (i.e., like a thermometer that measures 387.6 K at the same state-point)? (This is a *rhetorical* question; there is, of course, no such device.) What we can calculate (and measure, albeit indirectly) is the *change* in entropy between states 2 and 3; that is, $\Delta s = s_3 - s_2 = -0.3215 - (-0.0612) = -0.2603$ kJ/kg-K (i.e., entropy decreases). Let us explore this thought a little more.

In Section 3.1, combining the first and second laws of thermodynamics, we obtained the following *Tds* equation:

$$dh = Tds + vdP \qquad (3.22)$$

Let us assume that the gas in question obeys the ideal gas law so that the equation of state is $Pv = RT$. Furthermore, by the definition of specific heat at constant pressure, we know that $c_p = dh/dT$ (no *partial* derivative because ideal gas enthalpy is a function of temperature only, i.e., $h = f(T)$). Equation 3.22 is an exact differential; that is, *the partial derivative of enthalpy with respect to entropy while holding pressure constant* is given by

$$\left(\frac{\partial h}{\partial s}\right)_P = T \rightarrow \left(\frac{\partial s}{\partial h}\right)_P = \frac{1}{T}, \qquad (3.41a)$$

$$\left(\frac{\partial s}{\partial h}\right)_P = \left(\frac{\partial s}{\partial T}\right)_P \frac{dT}{dh} = \left(\frac{\partial s}{\partial T}\right)_P \frac{1}{c_p} = \frac{1}{T}. \qquad (3.41b)$$

Thus, integrating Equation 3.41b, we can calculate the change in entropy between state-points 2 and 3 while (i) holding pressure constant and (ii) assuming that c_p is constant (i.e., *perfect* gas) as follows:

$$\left(s_3 - s_2\right)_P = \int_2^3 \frac{c_p}{T} dT = c_p \ln\left(\frac{T_3}{T_2}\right). \qquad (3.42)$$

(Note that the assumption of constant specific heat is not necessary at all. It just simplifies the formulae.)

Now let us rewrite Equation 3.22 as follows

$$ds = \frac{1}{T} dh - \frac{v}{T} dP, \qquad (3.43)$$

which tells us that (since, for an ideal gas, constant h is equivalent to constant T)

$$\left(\frac{\partial s}{\partial P}\right)_h = \left(\frac{\partial s}{\partial P}\right)_T = -\frac{v}{T} = -\frac{R}{P}. \qquad (3.44)$$

Integrating Equation 3.44, the result is

$$\left(s_3 - s_2\right)_T = \int_2^3 -\frac{R}{P}\, dP = -R\ln\left(\frac{P_3}{P_2}\right). \tag{3.45}$$

Combining Equations 3.42 and 3.45, the change in entropy between state-points 2 and 3 is given by

$$s_3 - s_2 = c_p \ln\left(\frac{T_3}{T_2}\right) - R\ln\left(\frac{P_3}{P_2}\right). \tag{3.46}$$

Let us assume that c_p = 1.040 kJ/kg-K (a readily measurable parameter) and cal-culate that R = 8.314/28 = 0.297 kJ/kg-K (with R_{unv} = 8.314 kJ/kg-K). Then from Equation 3.46, using the data in Table 3.2, we calculate that

$$s_3 - s_2 = 1.040\ln\left(\frac{300}{387.6}\right) - 0.297\ln\left(\frac{3.06}{3.121}\right) = -0.2606\,\text{kJ/kg-K},$$

which is almost exactly equal to the value obtained using the data in Table 3.2 (from a bona fide equation of state, i.e., REFPROP package).

It is important to recognize that Equation 3.46, enabling us to calculate the change in entropy between two *equilibrium* states (which is the only thing of primary inter-est in engineering calculations), is exclusively derived from a macroscopic analysis of first principles without resorting to any *pseudo*-philosophical and *pseudo*-scientific sleight of hand. It does not tell us anything about system order or disorder, arrow of time, or any such nebulous concepts. To be perfectly frank, it does not tell us anything of practical value, à la temperature, pressure, or volume (density), at all. Its most significant contribution to the engineering analysis is quantification of lost work in work machines as discussed earlier (measured by the concept or key performance indicator (KPI) of *isentropic efficiency*).

Furthermore, the absolute value of entropy (e.g., s_3 = −0.3215 kJ/kg-K) is meaning-less in and of itself. It is true that the *third law of thermodynamics* assigns a value of zero to entropy at 0 K (see Section 2.2), but it is not of any practical interest to engi-neers. Numerically, one can assign a *reference value* (e.g., 0 kJ/kg-K) to the entropy of a fluid at an (arbitrarily) specified temperature and pressure, say, 200 K (a very low temperature – only possible in cryogenic systems) and 1 mbar (almost complete vac-uum). Thus, using Equation 3.46, with T in K and P in bara, the following formula,

$$s(P,T) = c_p \ln\left(\frac{T}{200}\right) - R\ln\left(\frac{P}{0.001}\right), \tag{3.47}$$

returns the entropy of an ideal gas in kJ/kg-K. Entropy values calculated with Equation 3.47 for state-points in Figure 3.5 are listed in Table 3.5 along with the

TABLE 3.5

Entropy Calculation for State-Points in Figure 3.5

State-Point	P, bara	T, K	s, kJ/kg-K (REFPROP)	s, kJ/kg-K Equation 3.47
1	1.0	270.0	−0.0988	−1.7395
2	3.121	387.6	−0.0612	−1.7015
3	3.06	300.0	−0.3215	−1.9621
4	1.02	226.0	−0.2889	−1.9304

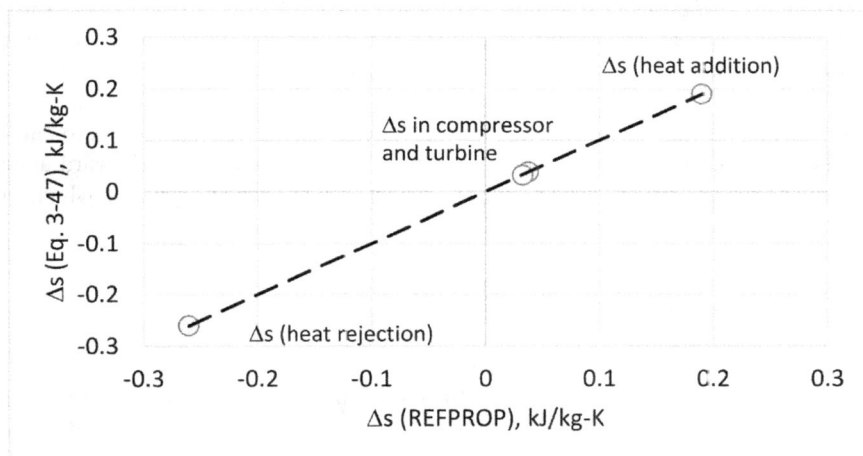

FIGURE 3.8 Entropy change comparison.

entropy values in Table 3.2. Entropy changes in reverse Brayton cycle processes in Figure 3.5, as calculated by each method, are plotted in Figure 3.8. Clearly, while significantly different in magnitude, both methods return negative values for entropy of each state-point, equally meaningless on their own. However, the agreement in entropy deltas from each method is perfect for all intents and purposes, and after all, this is what really counts.

Another way to write Equation 3.47 is as follows

$$s(P,T) = c_p \ln(T) - R \ln(P) + s_0, \tag{3.48}$$

where the s_0 is the (additive) *entropy constant* (in this case, of an ideal gas). By choosing an arbitrary reference state as discussed earlier, the implied value of the entropy constant is

$$s_0 = c_p \ln(T_0) - R \ln(P_0), \tag{3.49}$$

whose exact value is a function of units of temperature and pressure. However, it is important to realize that one can assign any arbitrary numerical value to s_0. (We will just do that later.)

We cannot apply the third law (i.e., $s = 0$ at $T = 0$ K, also known as *Nernst's theorem*) directly in this case, either, because $ln(T \to 0) \to -\infty$. This finding can be traced to a failure of the assumption of *constant c_p* (perfect gas) at such low temperatures rather than to a failure of the third law, which states that the entropy of a *perfect crystal of a pure substance* approaches zero as the temperature approaches zero. In other words, Equation 3.47 or 3.48, specifically derived for a *perfect gas* (i.e., an *ideal gas* with constant c_p) does not really apply at very low temperatures, especially about $T = 0$ K. (In fact, all *real* gases condense at sufficiently low temperatures.)

Quantum mechanics brings more clarity to this problem by showing that, for an ideal gas, modeled as comprising molecules as point masses with no intermolecular forces, specific heat is *not* constant, decreasing with decreasing temperature such that $c_p(T \to 0) \to 0$. This ensures that the term $c_p(T \to 0) \times ln(T \to 0)$ does not diverge.

Statistical mechanics (or rather, statistical thermodynamics) enables us to calculate the value of entropy as a function T and P along with a straightforward application of the third law. The simplest case is that of a monatomic gas, for which the entropy in cal/mol-K is given by the *Sackur-Tetrode* equation [4],

$$s(P,T) = R_{unv}\left(\frac{3}{2}\ln(MW) + \frac{5}{2}\ln(T) - \ln(P)\right) - 2.3141, \tag{3.50}$$

where R_{unv} = 1.98718 cal/mol-K, MW is the molecular weight in g/mol, T is in K, and p is in atm (1 calorie (cal) = 4.184 J). The additive constant on the right-hand side of Equation 3.50 is a combination of R_{unv}; Planck's constant, $\hbar$ (6.62607004 × 10^{-34} m²/kg-s); and Avogadro's number, N_A. Equation 3.50 and Equation 3.48 (with $s_0 = 2.25$) are plotted in Figure 3.9 ($MW = 28.014$ g/mol for nitrogen, $P = 3$ atm). Also included

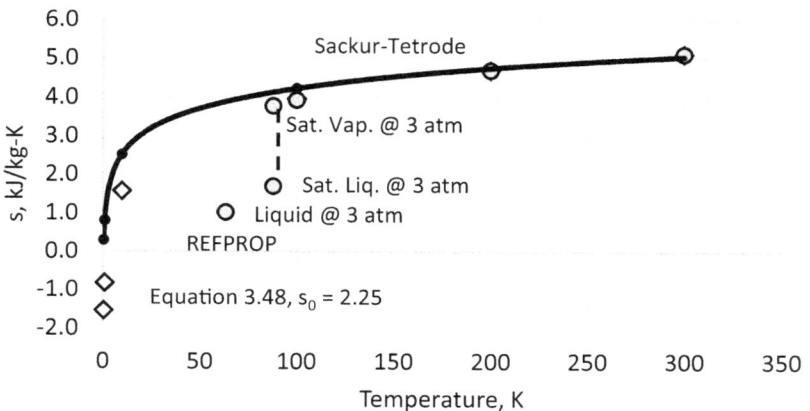

FIGURE 3.9 Ideal gas entropy comparison ($P = 3$ atm)

in the graph are entropy values for N_2 from the REFPROP package with an additive constant of 5.43 to bring it to the same basis as the other two formulations. Note that, at 3 atm pressure, nitrogen starts condensing at 88 K and the REFPROP package's lowest applicability limit is 63.35 K.

Let us now consider the entropy change between liquid N_2 at 63.35 K (3 atm) and gaseous N_2 at 300 K, which has three components; that is,

$$\Delta s = \Delta s_A + \Delta s_B + \Delta s_C,$$

with

$$\Delta s_A = c \ln\left(\frac{T_{sat}}{T_1}\right), \tag{3.51}$$

$$\Delta s_B = \frac{L(T_{sat})}{T_{sat}}, \text{ and} \tag{3.52}$$

$$\Delta s_C = c_p \ln\left(\frac{T_2}{T_{sat}}\right), \tag{3.53}$$

where $c = 2.040$ kJ/kg-K is the specific heat of liquid N_2, $L = 184$ kJ/kg is the latent heat of evaporation of N_2 at $T_{sat}(P = 3 \text{ atm}) = 88$ K, $T_1 = 63.35$ K, and $T_2 = 300$ K. Substituting these values into Equations 3.51–3.53 (with $c_p = 1.040$ kJ/kg-K), we obtain 0.67, 2.09, and 1.28 kJ/kg-K for Δs_A, Δs_B, and Δs_C, respectively, for total $\Delta s = 4.04$ kJ/kg-K. This result is very close to the "correct" value from REFPROP, 4.08 kJ/kg-K.

Finally, it should be noted that the Sackur-Tetrode equation can also be written as

$$s(V,T) = R_{unv}\left(\frac{3}{2}\ln(MW) + \frac{3}{2}\ln(T) + \ln(V)\right) + 2.6546, \tag{3.54}$$

with volume, V, in liters. Its counterpart is given by

$$s(V,T) = c_v \ln(T) + R\ln(V) + s_0', \tag{3.55}$$

with specific heat, $c_v = du/dT$ (no *partial* derivative because ideal gas internal energy is a function of temperature only, i.e., $u = f(T)$), and can be derived from the *Gibbs equation*, $du = Tds - Pdv$, by rewriting it first as

$$ds = \frac{1}{T}du + \frac{P}{T}dv. \tag{3.56}$$

The derivation is left as an exercise for the reader.

3.5 RECAP

Let us recap our learnings and, while doing that, separate the proverbial wheat from the chaff. First, let us lay down the law:

- The second law of thermodynamics has no single equation defining it. There is no single sentence or short paragraph that definitively lays it down. It can be summed up by two *negative* statements:
  - *Kelvin-Planck (K-P) statement*, which simply says that there is no perfect heat engine (i.e., 100% efficiency)
  - *Clausius statement*, which simply says that there is no perfect refrigerator (i.e., infinite *coefficient of performance*, COP)
- In mathematical logic, negative statements cannot be proven *formally*. They can only be *disproven* by finding conclusive evidence that negates the original statement (e.g., a heat engine that indeed converts heat, Q_H, into work, $W = Q_H$, without rejecting any heat to the surroundings). Thus, these two statements (and the second law) should be accepted as given. In other words, there is no point to asking why. They are true until there is an experiment that definitively proves otherwise. As of today, as far as this author can tell, on our planet, the Earth, no one has successfully conducted such an experiment.
- For practical heat engine analysis, the corollary of the K-P statement that provides us with an equation that we can use is the *Carnot cycle*. As demonstrated earlier, there can be no heat engine that operates in a thermodynamic cycle between two temperature reservoirs and has an efficiency higher than that of a Carnot engine operating in a Carnot cycle between the same two temperature reservoirs. The efficiency in question is given by Equation 3.2.
- Entropy as a thermodynamic property is inferred from the analysis of a reversible cyclic process taking place between two temperature reservoirs.
- Its change is defined by the generalized *Clausius inequality*, Equation 3.15b, of which a special case, for a cyclic process, is the *Clausius theorem*, Equation 3.18.
- The entropy formulation of the 2LT is given by Equation 3.14 and repeated subsequently for convenience

$$\Delta S \geq 0,$$

which is universally recognized as *the* second law, is a *special case* that applies to an *isolated* or (E, V, N) system (see Chapter 4). This is the *entropy postulate* of 2LT and can be verbalized as follows:

There is an extensive property of a system called entropy, S, and for an isolated system, S is either constant (reversible process) or increases (irreversible process) but can never decrease.

This is the wheat. Let us get rid of the chaff.

- Entropy is defined only for an *equilibrium* state. By itself, entropy does not tend to increase. It can increase or decrease depending on the system and/or process. If there is no process imposed on the system, entropy of the system does not change on its own spontaneously.
- Entropy of the universe is not defined. *Entropy of the universe* is as meaningless a term as *temperature of the universe*. If one uses the term *universe* very loosely for the ultimate isolated system (after all, what exactly is the *surroundings of the universe*?), that is one thing, but it is a useless construction that cannot and does not serve any useful purpose. The correct statement can be encapsulated by the *entropy postulate* as follows:

$$\left(\Delta S \big|_{\text{System}} + \Delta S \big|_{\text{Surroundings}} \right) \geq 0$$

- The entropy has nothing to do with time. There was no reference made to time as a variable or time derivatives of other variables in the earlier discussion. Granted, irreversibility and reversibility have *vague* connotations of passage of time (e.g., one cannot turn the clock back and the arrow of time points only in one direction). However, this is a misconception that will be explored in more detail in Chapter 4 when we dabble in *statistical mechanics*.
- There is the case of the *black hole thermodynamics* (briefly mentioned earlier), which originated in parallels between the area of a black hole's event horizon and the thermodynamic entropy established by Bekenstein and Hawking in the early 1970s. This led to the formulation of *analogies* to laws of classical thermodynamics for the black holes. The arrow of time characteristic of entropy, alluded to by Dr. Hawking, can most likely be traced to the black hole thermodynamics, which has its detractors as well as *zealous* supporters. In that sense, one should also point to other analogies between thermodynamics and other fields of science. One example is in the field of electricity. If one equates electrical current and voltage to heat flow and temperature difference, respectively, definitions of electrical charge and power can be shown to be analogous to heat and thermodynamic power, respectively. Consequently, Kirchhoff's laws, for instance, can be seen to have formal features similar to the laws of thermodynamics. (Another, even shakier, example is thermo-economics, which seems to finally have gone out of fashion.) However, they do not go beyond mere similarities and do not point to universal thermodynamic principles at work.
- Entropy of an *isolated* system does *NOT* always increase until it reaches a *maximum*. As the inequality in Equation 3.14 suggests, there are two possibilities, and one of them is that nothing will happen. At any given point when the system is in *equilibrium* (remember that, otherwise, entropy could not be defined), its entropy is already at a *maximum*.
- Entropy has *something* to do with randomness or disorder, but it is not a *measure* of them per se. Again, it is worth noting that there was no reference

made to *order* or *disorder* in the earlier discussion. Granted, this was an exercise in *macroscopic* thermodynamics. When one looks at the system at a *microscopic* level when billions of gas molecules dart here and there frantically, one may get the feeling that this may change (yes, we will discuss Boltzmann's famous equation later in the book). This, however, is not going to be the case as it will be made clear (hopefully) in Chapter 4.

In closing this chapter, the author would like to recommend a book that was an inspiration for him to write the present book: *Entropy – The Greatest Blunder in the History of Science* by Arieh Ben-Naim [5]. As Dr. Ben-Naim states in the Preface, "[his] book tells the incredible story of a simple, well-defined quantity, having well-defined limits of applicability, which had evolved to reach monstrous proportions, embracing and controlling everything that happens and everything that is unexplainable." The author could not have said it better.

NOTES

1 The other one is the famous Gibbs equation: $du = Tds - Pdv$ (Equation 2.1).
2 www.quantamagazine.org/craig-callender-are-we-all-wrong-about-black-holes-20190905/ (retrieved on October 6, 2022).

REFERENCES

1. Sultanian, B.K., *Gas Turbines – Internal Flow Systems Modeling*, 1st Edition, Cambridge: Cambridge University Press, 2018.
2. Young, J.B., and Horlock, J.H., Defining the Efficiency of a Cooled Turbine, *International Journal of Turbomachinery*, 128, pp. 658–667, 2006.
3. Young, J.B., and Wilcock, R.C., Modelling the Air-Cooled Gas Turbine: Part 1, General Thermodynamics; Part 2, Coolant Flows and Losses, *International Journal of Turbomachinery*, 124, pp. 207–213, 2002.
4. Hirschfelder, J.O., Curtiss, C.F., and Bird, R.B., *Molecular Theory of Gases and Liquids*, New York: John Wiley & Sons, Inc., 1964.
5. Ben-Naim, A., *Entropy—The Greatest Blunder in the History of Science*, eBookPro Publishing, 2020. www.ebook-pro.com.

4 Second Law – Microscopic Perspective

Ludwig Boltzmann, who spent much of his life studying statistical mechanics, died in 1906, by his own hand. Paul Ehrenfest, carrying on his work, died similarly in 1933. Now it is our turn to study statistical mechanics. Perhaps it will be wise to approach the subject cautiously.

—David L. Goodstein, *States of Matter*

In Chapter 3, we looked at the second law of thermodynamics (2LT) from a macroscopic system perspective. Starting from one of the two key statements of the 2LT (i.e., the Kelvin-Planck statement), we arrived at the Carnot corollary, which, for all practical purposes, is the 2LT for thermal engineering, specifically with a focus on the heat engines; that is, gas turbines, steam turbines, and others, and basic heat pumps (refrigerators).

The Carnot corollary led us to the Carnot efficiency and, from there, to the two important metrics (i.e., Carnot and technology factors), which enable us to assess the worthiness of any heat engine cycle and state of the art in engine technology based on a given thermodynamic cycle. (The *Clausius* statement sets the bar for heat pumps and refrigerators via the *Carnot COP* – see Section 3.2.2.)

At this point, the reader may have already realized that we did *not* need entropy to define the 2LT via Kelvin-Planck and Clausius statements. Strictly speaking, even the ideal Carnot cycle construct does not require a priori definition of entropy. It is defined as a reversible cycle comprising two adiabatic and two isothermal processes and can be depicted on a *pressure-volume* (*P-V*) diagram [1]. In fact, in Chapter 3, it was shown that entropy emerged as a thermodynamic property from the Carnot cycle following a logical deduction process that one can describe as the duck test; that is, *if it looks like a duck, swims like a duck, and quacks like a duck, then it probably is a duck.* To be precise, what really emerged from the said deductive reasoning was not entropy itself per se but the *change in a property* that we called entropy. This finding led to the conclusion that a reversible *and* adiabatic process had to be *isentropic* ($\Delta S = 0$), which is, of course, the basis of the most common embodiment of the Carnot cycle for heat engine/pump analysis: two isentropic and two isothermal processes depicted on a *temperature-entropy* (*T-s*) diagram.

Conversely, the 2LT is not necessary to define entropy, either. To demonstrate the validity of this assertion, however, we must look at the behavior of atoms and molecules forming the system rather than the (macroscopic) system itself. There are three levels at which one can look at the behavior of thermodynamic systems at a microscopic level:

DOI: 10.1201/9781003247418-5

1. Kinetic theory of gases (difficult)
2. Classical statistical mechanics (quite difficult)
3. Quantum statistical mechanics (very difficult)

The system under consideration in kinetic theory of gases is a *dilute gas* of N molecules in a container (box) of volume V. The temperature of the gas in the box is a measure of the kinetic energy of the molecules. The pressure of the gas in the box is a measure of the force (momentum) imparted on the walls of the box via molecular collisions. The mass density of the gas in the box is proportional to the ratio N/V. These are all easy-to-understand concepts. In other words, it is not that difficult to have a *mental picture* of a huge number of N microscopic billiard balls moving around randomly at high speeds in a box of volume V and colliding with each other and the walls of the box. If the box is *heated*, energy transfer *to* the billiard balls manifests itself in *higher* speeds and *increased* number of collisions. If the box is *cooled*, energy transfer *from* the billiard balls manifests itself in *lower* speeds and *reduced* number of collisions. This, in turn, manifests itself in higher or lower temperature and pressure (i.e., properties that we can readily measure). (For calculation of pressure using kinetic theory, for example, the reader is referred to the treatise by Pauli [1].)

It is, however, not so easy to form a mental picture of entropy. This is not so surprising when one considers that entropy is *not* a directly measurable property. In classical, macroscopic thermodynamics, it is derived from fundamental thermodynamic relationships and a known equation of state – that is, $P = f(P, V)$, as discussed in detail in Section 3.4. As will be shown in the paragraphs later, even at the molecular or microscopic level, entropy, a *non-mechanical* property, is arrived at via a careful thought process based on *statistical* analysis of *mechanical* properties. Even then, what exactly entropy does quantify (or represent) does not jump at one from the page in a proverbial aha moment. This is, after all, why there are so many pseudo-scientific explanations of entropy (and the second law) in popular science literature.

Strictly speaking, derivation of entropy from the statistical analysis of microscopic *particles* (idealized representations of atoms or molecules forming the matter) using Newton's laws of motion in a deductive process based on a few postulates is not necessary from a purely utilitarian perspective. For practical application of the second law to the analysis of engineering systems of interest, methodology and tools rigorously outlined in Chapter 3 are sufficient. In other words, skipping the present chapter will not hamper one's understanding of the rest of the book. Nevertheless, providing the reader with a mechanical explanation (less abstract and easier to grasp) of important, non-mechanical thermodynamic concepts (more abstract and much harder to grasp) is deemed important from a pedagogical perspective. The author will try to do this with minimum computational baggage in a colloquial language readily understandable by anyone with four years of undergraduate education in a technical field.

There are many renowned treatises available in the literature on statistical mechanics (including kinetic theory of gases), statistical thermodynamics, and quantum mechanics, all written by the giants in the field. In addition to Pauli's short volume on thermodynamics cited earlier (with a short and very readable section on the kinetic theory), the reader is pointed to the works by Schrödinger [2], Tolman [3], Hill [4, 5], and Wannier [6].

For the reader interested in self-learning, the best starting point is going through the short volumes by two Nobel laureates, Pauli (in 1945) and Schrödinger (in 1933). Hill's introduction on statistical thermodynamics is probably the best choice for getting acquainted with the ensemble method of Gibbs [4]. Tolman and Wannier's books are acknowledged classics covering quantum and classical statistical mechanics, but they can be overwhelming for the uninitiated. It should be noted that these books are quite dated (i.e., going back to 1930s, 1950s, and 1960s). Since then, especially in the field of non-equilibrium physics, there has undoubtedly been many developments published in the academic literature. A classic in this area is the book by Prigogine, which is extremely heavy in mathematics [7]. Nevertheless, it may be a starting point for those interested in the subject if nothing else to appreciate the complexity of the subject matter. A much more accessible yet in-depth and rigorous treatment is the monograph by Hollinger and Zenzen [8], which is highly recommended by the author for those interested in non-equilibrium and irreversible thermodynamics.[1] However, those are beyond the scope of the current book, which is a polite way of saying that they are well beyond the expertise and/or knowledge of the author. What is of interest herein is to lay out the theoretical foundation that leads one from fundamental mechanical principles to the concept of entropy. In that sense, it can be safely stated that absolutely nothing has changed in the past century.

4.1 DEFINING THE PROBLEM

Quantum mechanics teaches that atoms and molecules do not exactly behave as microscopically small billiard balls. They display the (admittedly, rather weird) characteristic referred to as the *wave-particle duality*. If the temperature is high enough (but not so high that their speed is getting closer to the speed of light, c, and they become *relativistic*) and the density is low enough, one can assume that the molecules are localized *wave packets*; that is, their wavelength (from the *de Broglie hypothesis* that matter behaves like wave) is small compared to the average intermolecular distance. In other words, the average *de Broglie wavelength*, λ, is much smaller than the average interparticle separation; that is,

$$\lambda = \frac{\hbar}{\sqrt{2mkT}} \left(\frac{V}{N} \right)^{1/3},$$ (4.1)

where $\hbar = \dfrac{h}{\sqrt{\pi}}$, h is the Planck constant, $6.62607004 \times 10^{-34}$ m²-kg/s; k is the Boltzmann constant, 1.380649×10^{-23} J/K; and m is the mass of a microscopic particle (i.e., molecule) in kg. The derivation of λ by de Broglie starting from Einstein's famous formula, $E = mc^2$, and using Planck's postulate that the energy of light quanta is proportional to their wave frequency, is straightforward and can be found in many textbooks or online. (A derivation of Equation 4.1 from fundamental arguments will be presented in Section 4.2.2.)

In Equation 4.1, the denominator in the definition of λ is the particle momentum; that is,

$$p = mv_{th} = \sqrt{2mkT}, \text{ so that} \tag{4.2}$$

$$v_{th} = \sqrt{\frac{2kT}{m}}, \text{ or} \tag{4.3}$$

$$kT = \frac{1}{2}mv_{th}^2. \tag{4.4}$$

Equation 4.4 gives the temperature from the kinetic theory, which is simply the kinetic energy of a particle moving at the most probable speed, v_{th} (*thermal* velocity), in the Maxwell-Boltzmann velocity distribution given by

$$f\left(v^2\right)dv = 4\pi v^2 \left(\frac{m}{2\pi kT}\right)^{\frac{3}{2}} \exp\left(-\frac{\frac{1}{2}mv^2}{kT}\right)dv, \tag{4.5}$$

which is the *probability density function* for the number of molecules with velocities between v and $v + dv$. Equation 4.5 is normalized (i.e., $\frac{dN}{N} = f(v^2)dv$) so that

$$\int_0^\infty f\left(v^2\right)dv = 1.$$

If the assumption formulated in Equation 4.1 holds, each molecule can be considered a *classical* particle (i.e., obeying Newton's laws of motion) with a well-defined position and momentum. Furthermore, any two molecules are distinguishable from each other. The molecules interact with each other and with the walls of the container via collisions. The walls are assumed to be idealized surfaces (i.e., their atomic structure is ignored) so that the collisions between the molecules and the walls are *elastic*. This simple model, forming the basis of kinetic theory of gases and classical statistical mechanics, is illustrated in Figure 4.1.

Well, one might think, the problem is easy now since even a first-year undergraduate student in engineering can carry out the calculations in Newtonian mechanics. This, of course, is easier said than done considering that just one mole of a gas (a rather small amount) contains $N_A = 6.02214076 \times 10^{23}$ molecules (the *Avogadro's number*). It is impossible to keep track of the position and momentum, $p = mv$, of each molecule in the box in three dimensions; that is, $\{x, y, z\}$ and $\{v_x, v_y, v_z\}$, with

$$m[\text{in kg}] = MW / \left(N_A \times 1000\right)$$

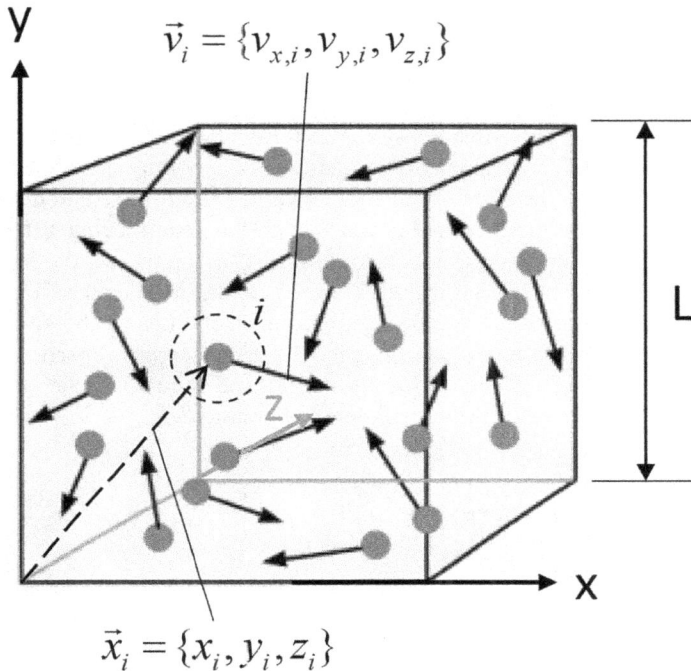

$$\vec{v}_i = \{v_{x,i}, v_{y,i}, v_{z,i}\}$$

$$\vec{x}_i = \{x_i, y_i, z_i\}$$

FIGURE 4.1   Simple molecular model used in kinetic theory of gases.

where *MW* is the molecular weight of the gas in g/mol or kg/kmol. This is when smart scientists of the nineteenth century decided to resort to statistical methods. Here is a brief description of how they went about it:

- We cannot determine the kinetic energy, position, and momentum of each *particle* (from now on, we will refer to the molecules as particles) in the box at any given time, t, and keep track of it for any length of time, $t = T$.
- What we can do (and really need) is a *time average* over a period of T long enough to smooth out fluctuations and independent of the starting time, $t = t_0$.
- But even this is impossible to do for the astronomically large number of particles.

At this juncture in the thought process, some of them resorted to statistical analysis, which resulted in the kinetic theory of gases and the Maxwell-Boltzmann velocity distribution given by Equation 4.5. Gibbs, in contrast, came up with the *ensemble* method, which is the cornerstone of classical statistical mechanics. Before continuing, a few remarks on the distinction between the *kinetic theory* and *statistical mechanics* are in order. Classical statistical mechanics, at least as it was originally

conceived, deals with a system in equilibrium, from macroscopic as well as microscopic perspectives. What this means is that, on average, except *fluctuations* around
the said average, there is no time dependence. Consider the closed system shown in
Figure 4.1 with rigid walls (i.e., constant volume, V, and number of particles, N) but
no insulation. Let us assume that the system is initially at time, t_0; at an energy state,
E_0; with temperature, T_0, which is higher than the temperature of the surroundings,
T. As time progresses, the system will cool down; that is, $T_0 \rightarrow T$ and $E_0 \rightarrow E$, with
$E < E_0$, until equilibrium is reached at time t with the system being at temperature
T and energy E. Kinetic theory does two things: (i) it determines the macroscopic
properties of the system in equilibrium at time t_0 and t, and (ii) describes how the
equilibrium is reached (i.e., how the macroscopic properties of the system evolve
with time). Statistical mechanics can describe the macroscopic properties of the system only at equilibrium; that is, either at time t_0 (if the system is isolated) or at time
t when the system is in equilibrium with its surroundings.

4.2 SOLVING THE PROBLEM

4.2.1 KINETIC THEORY

The starting point of the kinetic theory is the fraction of molecules in the box in
Figure 4.1 with velocities between $\{v_x, v_y, v_z\}$ and $\{v_x + dv_x, v_y + dv_y, v_z + dv_z\}$ (curly
brackets denote a *vector* in three dimensions). Assuming *isotropic* velocity distribution (i.e., there are no special or preferred directions in the system), one can work with
speeds of molecules (i.e., $v = |\vec{v}|$), which can be found from the vector relationship

$$v^2 = \left|\vec{v}_x\right|^2 + \left|\vec{v}_y\right|^2 + \left|\vec{v}_z\right|^2.$$

(Note that *velocity* is a vector, i.e., with a direction *and* magnitude, whereas *speed* is
the magnitude of that vector.)
    Maxwell postulated that the distribution of each component of the velocity is
independent of the other two so that

$$f\left(v^2\right) = g\left(v_x^2\right)g\left(v_y^2\right)g\left(v_z^2\right) \tag{4.6}$$

where all three distributions are the same because of *isotropy* and depend only on
squares of velocity components. The second condition follows from the assumption
of a *symmetrical distribution* (i.e., invariance for $\vec{v} \rightarrow -\vec{v}$). At this point, Maxwell
introduced two new functions such that

$$\varphi\left(v_i^2\right) \equiv \ln g\left(v_i^2\right), i = x,y,z, \text{ and} \tag{4.7}$$

$$\psi\left(v^2\right) \equiv \ln f\left(v^2\right). \tag{4.8}$$

Substituting Equations 4.7 and 4.8 into Equation 4.6, the result is

$$\psi\left(v^2\right) = \psi\left(v_x^2 + v_y^2 + v_z^2\right) = \varphi\left(v_x^2\right) + \varphi\left(v_y^2\right) + \varphi\left(v_z^2\right).$$

It can be shown that Equation 4.9 can only be satisfied if φ and ψ are *linear* functions of their arguments; that is,

$$\varphi\left(v_i^2\right) = -\alpha v_i^2 + \beta, \text{ and} \tag{4.9}$$

$$\psi\left(v^2\right) = -\alpha v^2 + 3\beta. \tag{4.10}$$

Combining Equations 4.8 and 4.11, we obtain that

$$f\left(v^2\right) = \exp\left(-\alpha v^2 + 3\beta\right) = \exp(3\beta)\exp\left(-\alpha v^2\right), \text{ or} \tag{4.11}$$

$$f\left(v^2\right) = C\exp\left(-\alpha v^2\right), C = \exp(3\beta). \tag{4.12}$$

From the boundary condition that $\int_{-\infty}^{+\infty} f\left(v^2\right)dv = 1$, it can be shown that $C = \left(\dfrac{\alpha}{\pi}\right)^{3/2}$ so that

$$f\left(v^2\right) = \left(\frac{\alpha}{\pi}\right)^{3/2} \exp\left(-\alpha v^2\right). \tag{4.13}$$

This is the derivation of the *probability density function* (PDF) of molecular speeds starting from a postulate and two assumptions (i.e., isotropy and symmetry) with an unknown constant α, which can be readily determined from the definition of the *mean-square speed*, $\overline{v^2}$ as. Finding C and α as $\alpha = \dfrac{m}{2kT}$ described earlier is left as an exercise for the reader. (If stuck, you can refer to Pauli [1], pp. 100–101). Substituting for α in Equation 4.13, the fraction of molecules with velocities between $\{v_x, v_y, v_z\}$ and $\{v_x + dv_x, v_y + dv_y, v_z + dv_z\}$ (or with speeds between v and $v + dv$) is given by

$$\frac{dN}{N} = f\left(v^2\right)d^3v = \left(\frac{m}{2\pi kT}\right)^{3/2} \exp\left(-\frac{m}{2kT}v^2\right)d^3v. \tag{4.14}$$

Equation 4.14 represents the integrand for integration over three directions or a volume integral; that is,

$$\int f\left(v^2\right)d^3v = \iiint f\left(v^2\right)dv_x dv_y dv_z, \tag{4.15}$$

which, in polar coordinates, becomes

$$\int f\left(v^2\right)d^3v = \int f\left(v^2\right)v^2 dv\, d\theta, \tag{4.16}$$

$$\int f\left(v^2\right)d^3v = 4\pi \int f\left(v^2\right)v^2 dv. \tag{4.17}$$

(In other words, $4\pi v^2 dv$ is the volume of the spherical shell bounded by the surfaces with radii v and $v + dv$.) Thus, combining Equations 4.14 and 4.17, we arrive at Equation 4.5 introduced earlier

$$f\left(v^2\right)dv = 4\pi v^2 \left(\frac{m}{2\pi kT}\right)^{\frac{3}{2}} \exp\left(-\frac{\frac{1}{2}mv^2}{kT}\right)dv. \tag{4.5}$$

Equation 4.5 is evaluated for a molecule with molecular weight of $MW = 28$ g/mol, idealized as a microscopic particle with the mass, $m = MW/(N_A \times 1{,}000) = 4.65E{-}26$ kg, at a temperature of $T = 300$ K. Plots of the PDF and cumulative PDF are shown in Figure 4.2.

The most probable speed, which we earlier referred to as the *thermal speed*, is the speed most likely to be possessed by any molecule in the system and corresponds to the maximum value or the *mode* of $f(v^2)$. To find it, we take the derivative of $f(v^2)$ given by Equation 4.5 and set it to zero. Solving the resulting equation for v, the result is given by Equation 4.3 introduced earlier; that is,

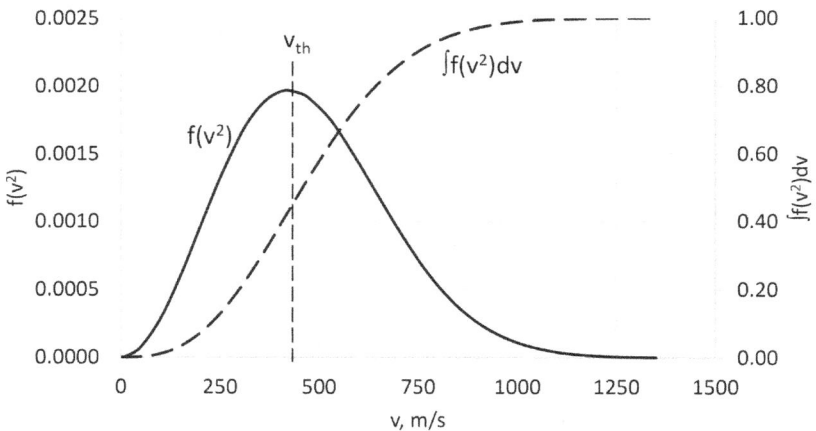

FIGURE 4.2   Maxwell-Boltzmann velocity distribution.

$$v_{th} = \sqrt{\frac{2kT}{m}}.$$ (4.3)

There are other representative values that can be calculated from Equation 4.5; that is,

$$\text{Mean speed}: \overline{v} = \int_0^\infty vf\left(v^2\right)dv = \sqrt{\frac{8kT}{\pi m}} = \frac{2}{\sqrt{\pi}} v_{th},$$ (4.18)

$$\text{Mean-square speed}: \overline{v^2} = \int_0^\infty v^2 f\left(v^2\right)dv = \frac{3kT}{m} = \frac{3}{2}v_{th}^2,$$ (4.19)

$$\text{Root-mean-square speed}: v_{rms} = \sqrt{\overline{v^2}} = \sqrt{\frac{3kT}{m}} = \sqrt{\frac{3}{2}}v_{th}.$$ (4.20)

For the numerical example that we used in Figure 4.2, we find that $v_{th} = 422.1$ m/s, $\overline{v} = 476.3$ m/s, and $v_{rms} = 517.0$ m/s. Noting that the gas constant for a particular gas is given by $R = R_{unv}/MW$ and $R_{unv} = k \cdot N_A$, with $m = MW/N_A$, $R = k/m$, Equation 4.20 can be written as $v_{rms} = \sqrt{3RT}$, which is analogous to the equation for the *speed of sound*, $c = \sqrt{\gamma RT}$. In other words, $c = \sqrt{\frac{\gamma}{3}} v_{rms}$. For a diatomic molecule (e.g., nitrogen, N_2), with $MW = 28$ g/mol and $\gamma = 1.4$ so that $c = 0.68313 \times v_{rms} = 353.2$ m/s.

From Equation 4.19, it is found that the average kinetic energy per molecule is equal to $\frac{3}{2}kT$ and has the same value for all gases at the same temperature (i.e., there is no parameter such as molecular weight that is unique to a particular gas). This finding is known as the *equipartition of energy*, which is formulated as follows: the average kinetic energy of molecules or other particles in the region where classical statistics apply is $\frac{1}{2}kT$ per *degree of freedom* (e.g., see Pauling [8], p. 326). In this case, with *three* degrees of freedom of translational motion (i.e., along the x, y, and z coordinates), we have $\frac{3}{2}kT$.

In kinetic theory, the energy of particles (i.e., molecules or atoms) is the translational kinetic energy (i.e., for each individual particle, i, $\mu_i = \frac{1}{2}mv_i^2$). Thus, for the fraction of molecules with values of energy between ε and $\varepsilon + d\varepsilon$, Equation 4.5 can be rewritten as (dropping the subscript i)

$$f(\varepsilon)d\varepsilon = \frac{2\pi}{\left(\pi kT\right)^{\frac{3}{2}}} \sqrt{\varepsilon} \exp\left(-\frac{\varepsilon}{kT}\right)d\varepsilon.$$ (4.21)

Note that the term $exp(-\varepsilon/kT)$ is referred to as the *Boltzmann factor*. Equation 4.21 is the Maxwell-Boltzmann *energy* distribution from the kinetic theory (i.e., *classical statistics*) and can be derived in the same form from *quantum statistics*. Doing this requires the introduction of the concept of *partition function*, usually denoted by the letter Q or Z (from *Zustandsumme*, German for sum of states). Herein, we will go with Q, which was used by Hill [4], whose treatment of the subject is loosely followed subsequently. For a system of gas with N molecules (e.g., see Figure 4.1), the partition function is given as

$$Q = \frac{q^N}{N!},\qquad(4.22)$$

where $q = \sum_j \exp\left(-\frac{\varepsilon_j}{kT}\right)$ is the partition function for energy states ε_j for a molecule (in a system) or a subsystem (in an *ensemble* of systems). For N subsystems in an ensemble, for example, one would expect that $Q = q^N$, and that is indeed the case for a *crystalline* system. In a crystal lattice, each atom's motion is confined to the immediate neighborhood of its place in the lattice (at least in the lattice *model* if not all the time in real crystals). In other words, the atoms are *distinguishable*; that is, if there are N atoms in the system (lattice), they can be numbered as 1, 2, 3, and so on up to N and identified individually. In a system comprising a gas, however, identical and independent molecules moving randomly all over the system volume are *indistinguishable*. The correction for indistinguishability is achieved via division by $N!$ (for a proof of that assertion, see Hill [4], pp. 62–63, or Wannier [6], pp. 167–170), which results in Equation 4.22.

In kinetic theory, we only considered the translational kinetic energy for individual molecules (particles) obeying Newton's laws of motion. (This is how, for example, pressure is determined.) In general, one can consider other forms of energy as well (i.e., rotational, vibrational, and electronic). For example, in a crystal lattice, instead of translational motion, one must consider vibrational and rotational motions. A different partition function can be written for each energy type (e.g., $q_{rot} = \sum_j \exp\left(-\frac{\varepsilon_{rot,j}}{kT}\right)$ for rotational energy). Thus, the general form of Equation 4.22 is

$$Q = \frac{q_{trans}^N q_{rot}^N q_{vib}^N q_{elec}^N}{N!} = \frac{\left(q_{trans} q_{rot} q_{vib} q_{elec}\right)^N}{N!}.\qquad(4.23)$$

For our purposes, Equation 4.22 is adequate, and it is the partition function of a system of N identical, indistinguishable molecules with the number of available energy states much greater than N. In other words, there are so many choices of energy states available to the molecules that any two of them rarely find themselves in the same state. Thus, Equation 4.22 is a limiting form of quantum statistics and is referred to as classical or Boltzmann statistics.

4.2.2   From Quantum Mechanics to Kinetic Theory

The origins of the idea of considering molecular motion to describe the behavior of a gaseous substance can be traced back to Daniel Bernoulli (1738). Starting with August Kronig's simple gas-kinetic model (1856), the idea was fully developed into a detailed theory by the works of Clausius (1857), Maxwell (1859), and Boltzmann (1871). An experimental verification of the Maxwell-Boltzmann velocity distribution was provided by Otto Stern via deflection of a molecular beam by gravity (1920). Before that, Einstein surmised that, although molecules cannot be visually observed moving randomly, momentum imparted by them to larger particles could be observed under the microscope. This, in turn, would be a directly observable effect that would corroborate the kinetic theory. Starting from this premise, Einstein developed his quantitative theory of *Brownian motion* and published it in a landmark paper (1905).

Clausius, Maxwell, and Boltzmann used principles of classical mechanics (i.e., Newton's laws of motion) to develop the kinetic theory of gases. In developing a full picture of the thermodynamic state of the gas, they were assisted by the *ideal gas equation of state*; that is, $PV = n(kN_A)T$ which combined three empirical laws – that is, the laws of Boyle (*PV* = constant at constant *T*), Charles and Gay-Lussac (*V* ∝ *T* at constant *P*), and Avogadro – into a single equation. (In the ideal gas equation, *n* is the number of moles of gas.) Around 1900, the quantum theory came to the fore with Planck's theory that light of a certain wavelength, λ, consists of packets of energy (referred to as *quanta* – for plural of *quantum* – or *photons*) quantified by the simple equation, $E = h\nu = h\dfrac{c}{\lambda}$ where ν (the Greek letter *nu*) is the frequency of the light wave, *c* is the speed of light, and *h* is Planck's constant. The mass of the photon is related to its energy by probably the most famous equation in physics (i.e., $E = mc^2$, the *Einstein equation*). This established the *wave-particle duality* of light in a simple equation; that is,

$$E = mc^2 = h\nu = h\frac{c}{\lambda}, \text{ so that}$$

$$\lambda = \frac{h}{mc}.$$

De Broglie realized that the same equation can be applied to an electron by replacing *c* with the speed of the electron, *v*, so that, $\lambda = \dfrac{h}{mv}$ where λ is the *de Broglie wavelength*.

In classical mechanics and kinetic theory, velocity of a particle and its energy are continuous; that is, Δv or $\Delta\varepsilon$ separating *v* and $v + \Delta v$ or ε and $\varepsilon + \Delta\varepsilon$, respectively, can become infinitesimally small. In quantum mechanics, particle velocity and energy are quantized in analogy to the photons of light. In other words, there are *discrete* values of velocities and energies, which are referred to as *microstates*. For the cubic volume in Figure 4.1 with a volume of $V = L^3$, the quantized states of the particles in that space are given by

$$p_{x,y,z} = m v_{x,y,z} = \frac{n_{x,y,z} h}{2L}, \tag{4.24}$$

where $n_{x,y,z} = 1, 2, 3, \ldots, n$, represents the *quantum numbers*. The corresponding wave function is given by

$$\Psi\left(n_x, n_y, n_z\right) = \left(\frac{2}{L}\right)^{\frac{3}{2}} \sin\left(\pi n_x \frac{x}{L}\right) \sin\left(\pi n_y \frac{y}{L}\right) \sin\left(\pi n_z \frac{xz}{L}\right), \tag{4.25}$$

For an elegant (and readily digestible) derivation of Equations 4.24 and 4.25, the reader is referred to Pauling [8] (pp. 317–318). The energy of the quantized states is given by

$$\varepsilon\left(n_x, n_y, n_z\right) = \frac{1}{2} m \left(v_x^2 + v_y^2 + v_z^2\right) = \frac{1}{2} m v^2, \tag{4.26}$$

and making use of Equation 4.24, it becomes

$$\varepsilon\left(n_x, n_y, n_z\right) = \frac{h^2}{8mL^2}\left(n_x^2 + n_y^2 + n_z^2\right). \tag{4.27}$$

In a three-dimensional space of the quantum numbers $n_{x,y,z} = 1, 2, 3, \ldots, n$, each possible quantum state, say, $(n_x, n_y, n_z) = (1,2,1)$, corresponds to a point in this space with positive integers as coordinates (see Figure 4.3). In other words, there is one quantum state per unit volume of this space.

Using Equation 4.27, the *octant* (one eighth) of a sphere (which contains nonzero positive integers for the quantum numbers) with radius R in that space is defined by the following relationship:

$$R^2 = \frac{8mV^{\frac{2}{3}}}{h^2} \varepsilon. \tag{4.28}$$

Equation 4.28 directly leads one to the number of quantum states with energy less than ε with the assumption is that the said number is so big that the edge effects are neglected. That number is given by the octant volume; that is,

$$N(\varepsilon) = \frac{\pi}{6}\left(\frac{8m\varepsilon}{h^2}\right)^{\frac{3}{2}} V. \tag{4.29}$$

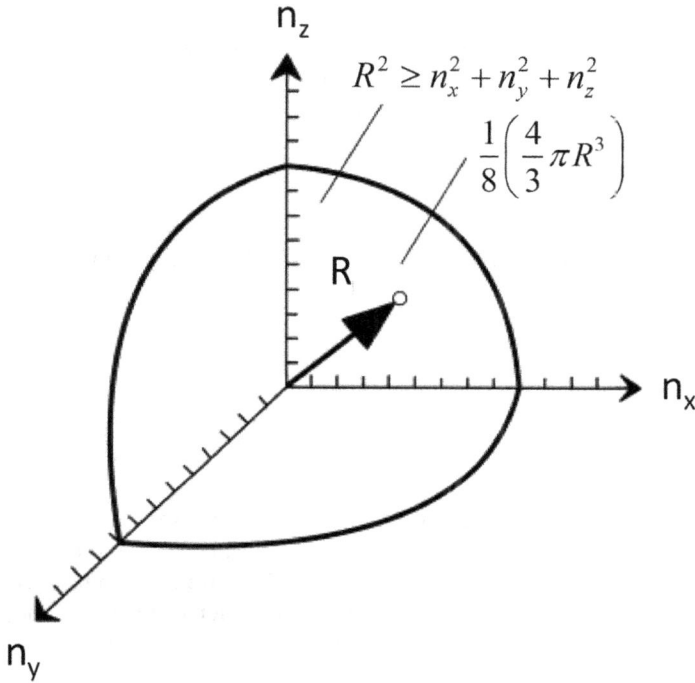

FIGURE 4.3 Octant of a sphere corresponding to the constant energy surface (quantum numbers are nonzero positive integers).

If the system in question is a dilute gas with independent and indistinguishable molecules, the number of quantum states, $N(\varepsilon)$, is much larger than the number of molecules, N (i.e., $N(\varepsilon) \gg N$). In that case, energy states with $\varepsilon \gg O(kT)$ (read as "of the order of kT") would be sparsely populated (i.e., effectively unavailable). Consequently, the partition function given by Equation 4.22 with

$$q = \sum_{n_{x,y,z}=1}^{\infty} \exp\left(-\frac{\varepsilon\left(n_x, n_y, n_z\right)}{kT}\right)$$

would be valid. Rewriting Equation 4.29 as

$$N(\varepsilon) = \frac{4}{3}\sqrt{\pi}\left(\frac{2\pi m \varepsilon}{h^2}\right)^{\frac{3}{2}} V,$$

this assumption can be expressed as

$$N(\varepsilon) \cong \left(\frac{2\pi mkT}{h^2}\right)^{\frac{3}{2}} V >> N, \text{ or}$$

$$\frac{\lambda^3 N}{V} << 1,$$

which, of course, is identical to Equation 4.1 introduced earlier in the chapter (with $\hbar = \dfrac{h}{\sqrt{\pi}}$),

$$\lambda = \frac{\hbar}{\sqrt{2mkT}} << \left(\frac{V}{N}\right)^{1/3}. \tag{4.1}$$

Strictly speaking, this condition is valid for monatomic gases with only translational kinetic energy states. However, it would be applicable to diatomic or polyatomic molecules as well (with rotational energy states) because the number density of the quantum states is dominated by the density of translational states (i.e., they are more closely spaced, i.e., $\Delta\varepsilon << kT$). If this is indeed the case, then for the partition function, we would have

$$q = \sum_{n_{x,y,z}=1}^{\infty} \exp\left(-\frac{\varepsilon\left(n_x,n_y,n_z\right)}{kT}\right) \rightarrow \iiint_{\varepsilon_{x,y,z}} \exp\left(-\frac{\varepsilon\left(n_x,n_y,n_z\right)}{kT}\right) d\varepsilon_x d\varepsilon_y d\varepsilon_z. \tag{4.30}$$

Equation 4.30 is the mathematical expression of the transition from quantum states or statistics (discrete) to classical states or statistics (continuous). To show that $\Delta\varepsilon << kT$, consider that, from Equation 4.27, with some algebraic manipulation, it follows that $\Delta\varepsilon = O\left(\dfrac{h^2}{mV^{2/3}}\right)$ and

$$\frac{\Delta\varepsilon}{kT} = O\left(\frac{h^2}{mkTV^{2/3}}\right) = O\left(\frac{\lambda^2}{V^{2/3}}\right),$$

which, using Equation 4.1 and considering that $N = O(N_A)$, translates to $O(10^{-14})$. QED.

Making use of Equation 4.29, the number of molecular energy states (microstates) between ε and $\varepsilon + d\varepsilon$ is given by

$$\frac{dN(\varepsilon)}{d\varepsilon} d\varepsilon = \acute{E}(\varepsilon) d\varepsilon = \frac{\pi}{4}\left(\frac{8m}{h^2}\right)^{\frac{3}{2}} V\sqrt{\varepsilon}\, d\varepsilon, \tag{4.31}$$

which we can use to carry out the integration in Equation 4.30. Dropping the subscripts and triple integral notation, Equation 4.30 is rewritten as

$$q = \int_0^\infty \exp\left(-\frac{\varepsilon}{kT}\right) dN(\varepsilon).$$

(4.32)

so that, after substituting Equation 4.31 into Equation 4.32 and defining a new integration variable, $u = \varepsilon/kT$, we end up with

$$q = \frac{\pi}{4}\left(\frac{8m}{h^2}\right)^{\frac{3}{2}} V \int_0^\infty \sqrt{u}\, e^{-u} du,$$

$$q = \left(\frac{2\pi mkT}{h^2}\right)^{\frac{3}{2}} V = \frac{V}{\lambda^3}.$$

(4.33)

(Note that q is a function of V and T, i.e., $q = q\,(V,\,T)$, which means that it represents the partition function of a *canonical ensemble*, which will be discussed in more detail in the following section.)

The fraction of molecules with energies between ε and $\varepsilon + d\varepsilon$ is given by

$$\frac{dN(\varepsilon)}{N} = dN(\varepsilon)\frac{\exp\left(-\dfrac{\varepsilon}{kT}\right)}{q},$$

(4.34)

which, after substitution with Equations 4.31 and 4.33, reproduces Equation 4.21 (i.e., Maxwell-Boltzmann distribution), which is repeated subsequently for convenience:

$$f(\varepsilon)d\varepsilon = \frac{2\pi}{(\pi kT)^{\frac{3}{2}}} \sqrt{\varepsilon}\,\exp\left(-\frac{\varepsilon}{kT}\right) d\varepsilon.$$

(4.21)

Substituting Equation 4.33 into Equation 4.22 and taking the logarithm of both sides of the equation, we arrive at the canonical ensemble partition function in logarithmic form as

$$\ln Q = N \ln\left[\left(\frac{2\pi mkT}{h^2}\right)^{\frac{3}{2}} \frac{V}{N}\right] + N.$$

(4.35)

The characteristic function for the canonical ensemble (i.e., (N, V, T)) thermodynamic system) is *Helmholtz free energy*, F. (In literature, different notations are used for Helmholtz free energy, (e.g., A). In this book, A is used for the *extensive* property, exergy – for the alternative, and nowadays rarely used, US term *availability*.) For the derivation of the characteristic function, F, the reader is pointed to Schrödinger [2], pp. 11–12. In classical thermodynamics, Helmholtz free energy is one of the *four* fundamental thermodynamic functions. (The other three are internal energy, U; enthalpy, H; and *Gibbs free energy*, G; e.g., see Moran and Shapiro [1] in Chapter 2.) The function F is related to the internal energy and entropy via the fundamental thermodynamic relationship

$$F = U - TS. \tag{4.36}$$

Helmholtz free energy derived from the canonical ensemble partition function is given by $F = -kT \cdot lnQ$ (see Schrödinger [2], pp. 13–14) so that, using Equation 4.35, the result is

$$F = -NkT \ln\left[\left(\frac{2\pi m\, kT}{h^2} \right)^{\frac{3}{2}} \frac{V}{N} \right] - NkT. \tag{4.37}$$

Before deriving other thermodynamic properties from the partition function, let us combine Equation 4.33 with Equation 4.22 to obtain

$$\ln Q = -N\left(\ln N - 1 - \ln q \right). \tag{4.38}$$

Using Equations 4.33 and 4.38, system pressure is found as

$$P = kT\left(\frac{\partial \ln Q}{\partial V} \right)_{N,T} = NkT\left(\frac{\partial \ln q}{\partial V} \right)_{N,T} = \frac{NkT}{V}, \tag{4.39}$$

which is immediately recognized as the *ideal gas equation of state*. System energy, denoted by U, is given by

$$U = kT^2\left(\frac{\partial \ln Q}{\partial T} \right)_{N,V} = \frac{3}{2} NkT. \tag{4.40}$$

Finally, combining Equations 4.36, 4.37 and 4.40, entropy is found as

$$S = Nk \ln\left[\left(\frac{2\pi mkT}{h^2} \right)^{\frac{3}{2}} \frac{V}{N} \right] + \frac{5}{2} Nk. \tag{4.41}$$

Recalling that the definitions of the universal gas constant, $R_{unv} = k \cdot N_A$; and molecular mass, $m = MW/N_A$; and setting $N = N_A$; Equation 4.41 becomes, per unit mole (so that the letter s is used instead of S to denote an *intensive* property),

$$s = R_{unv}\left[\frac{3}{2}\ln\left(\frac{2\pi k}{h^2}\right) + \ln\left(\frac{MW \cdot T}{N_A}\right)^{\frac{3}{2}} + \ln\left(\frac{V}{N_A}\right) + \frac{5}{2}\right],$$

(4.42)

which, after combining all the terms containing π, k, h, N_A, R_{unv}, and numerical constants into a single constant term, c, becomes

$$s = R_{unv}\left[\frac{3}{2}\ln(MW) + \frac{3}{2}\ln(T) + \ln(V)\right] + c.$$

(4.43)

Equation 4.42 is the *Sackur-Tetrode equation* introduced earlier in Section 3.4, Equation 3.54, which is repeated subsequently for convenience:

$$s(V,T) = R_{unv}\left(\frac{3}{2}\ln(MW) + \frac{3}{2}\ln(T) + \ln(V)\right) + 2.6546,$$

(4.54)

Using the classical statistical description of the Helmholtz free energy, $F = -kT \cdot \ln Q$, Equation 4.36 is rewritten as

$$-kT \cdot \ln Q = U - TS.$$

(4.43)

from which the classical statistical description of entropy for a canonical ensemble is found as

$$S = k\ln Q + \frac{U}{T}.$$

(4.44)

The probability that the system is in energy state ε_i is given by

$$P_i = \frac{\exp\left(-\dfrac{\varepsilon_i}{kT}\right)}{Q}.$$

(4.45)

Using the thermodynamic relationship between internal energy and Helmholtz free energy

$$U = -T^2\frac{\partial}{\partial T}\left(\frac{F}{T}\right)_{N,V} = kT^2\left(\frac{\partial \ln Q}{\partial T}\right)_{N,V}.$$

(4.46)

and substituting it into Equation 4.45, we obtain

$$S = k \ln Q + kT \left(\frac{\partial \ln Q}{\partial T} \right)_{N,V},$$ (4.47)

which, when combined with Equation 4.45, can be shown to result in the following relationship:

$$S = -k \sum_i P_i \ln P_i.$$ (4.48)

(For a proof, start with Equation 4.48 and evaluate it via substituting Equation 4.45 and noting that $\sum_i P_i = 1$.)

Equation 4.48 is the definition of the *Gibbs entropy*, which can be shown to lead to the famous *Boltzmann entropy*, $S = k \ln Q$ for equal probability of energy microstates (i.e. $P_i = \frac{1}{Q}$). Furthermore, Gibbs entropy is analogous to *Shannon's entropy* in information theory defined as

$$H(x) = -\sum_{i=1}^n P(x_i) \log P(x_i),$$ (4.49)

which is a quantification of the average level of information inherent to the possible outcomes $\{x_1, x_2, \ldots, x_n\}$, with probabilities of occurrence of $\{P(x_1), P(x_2), \ldots, P(x_n)\}$, of the discrete random variable x [9]. (The symbol H was chosen in reference to Boltzmann's *H-theorem*.) The base of the logarithm in Equation 4.49 can be 2, e (i.e., 2.718282), or 10, depending on the application.

As an example, consider a dice with six faces numbered from 1 to 6. If the dice is straight, the probability of rolling any of the six numbers is 1/6 or 0.167 (16.7%). From Equation 4.49, with 10-base logarithm, $H(x)$, with $x = \{1, 2, 3, 4, 5, 6\}$, is calculated as 0.7782. If the dice is manipulated so that the probability of rolling 1, $P(1)$, can have any value within the range of 0 to 1 with rolling the rest (i.e., from 2 to 6), having equal probabilities, the variation in $H(x)$ is plotted in Figure 4.4. As shown in the figure, $H(x)$ has a maximum at $P(1) = 0.167$, and it is 0 at $P(1) = 1.0$. As $P(1)$ approaches 0, $H(x)$ approaches a limiting value of about 0.7.

The interpretation of the plot in Figure 4.4 is as follows. If $P(1) = 1.0$, there is no uncertainty or surprise in rolling the dice. The outcome is known a priori; that is, the information content of the probability distribution is zero. If the dice is straight with each outcome having the same probability of 1/6, the uncertainty is at its maximum; that is, the information content of the probability distribution (uniform) is also maximum. As $P(1)$ approaches 0, one outcome (out of the six) is eliminated so that we are left with five possible outcomes with the same probability; that is, 1/5 = 0.2 (20%). There is a slight reduction in the uncertainty of the outcome of the roll (i.e., one out

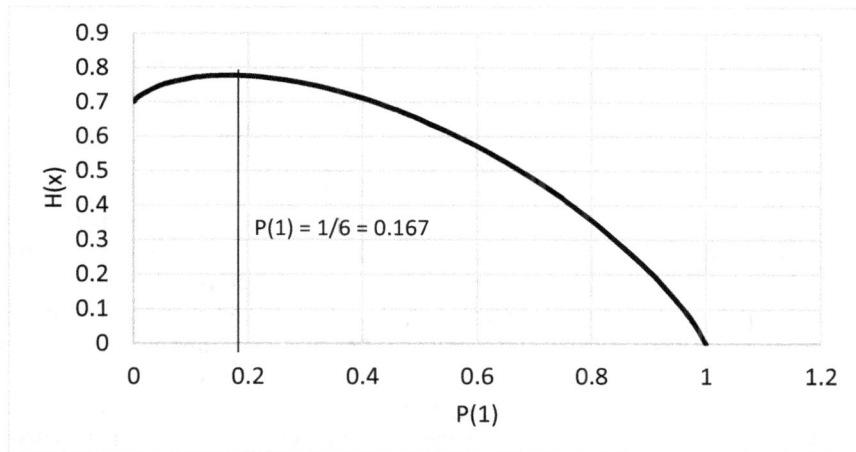

FIGURE 4.4 Shannon's entropy for a loaded dice.

of five instead of one out of six) so that the information content of the probability distribution is somewhat lower than its maximum.

4.2.3 ENSEMBLE METHOD

What is an ensemble? An ensemble is a mental collection of systems that represent a *microstate* of the system being considered with a certain *macrostate*. All systems in the ensemble have the *same macrostate*; that is, they are *replicas* of the system under investigation, but each has a *different microstate*. Of course, now we must define what we mean by microstate and macrostate. This will be done in the following paragraphs.

Let us start by considering the system shown in Figure 4.1 with N particles and having energy of E and volume of V. Both E and V are measurable macroscopic quantities and define the *macrostate* of the system. If the system is in equilibrium, if you measure E_1 and V_1 at time t and E_2 and V_2 at time $t + \Delta t$, you will find out that $E_2 = E_1$ and $V_2 = V_1$. However, at time t, N particles constituting the system have coordinates x_i, y_i, and z_i for each N_i and velocities v_{xi}, v_{yi}, and v_{zi} for each N_i, which will be different at time $t + \Delta t$. Each unique combination of those six variables for N particles represent a *microstate* of the system. In other words, there is an astronomically large number of microstates (of the order of the Avogadro's number, i.e., 10^{23}) corresponding to the same macrostate. Furthermore, at any given time, for a system in equilibrium, we know that the system is in one of those microstates, but we do not know which one.

Before continuing, as already noted earlier, in statistical and quantum mechanics, it is customary to deal with *momenta* instead of velocities; that is, for a particle with mass of m, $p_x = m \cdot v_x$, $p_y = m \cdot v_y$, and $p_z = m \cdot v_z$, or, in vector notation, $\vec{p} = m \cdot \vec{v}$ where p denotes the *momentum* of an individual particle. Consequently, the energy of each particle is given by

$$\varepsilon_i = \frac{p_x^2 + p_y^2 + p_z^2}{2m} = \frac{|\vec{p}|^2}{2m}. \tag{4.50}$$

By the way, this is the origin of the term *statistical mechanics*. To investigate the system at a microscopic level, we use *mechanical* properties (i.e., energy, volume, and pressure), all of which can be defined in purely mechanical terms (i.e., velocities, momenta, and kinetic, translational, and rotational energy). (This is the same for quantum mechanics as well.) Transition from statistical mechanics to *statistical thermodynamics* will take place via introduction of temperature, which will lead to other *non-mechanical* variables (e.g., entropy, free energy, and chemical potential).

There are two ways of looking at this problem. The first one is time averaging, which, as already stated earlier, is practically impossible. The other ingenious way of looking at the problem was proposed by Gibbs, who postulated that the time average of a mechanical variable of the system of interest is equal to the ensemble average of that variable in the limit as $M \to \infty$, where M is the number of systems in the ensemble, provided that the systems of the ensemble replicate the macrostate and environment of the actual system.

A thermodynamic system can be defined in myriad ways. Three most important ones are as follows:

1. Isolated system with constant E, V, and N, represented by a *microcanonical* ensemble.
2. Closed, isothermal system with constant T, V, and N, represented by a *canonical* ensemble.
3. Open, isothermal system with constant chemical potential $\{\mu_1, \mu_2, \ldots\}$, V, and T with components $\{N_1, N_2, \ldots\}$, represented by a *grand canonical* ensemble.

There is a second postulate, specifically applying to a microcanonical ensemble for an isolated system (E, V, N), which states that the M systems in the ensemble $(M \to \infty)$ are distributed uniformly (i.e., with equal probability) over the possible microstates. In other words, if one selects a system at random from the ensemble, the probability that it will be in a particular microstate is the same for all the possible microstates. (In layman's terms, there are no *preferred* microstates.) The combination of the two postulates tells us this: The actual isothermal system of interest (E, V, N), which is the prototype of M systems forming the microcanonical ensemble $(M \to \infty)$, spends equal amounts of time over a long period of time in each of the available microstates. (This statement is known as the *ergodic hypothesis*.)

Canonical ensemble approach will be covered herein in detail because, starting from a few basic premises, it directly leads to Equation 4.48 for the entropy derived in the preceding section. The mathematical derivation is presented in Section 14.3. We pick the tread from the *Boltzmann equation*, Equation 14.46, reproduced here for convenience,

$$S = k \ln \Omega, \tag{4.51}$$

where Ω represents the number of quantum states with energy E^* (the most probable energy, which is practically equal to the average value $\bar{E}$) This unique case (corresponding to a *microcanonical* ensemble in statistical mechanics) is the origin of the qualitative (to the point of being almost folkloric) statements correlating entropy with concepts such as uncertainty (probability) – that is, akin to the Shannon's entropy in information theory, randomness, and disorder. While it provides a convenient mental picture of what entropy is, it should be remembered that it strictly applies to an isolated or isothermal system.

Nevertheless, Equation 4.51 has general validity; that is, it can be used as a general correlation between the magnitude of S and the number of available quantum states Ω. For different systems (ensembles), different distributions (partition functions) and corresponding counting of microstates (Ω) will be applicable. As such, it is the starting point for the statistical mechanical version of the 2LT (i.e., for spontaneous processes, $\Delta S > 0$). The famous example used for demonstration of the 2LT is the gas confined to one side of an isolated container by a removable barrier. The system (gas) is characterized by N, V, and E with Ω quantum states accessible to it. All of a sudden, the barrier is removed, and the gas expands so that $V' > V$. It follows that the number of quantum states available to it, Ω', is larger (i.e., $\Omega' > \Omega$). Consequently,

$$\Delta S = S' - S = k \ln \frac{\Omega'}{\Omega} > 0. \tag{4.52}$$

Let us now consider a spontaneous process that goes in the reverse direction. Initially, the gas occupies the entire container with Ω' quantum states accessible to it. All of a sudden, one half of the container is empty; that is, all the gas molecules occupy the other half with $\Omega < \Omega'$ accessible quantum states. From Equation 4.52

$$\frac{\Omega}{\Omega'} = e^{-\frac{\Delta S}{k}}. \tag{4.53}$$

From Equation 4.41, we can calculate that $\Delta S = Nk \ln 2$ (assuming that the removable barrier was dividing the container exactly into two halves). Thus, the probability of observing a violation of the second law is

$$\frac{\Omega}{\Omega'} = e^{-\frac{Nk \ln 2}{k}} = \frac{1}{2^N}, \tag{4.54}$$

which is a very small number (i.e., of the order of 10^{-20}). In conclusion, the possibility of a spontaneous process with a negative entropy change, while not exactly zero, is so low that it can be considered to be practically impossible.

Equation 4.51 is also the source of the association between entropy and disorder or randomness. The greater the degree of randomness or disorder in the system, the

larger Ω and, therefore, S. Indeed, there is ample qualitative evidence for this assertion. For a pure substance, for example, as the phase changes from solid (crystalline
structure) to liquid to gas at constant pressure, the entropy increases. However, there
are also counter-examples; for example, the spontaneous crystallization of a super-
cooled melt (i.e., a *metastable* state of a pure substance as a liquid below its melting
temperature). If a super-cooled melt is allowed to crystallize under adiabatic conditions, the entropy of the system increases. As an example, consider the experiment
where 1-chlorine-2-nitrobenzene (1C2NB) is melted in a test tube equipped with
a thermocouple.[2] The melting (or solidification) point of 1C2NB is 35.4°C. In the
experiment, this temperature is reached without solidification. Upon further cooling,
at 28.8°C, there is still no sign of solidification, i.e., 1C2NB in the test tube is still
supercooled. At 23.2°C (i.e., 12.2°C below the actual solidification point), the first
crystals form. (Even the smallest amount of impurities present can act as nucleation
sites for crystals to form. In some cases, just physically jarring the substance can lead
to a release of the melting enthalpy and the substance spontaneously crystallizes.)
The heat energy released as the substance crystallizes causes the temperature to rise
to 31.2°C with a concomitant increase of the entropy.

Another counter-example is the phase transformation of solid helium to liquid
helium II, which is not accompanied by release of energy (latent heat). Helium II
is one of the two liquid forms of the helium isotope, ^{4}He or helium 4. (Helium 4
makes up over 99% of naturally occurring helium.) Helium I acts more or less like
a conventional liquid, whereas Helium II has some strange properties. The lack of
latent heat release during phase transformation means that the entropy stays constant
during the process with the implication of the same level of disorder in solid and in
liquid phases. The transition temperature is 2.17 K. (It varies slightly with pressure.)
This was first recognized by the Soviet scientist Shubnikov in 1936, who referred to
the liquid crystal nature of helium II [10].[3]

4.2.4 RECAP

In this chapter, we covered the derivation of the entropy starting from the examination of the microscopic particles (e.g., individual molecules) comprising the macroscopic systems (e.g., a volume of gas). At this point, at least two things must be clear
to the reader:

1. Definition of entropy has nothing to do with time.
2. Entropy can only be defined for a *precisely defined* system at *equilibrium*.

Not only does entropy *not* change with time, but it does also *not* strive to a maximum. As the derivation of the Gibbs's entropy (Equation 4.48 earlier), using the
canonical ensemble method in Section 14.4 indicates, *at the given equilibrium state,
it is already at a maximum*. The easiest way to explain this observation is to use the
analogy between Gibbs's entropy and Shannon's "entropy" (Shannon's "measure of
information" is a better term – see Ref. [5] in Chapter 3 for an in-depth discussion

of this) and refer to the graph in Figure 4.4 for a *loaded dice* as an example. The maximum occurs at the point of equal probabilities for each outcome (i.e., maximum uncertainty). If there is a preference for a particular outcome (i.e., reduced uncertainty with increased amount of information coming from the *loaded* dice), the entropy (uncertainty) decreases.

Similarly, for a closed, isothermal thermodynamic system (E, V, N), represented by the canonical ensemble, there are no preferred microstates. Probability distribution of the energy states can be shown to be a normal distribution with a mean value of $\bar{E}$ and standard deviation of the order of $10^{-10} \cdot \bar{E}$. In other words, the distribution is practically a Dirac δ-function at $E = \bar{E}$ (For an elegant mathematical derivation of this statement, refer to pp. 33–35 in Hill [4].) Furthermore, the average value $\bar{E}$ is practically identical to the *most probable* value E^* with the highest probability, $P^* = 1/\Omega$. From these two observations, it logically follows that the entropy calculated via Equation 4.48 is a maximum, too.

Coming back to the question of what entropy really *is* (i.e., what it really *measures*), what we learned in this chapter does not really support the common interpretation that entropy measures the disorder of a system. As shown at the end of the preceding section, there are well-known experimental observations contradicting that view. Furthermore, the emergence of the 2LT on its own from the microscopic analysis of the system (i.e., independently) is not exactly the case. As shown at the end of Section 14.3, the expression for entropy is derived from the first law of thermodynamics expressed by the Gibbs formula, $dU = TdS - pdV$, by *associating* it with the statistical mechanical result expressed by Equation 14.41. In other words, the origins of the statistical mechanical formula for entropy (i.e., Equations 14.44 through 14.46) go back to the changes in the internal energy of the system and the first law for a closed system (i.e., $dU = \delta Q - \delta W$), which establishes the Clausius *equality*, $\delta Q = TdS$. This, of course, is Equation 3.10, which was derived based entirely on macroscopic considerations. In fact, as stated in Ref. [8] (p. 221), "entropy is a macroscopic property with no microscopic counterpart."

The preceding discussion leads one to the somewhat disturbing conclusion that the second law of thermodynamics is not an independent law after all. It is derived from the first law using the theoretical construct of a *reversible heat engine* (i.e., the *Carnot* engine). To be comfortable with this assertion, the reader is encouraged to follow the discussion in Chapter 3 around Equations 3.1 through 3.6. Entropy, as it emerges from that discussion, *does not measure anything*. It is simply an *invented quantity* for computational convenience to quantify the equivalence between *heat* and *work*. It is defined strictly for a system in equilibrium, and as such, it is a *state variable*. Once it is defined, however, it is amenable to computation from other thermodynamic state variables, such as pressure, volume, or temperature (any combination of two of them), which actually *do measure something real*.

In Chapter 3, it was emphasized that the second law could not be reduced to a simple statement. It is expressed through its two corollaries, which basically state that there are no perfect heat engines or refrigerators. The perfectness in question rests upon the theoretical construct of a *reversible process*, which is a continuous string of consecutive equilibrium states. (Remember that entropy and other thermodynamic variables can only be defined for an equilibrium state.) If all the constituent processes in a heat engine are reversible, its efficiency represents the *theoretical maximum* (i.e., the Carnot engine and Carnot efficiency). Any real engine comprising a combination of real *irreversible* processes is bound to have a lower efficiency. This assertion leads one to the *Clausius inequality*, Equation 3.15, which is the commonly identified formulaic version of the 2LT. It is imperative for the reader to be cognizant of the fact that none of the theoretical considerations leading to the Clausius inequality necessitate the definition of entropy and its unique properties. Its introduction, as stated earlier, is a mathematical trick to ensure simplification of formulae resulting from the said theoretical considerations.

As stated in the beginning, looking at the system at a microscopic level did not directly lead to entropy either. Its identification required introduction of the first law of thermodynamics. The second law associated with the statistical mechanical entropy quantifying the information content via probabilities of energy states is formulated such that, in an isolated system, *spontaneous processes* with decreasing entropy are *extremely unlikely* (but *not impossible*) to take place. This finding is the basis of all the pseudo-scientific explanations of the 2LT. At this point, it is hoped that the reader can see such explanations are not to be taken seriously. As far as a connection between the microscopic considerations and their macroscopic ramifications, this will be undertaken in Section 5.1.2.

Before moving on, however, let us take a closer look at the reversible-irreversible dichotomy. As pointed out by Hollinger and Zennen [8] (p. 69), philosophically, two types of irreversibilities can be identified: *nomological*[4] (i.e., as dictated by a law of nature) and *de facto*. A well-known example of nomological irreversibility is shattering of a porcelain cup. A process is de facto irreversible when its temporal inverse (i.e., its going back to its initial state) never (or hardly ever) occurs with increasing time. (The time in question here is the Newtonian, i.e., *absolute*, time.) There is no law prohibiting the reversal of the process, but it hardly ever occurs. As an example, consider a room in a vacuum except a small corner in the ceiling with a balloon filled with air molecules. When the balloon bursts, the entire room is filled with air. However, the reversal of that process – that is, all the air molecules in the room going to the same corner in the ceiling and leaving the rest of the room in a vacuum – will practically never happen. The probability of this happening, while statistically speaking nonzero, is so small that an immortal observer would not see it even after passing of centuries.

The distinction between nomological and de facto irreversibility is somewhat fuzzy. Clearly, a porcelain cup reconstituting itself from its shards (or a cigarette constituting itself from its ashes) is impossible. But so are the air molecules in a room gathering together in a corner in the ceiling. In the latter case, however, the irreversibility criterion is *probabilistic* and nonentropic. This goes back to the statement made earlier that the entropy has no microscopic counterpart.

Hollinger and Zenner [8] defined a "thermodynamic reversibility" by identifying three criteria of irreversibility: inertial, temporal, and exclusion (pp. 57–66 in the cited work). The reader is encouraged to consult that reference. Herein, we use an example to illustrate the concept by considering a process in which heat is flowing from a hot body A to a cold body B; that is,

$$A(hot) \rightarrow B(cold).$$

According to the *exclusion* criterion, the process is irreversible; that is,

$$A(hot) \leftarrow B(cold).$$

is impossible and *increases* entropy. However, if A and B are near thermal equilibrium (i.e., the temperature difference is *infinitesimally small*), according to the *inertial* criterion, the second process is also possible (i.e., the first process is reversible); that is, it can be reversed in a *short time* with a *slight change* of external influences with no change in entropy. In other words, the system has no *thermal inertia*, quantified by the temperature difference. (According to the *temporal* criterion, the process is always reversible. For an explanation of this counterintuitive assertion, the reader is referred to the cited work.)

NOTES

1 Late Dr. Hollinger was the author's teacher in statistical mechanics at the Rensselaer Polytechnic Institute (RPI).
2 From https://home.alb42.de/versuche/english/v94-1.html#:~:text=A%20supercooled%20 melt%20is%20a,range%20below%20the%20solidification%20point. Last accessed on October 24, 2022.
3 During the Great Purge in 1937, the NKVD launched the Ukrainian Physics and Technology Institute Affair on the basis of falsified charges, and Shubnikov (along with several colleagues) was convicted and executed.
4 *Nomological* means relating to or denoting certain principles, such as laws of nature, that are neither logically necessary nor theoretically explicable but are simply taken as true.

REFERENCES

1. Pauli, W., *Thermodynamics and the Kinetic Theory of Gases: Volume 3 of Pauli Lectures on Physics*, Mineola, NY: Dover Publications, 2010.
2. Schrödinger, E., *Statistical Thermodynamics*, Mineola, NY: Dover Publications, 1989 – originally published in 1952.
3. Tolman, R.C., *The Principles of Statistical Mechanics*, New York: Dover Publications, 1979 – originally published in 1938.
4. Hill, T.L., *An Introduction of Statistical Thermodynamics*, Mineola, NY: Dover Publications, 1986 – originally published in 1962.
5. Hill, T.L., *Statistical Mechanics – Principles and Selected Applications*, Mineola, NY: Dover Publications, 1987 – originally published in 1956.
6. Wannier, G.H., *Statistical Physics*, Mineola, NY: Dover Publications, 1987 – originally published in 1966.
7. Prigogine, I., *Non-Equilibrium Statistical Mechanics*, New York: Dover Publications, 2017 – This is an unabridged republication of the 1962 original.

8. Hollinger, H.B., and Zenzen, M.J., *The Nature of Irreversibility*, Dordrecht: D. Reidel Publishing Company, 1985.
9. Pauling, L., *General Chemistry*, New York: Dover Publications, 1988 – originally published in 1970.
10. Shannon, C.E., A Mathematical Theory of Communication, *Bell System Technical Journal*, 27 (3), pp. 379–423, 1948.
11. Shubnikov, L.W., and Kikoin, A.K., Optical Experiments on Liquid Helium II, *Nature*, 138 (3493), p. 641, https://doi.org/10.1038/138641a0.

Section II

Obeying the Law

5 Second Law in Practice

In theory, theory and practice are the same. In practice, they are not.
—Albert Einstein

In Chapter 3, the groundwork for understanding the second law was laid down using the key postulates of classical thermodynamics. In Chapter 4, the same exercise was repeated by making use of statistical mechanics. At this point, the assumption is that the reader has a firm grasp of what entropy really is (for all practical purposes, not more than a mathematical trick) and what the second law really tells us (hard to encapsulate in a simple statement). At the end of the day, however, the objective is to put that theoretical understanding to use in solving practical problems encountered in thermal power engineering.

This was already done in Chapter 3 by looking at a particular heat engine (i.e., the gas turbine and at a particular process in that heat engine, i.e., combustion) to illustrate the application of the second law using quantitative examples. Herein, the focus will be on the most important engineering tool provided to us by the second law, namely, the *exergy*. Exergy (in some older US textbooks, *availability*) is the *maximum theoretical work* obtainable from a system as it interacts with its environment to equilibrium. For an undergraduate level of discussion of exergy, the reader is referred to chapter 7 of Moran and Shapiro (Ref. [1] in Chapter 2). For a comprehensive treatise on application of exergy analysis to common problems in thermal engineering, the best resource is the monograph by Kotas (Ref. [1] in Chapter 1). For combination of exergy and economic analysis of engineering problems, the reader should consult the book by Bejan et al. (Ref. [4] in Chapter 2.)

5.1 EXERGY

5.1.1 THE THERMODYNAMIC PROPERTY

Herein, the concept will be introduced using a practical example (i.e., a condensing steam turbine rated at 60 MWe). Pertinent heat and mass balance data from the THERMOFLEX model of the turbine is shown in Figure 5.1.

The source of power generated by the turbine in Figure 5.1 is *main steam* at 100 bara and 565°C (enthalpy, h_1 = 3,536.9 kJ/kg) with a flow rate of 50 kg/s. The question is this: What is the *maximum useful work* that one can generate utilizing this source of thermal energy? Theoretically, one can think of an unspecified process through which the steam (the *system*) is brought down from its high energy state to a low energy or *rest* state, when it is at equilibrium with its surroundings (the *environment*). What is meant by environment is another system, surrounding our system of interest, which is so large that it is at constant (and uniform) T_0 and P_0 no matter what

DOI: 10.1201/9781003247418-7

$$\frac{\text{bar} \mid \text{C}}{\text{kg/s} \mid \text{kJ/kg}}$$

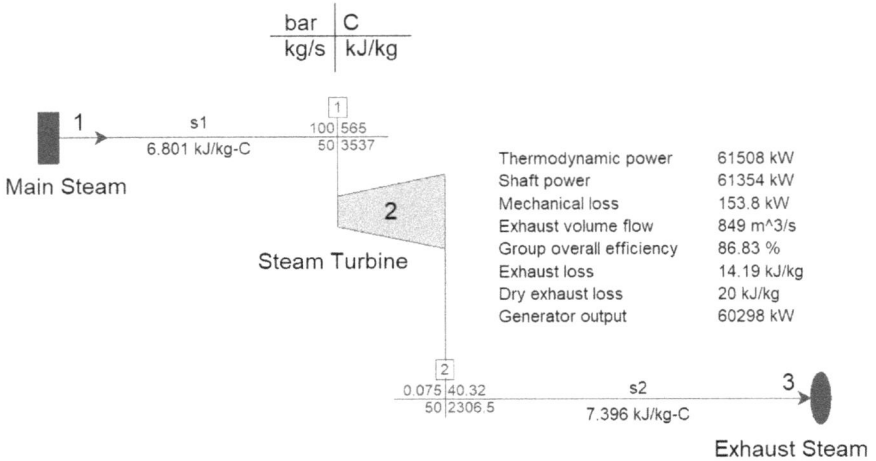

FIGURE 5.1   Steam turbine example for the exergy concept.

the interaction between itself and our system. The logical choice for defining the environment is the ambient conditions, say, at $P_0 = 1$ atm (1.01325 bar) and $T_0 = 15°C$ (288.15 K). (In passing, this reference (rest) state defining the environment is commonly referred to as the *dead state*.) Consequently, in this hypothetical process, we can envision that the enthalpy of steam is brought down to $h_0(P_0, T_0) = 63.0$ kJ/kg (at which point, the fluid is not steam anymore; it is compressed liquid). Thus, the heat given up by steam/water in this process is

$$q = h_1 - h_0 = 3,536.9 - 63.0 = 3,473.8 \text{ kJ / kg.}$$

(Enthalpies are from the *ASME Steam Tables* [1].) Now, let us consider a (hypothetical, of course) *Carnot engine* that is making use of this amount of heat to generate useful work. The hypothetical isothermal heat addition in this Carnot engine takes place at a temperature given by (noting that $s_0(P_0, T_0) = 0.2243$ kJ/kg-K from the *ASME Steam Tables*)

$$T_H = \frac{h_1 - h_0}{s_1 - s_0} = \frac{3536.9 - 63.0}{6.8010 - 0.2243} = 528.2 \text{ K.}$$

Furthermore, this Carnot engine rejects heat to the environment isothermally at $T_0 = 288.15$ K so that its efficiency is given by

$$\eta_{Carnot} = 1 - \frac{T_0}{T_H} = 1 - \frac{288.15}{528.2} = 0.4545 = 45.45\%.$$

Thus, work generated by this Carnot engine is found as

$$w_{Carnot} = 45.45\% \times 3,473.8 = 1,578.7 \text{kJ} / \text{kg}.$$

This is the *maximum theoretical work* that can be generated by utilizing steam at 100 bara and 565°C and, referring to the earlier discussion, can be formulated as

$$w_{max} = \left(1 - T_0 \frac{s_1 - s_0}{h_1 - h_0}\right)(h_1 - h_0)$$

$$w_{max} = (h_1 - h_0) - T_0 (s_1 - s_0). \tag{5.1}$$

Since the terms on the right-hand side (RHS) of Equation 5.1 are thermodynamic properties and since any combination of thermodynamic properties is a property itself, it follows that *maximum theoretical work* is a property of the fluid, in this case, steam. This property is referred to as *exergy*, denoted herein by the symbol a; that is,

$$a_1 = (h_1 - h_0) - T_0 (s_1 - s_0). \tag{5.2}$$

Strictly speaking, Equation 5.2 defines the *specific* exergy (in kJ/kg or Btu/lb) of a non-reacting fluid and should include kinetic and potential energy terms, too; that is,

$$a_1 = KE_1 + PE_1 + (h_1 - h_0) - T_0 (s_1 - s_0), \tag{5.3}$$

but they are too small and can be neglected with little or no loss of accuracy in most cases of interest. Total exergy (in kW) is given by

$$A_1 = \dot{m}_1 a_1, \tag{5.4}$$

which, for this example, results in $A_1 = 50 \times 1,578.5 = 78,926$ kW. Actual thermodynamic (expansion) power is

$$50 \times (3536.6 - 2306.5) = 61,507 \text{kW},$$

which is 78% of the theoretical maximum given by Equation 5.4 and quantifies the second law or *exergetic efficiency* of the steam turbine. It is lower than the *isentropic efficiency* of the turbine given by $\eta_s = \Delta h / \Delta h_s = 86.83\%$. The best possible design of a steam turbine is an isentropic machine; that is, with $\eta_s = 100\%$, and $\Delta h_s = (3,536.6 - 2,306.5)/0.8683 = 1,416.7$ kJ/kg. In other words, practically possible maximum exergetic efficiency of the steam turbine with given steam inlet conditions is $1,416.7/1,578.7 = 89.8\%$.

Before moving on, it should be noted that the similarity between a change in Gibbs free energy, g, and exergy for a *pure fluid* at P and T is unmistakable; that is, using Equation 5.2

$$a = [h - T_0 s] - [h_0 - T_0 s_0],$$

so that, comparing with Equation 2.26, we can see that

$$a = g' - g'_0,$$

which shows that the exergy of a pure fluid (ignoring kinetic and potential energy terms) is equal to the change in a *special form* of the Gibbs free energy of the fluid between the initial and dead states, that is, $g' = h - T_0 s$. (Kotas (ibid.) refers to g' as the *specific exergy function* and uses the symbol β.)

For the control volume of the steam turbine in Figure 5.1, the *steady-state, steady-flow* (SSSF) energy balance (i.e., the first law) can be written as

$$0 = \dot{Q} - \dot{W} + \dot{m}(h_1 - h_2), \tag{5.5}$$

and since the expansion in the turbine is adiabatic, Equation 5.5 becomes

$$0 = -\dot{W} + \dot{m}(h_1 - h_2). \tag{5.6}$$

A similar SSSF balance equation can be written for exergy as well. From Equation 5.2, exergy change during expansion is found as

$$a_1 - a_2 = (h_1 - h_2) - T_0(s_1 - s_2). \tag{5.7}$$

Substituting Equation 5.7 into Equation 5.6, one obtains

$$0 = -\dot{W} + \dot{m}(a_1 - a_2) + \dot{m}T_0(s_1 - s_2). \tag{5.8}$$

Equation 5.8 is the SSSF exergy balance for a control volume that is of most interest in practical engineering calculations. A more generalized version for systems including heat transfer between the system (control volume) and its surroundings can be derived as follows:

$$0 = \dot{Q}\left(1 - \frac{T_0}{\overline{T}}\right) - \dot{W} + \dot{m}(a_1 - a_2) - \dot{I}. \tag{5.9}$$

In Equation 5.9, $\dot{Q}$ is the heat transfer, which is a negative (positive) value if out of (into) the system, and $\overline{T}$ is the average temperature at which the heat transfer takes place. Thus, the four terms on the RHS of Equation 5.9 are as follows, from left to right:

- Reversible work (in a hypothetical Carnot engine operating between $\overline{T}$ and T_0) associated with heat transfer $\dot{Q}$ which quantifies exergy *transfer* with heat

- Useful work done by the system, $\dot{W}$ (positive) or on the system (negative), which quantifies exergy transfer with work
- Net exergy transfer into the system (control volume) with material flow
- Exergy destruction $\dot{I}$,

Equation 5.9 can be written in per unit mass flow rate terms as well; that is,

$$0 = q\left(1 - \frac{T_0}{T}\right) - w + (a_1 - a_2) - i. \tag{5.10}$$

The translation between first and second law equations can be seen by rewriting Equation 5.10 with the help of Equation 5.2; that is,

$$0 = \{q - w + (h_1 - h_2)\} - q\frac{T_0}{T} - T_0(s_1 - s_2) - i. \tag{5.11}$$

The first term on the RHS of Equation 5.11 is equal to zero via the first law so that

$$0 = -\frac{q}{T} - (s_1 - s_2) - \frac{i}{T_0}, \text{ or} \tag{5.12}$$

$$s_2 - s_1 = \frac{q}{T} + \sigma, \tag{5.13}$$

which, of course, is the second law equation that we derived earlier in Chapter 3 (e.g., see Equation 3.12b). Voila!

In Equation 5.13, $\sigma = i/T_0$ (in total terms, $\Sigma = \dot{I}/T_0$) represents entropy produced by internal irreversibilities, which, of course, is the same as exergy destruction scaled by a factor of $(1/T_0)$.

Coming back to the example herein, lost work is calculated as the difference between actual and isentropic work; that is, using Equation 2.28 for a turbine,

$$i = (h_1 - h_{2s})(1 - \eta_s) = 1,416.7(1 - 86.83\%) = 186.6 \text{ kJ/kg}.$$

Entropy increase during turbine expansion is given by

$$\Delta s = s_2 - s_1 = 7.3963 - 6.801 = 0.5953 \text{kJ/kg} - \text{K},$$

so that the scaling factor or reference temperature is calculated as

$$T_0 = \frac{i}{s_2 - s_1} = \frac{186.6}{0.595} = 313.4 \text{ K} = 40.3 °C.$$

The difference between results obtained with this true reference temperature and the arbitrarily selected reference temperature, 15°C, is illustrated in the comparison in Table 5.1. For given turbine inlet conditions, exhaust pressure and technology (represented by the isentropic efficiency), isentropic, actual, and lost work are fixed. Furthermore, for given turbine inlet conditions and exhaust pressure, theoretically possible maximum work is the isentropic work. This is equal to the exergy of steam at turbine inlet, a_1, when T_0 is set to 40.3°C.

For the arbitrary reference temperature (temperature of the *environment*), a_1 = 1,587.5 kJ/kg, which is higher than the isentropic work by 1,587.5 − 1,416.7 = 161.8 kJ/kg. This is the work that *could* be produced if steam at *isentropic* turbine exhaust conditions *could* go through another, *hypothetical* work-producing process that would bring it to an equilibrium with the environment (see later for more on this). This, of course, is a practically hopeless proposition (for one, steam pressure at the exhaust is *partial vacuum*). Leaving the mechanical impossibility aspect for now aside, to quantify the hopelessness of the proposition, consider that the mean-effective temperature, $\overline{T} = (h_2 - h_0)/(s_2 - s_0) = 312.8K$ which translates into a Carnot efficiency of $1 - T_0/\overline{T} = 7.9\%$. Exergy of steam at turbine exhaust conditions is calculated as a_2 = 176.9 kJ/kg. Even with this hypothetical Carnot engine, work potential is 7.9% × 176.9 ~ 14 kJ/kg, which, in real terms, translates into pretty much nothing.

Finally, in Table 5.1, we note that there is a difference of 15.1 kJ/kg between lost work and the exergy destruction term with the reference temperature, T_0 = 15°C. Taking a closer look at this difference, we have

$$\left(h_1 - h_{2s}\right) - \left(h_1 - h_2\right) - T_0\left(s_2 - s_1\right) = 15.1\,kJ/kg$$

$$h_1 - h_{2s} = a_1 - a_{2s}$$

$$\left(a_1 - a_{2s}\right) - \left(a_1 - a_2\right) = 15.1\,kJ/kg$$

TABLE 5.1
Second Law Calculations for the Steam Turbine in Figure 5.1

Reference temperature, °C	15.0	40.3
Reference enthalpy, kJ/kg	63.0	168.7
Reference entropy, kJ/kg-K	0.2243	0.5759
Actual work, kJ/kg	1,230.1	1,230.1
Isentropic work, kJ/kg	1,416.7	1,416.7
Lost work, kJ/kg	186.6	186.6
$s_2 - s_1$, kJ/kg-K	0.5953	0.5953
$T_0(s_2 - s_1)$, kJ/kg	171.5	186.6
a_1, kJ/kg	1,578.5	1,416.7

$$a_2 - a_{2s} = 15.1 \, kJ/kg$$

Consequently, since a_2 was calculated as 176.9 kJ/kg, a_{2s} = 161.8 kJ/kg, which proves the assertion made earlier.

5.1.2 COMPONENTS OF EXERGY

As stated earlier, Equation 5.2 defines the *specific* exergy (in kJ/kg or Btu/lb) of a non-reacting fluid. To be precise, it is the *physical* exergy of a material stream and is also known as the *flow* exergy. Although not readily apparent from Equation 5.2, physical exergy (ignoring the kinetic and potential energy terms) has two components: *thermal* and *pressure*. The difference between the two can be best understood when Equation 5.2 is written for a perfect gas (i.e., obeying the ideal gas law with constant specific heat). Consequently,

$$h_1 - h_0 = c_p \left(T_1 - T_0 \right), \tag{5.14}$$

$$s_1 - s_0 = c_p \ln \frac{T_1}{T_0} - R \ln \frac{P_1}{P_0}, \tag{5.15}$$

$$a_1 - a_0 = c_p \left(T_1 - T_0 \right) - T_0 \left(c_p \ln \frac{T_1}{T_0} - R \ln \frac{P_1}{P_0} \right), \tag{5.16}$$

$$\Delta a_{Th} = c_p \left(T_1 - T_0 \right) - T_0 c_p \ln \frac{T_1}{T_0}, \tag{5.17}$$

$$\Delta a_{Pr} = R T_0 \ln \frac{P_1}{P_0}, \tag{5.18}$$

$$a_1 - a_0 = \Delta a_{Th} + \Delta a_{Pr}. \tag{5.19}$$

Let us consider air as perfect gas with the following assumptions: molecular weight, MW = 29 kg/kmol; c_p = 1.005 kJ/kg; T_0 = 15°C; P_0 = 1 atm; and R_{unv} = 8.314 kJ/kg-K. Exergy terms appearing in Equations 5.14 through 5.19 are plotted in Figure 5.2 as a function of temperature. The pressure component of physical exergy is calculated for P_1 = 1.2 atm (i.e., $P_1 > P_0$) and P_1 = 0.8 atm (i.e., $P_1 < P_0$) as 15.1 kJ/kg and −18.1 kJ/kg, respectively.

The first observation from Figure 5.2 is that the thermal component of physical exergy is always positive (i.e., even when $T_1 < T_0$). (As expected, it is zero when $T_1 = T_0$.) This is somewhat difficult to grasp intuitively. When $T_1 > T_0$, rewriting

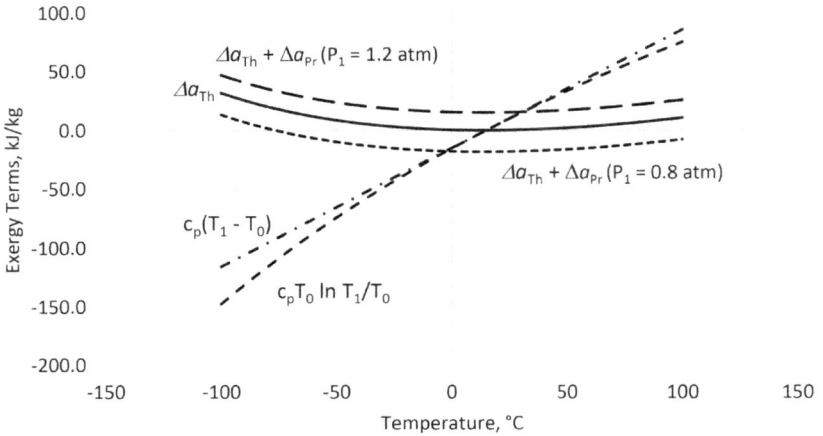

FIGURE 5.2 Physical exergy components.

Equation 5.17 as follows makes the physical significance of thermal energy eminently clear:

$$\Delta a_{Th} = c_p \left(T_1 - T_0 \right) \left(1 - \frac{T_0}{\left(T_1 - T_0 \right) \Big/ \ln \frac{T_1}{T_0}} \right) = q_{rev} \left(1 - \frac{T_0}{\overline{T}} \right). \quad (5.20)$$

According to Equation 5.20, the thermal component of physical exergy is the work produced in a Carnot engine receiving the heat quantified by the temperature difference between the fluid's extant temperature, T_1, and the dead state temperature, T_0, q_{rev}, reversibly and isothermally at the mean-effective (the logarithmic mean) temperature, $\overline{T}$ and rejecting heat to the environment reversibly and isothermally at the dead state temperature, T_0. In other words, it is a quantification of the Kelvin-Planck statement of the 2 LT for a specific fluid stream.

This perfectly sensible interpretation falls apart when $T_1 < T_0$. In that case, we have to look at the problem in a different way. Note that the work done by a Carnot engine receiving heat from a high temperature reservoir and rejecting heat to a low temperature reservoir can be expressed in two different ways. Namely,

$$w = q_H - q_L, \text{ i.e.} \quad (5.21)$$

$$w = q_H \left(1 - \frac{q_L}{q_H}\right) = q_H \left(1 - \frac{T_L}{T_H}\right), \text{ or} \tag{5.22}$$

$$w = q_L \left(\frac{q_H}{q_L} - 1\right) = q_L \left(\frac{T_H}{T_L} - 1\right). \tag{5.23}$$

Equation 5.22 is the conventional version with the term in parentheses on the RHS of the equation representing the *efficiency* of the Carnot engine (the numerical value less than one). In the alternative version, Equation 5.23, the Carnot engine work is referenced to the heat *rejected* from the engine, and the term in parentheses on the RHS of the equation represents a *coefficient of performance*, which is greater than one.

In order to demonstrate how this different interpretation works, let us rewrite Equation 5.17 as follows:

$$\Delta a_{\mathrm{Th}} = -c_p (T_0 - T_1) + T_0 c_p \ln \frac{T_0}{T_1}, \tag{5.24}$$

$$\Delta a_{\mathrm{Th}} = c_p (T_0 - T_1) \left(\frac{T_0}{(T_0 - T_1) \Big/ \ln \frac{T_0}{T_1}} - 1 \right) = q_{rev} \left(\frac{T_0}{\overline{T}} - 1 \right). \tag{5.25}$$

According to Equation 5.25, the thermal component of physical exergy when $T_1 < T_0$ is the work produced in a Carnot engine *rejecting* the heat quantified by the temperature difference between the fluid's extant temperature, T_1, and the dead state temperature, T_0, q_{rev}, reversibly and isothermally at the mean-effective (the logarithmic mean) temperature, and *receiving* heat from the environment reversibly and isothermally at the dead state temperature, T_0.

According to Equation 5.18, the pressure component of physical exergy when $P_1 > P_0$ quantifies the work *produced* in an isothermal *expander*. Thus, when $P_1 > P_0$, physical exergy is always greater than zero. Once again, this makes intuitively a lot of sense and does not require any more elaboration. Now, when $P_1 < P_0$, the pressure component of physical exergy is *negative*, and depending on its exact numerical value, the physical exergy itself can be a negative value. In other words, the pressure component of physical exergy quantifies the work *consumed* by an isothermal *compressor*. If this work is more than the work quantified by the thermal component (i.e., the work produced by a Carnot engine), the physical exergy is *negative*. If it is more than compensated by the thermal component, the physical exergy is *positive*.

In most practical problems of interest in power (heat engine) cycles, we do not have to worry about the possibility of $T_1 < T_0$ or $P_1 < P_0$. Furthermore, in exhaust gas exergy or in heat exchange irreversibility calculations, we can assume $P_1 \sim P_0$ with little loss in accuracy while simplifying the calculations considerably. However, one interesting case is presented by air liquefaction via Linde-Hampson (L-H) process in *cryogenic energy storage* (CES) applications. In particular, while the internal energy and enthalpy of the air is *reduced* in the cryogenic liquefaction process, its *exergy* (i.e., its ability to produce useful work) is *increased*. Why this is so can be easily understood by referring to the *T-s* diagram of the L-H process Figure 5.3.

If one writes the change in exergy of air between states 1 and 7 in the *T-s* diagram in Figure 5.3, the following relationship is obtained:

$$e_7 - e_1 = \left(h_7 - h_1\right) - T_0\left(s_7 - s_1\right) \tag{5.26}$$

$$\Delta e = \Delta h - T_0 \Delta s \tag{5.27}$$

Furthermore, also from the *T-s* diagram, one can *qualitatively* infer that

$$\Delta h < 0, \Delta s \left\langle 0 \,\text{and}\, |\Delta s| \right\rangle |\Delta h| \tag{5.28}$$

Thus, combining Equations 5.27 and 5.28, one can confirm that

$$\Delta e = -|\Delta h| + T_0 |\Delta s| > 0. \tag{5.28}$$

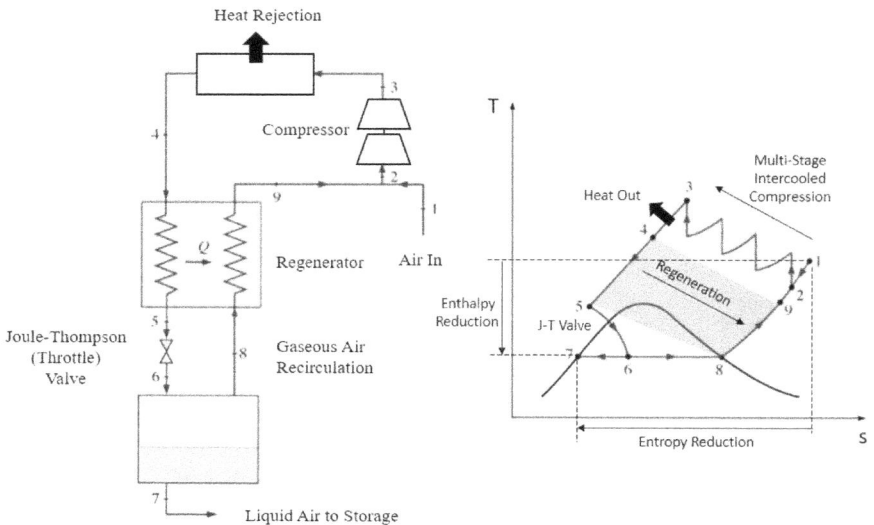

FIGURE 5.3 Schematic description of the basic Linde-Hampson air liquefaction process and its conceptual *T-s* diagram.

This example demonstrates that a reduction in entropy can indeed increase the exergy of the fluid in question, which is exactly what we observe in Equations 5.15 and 5.16. To investigate this observation further, let us consider a change in state from the dead state, (T_0, P_0), to a state at higher temperature and pressure, (T_1, P_1), which is depicted on an enthalpy-entropy (h-s) diagram in Figure 5.4. The change takes place in two steps:

- Isothermally (at T_0) from P_0 to P_1 ($0 \rightarrow a$)
- Isobarically (at P_1) from T_c to T_1 ($a \rightarrow 1$)

In the first step, the entropy *decreases* by the amount of $R \ln \dfrac{P_1}{P_0}$. In the second step, entropy *increases* by the amount of $c_p \ln \dfrac{T_1}{T_0}$ Thus, the total change in the entropy is

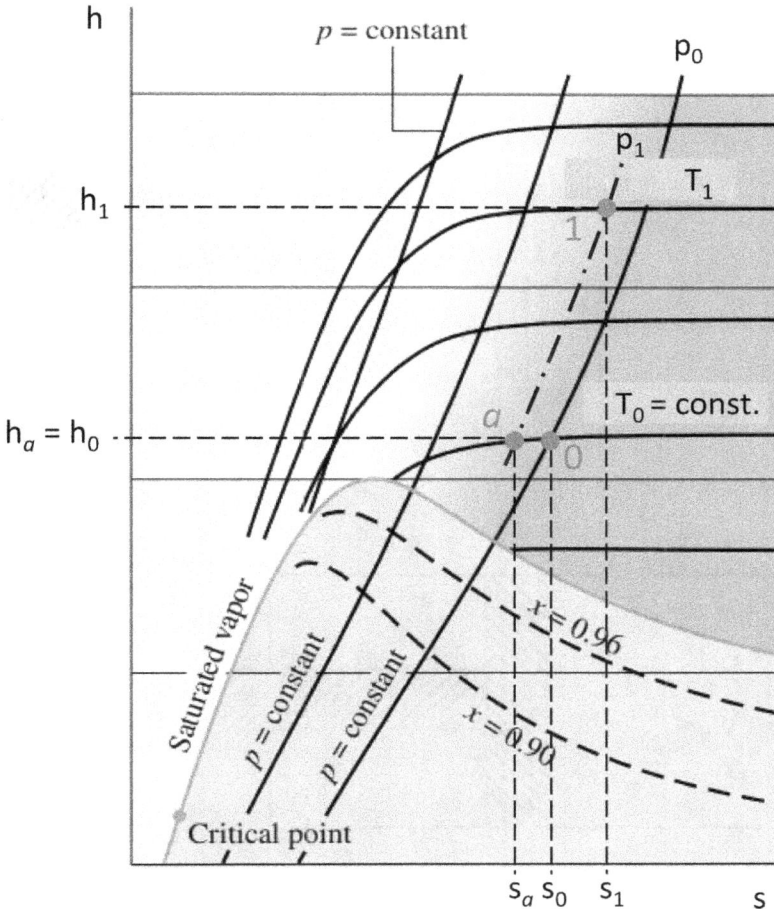

FIGURE 5.4   Exergy change on h-s diagram.

that given by Equation 5.15, which when substituted into the exergy formula, as seen in Equation 5.16, *increases* the exergy by the amount of $R \ln \frac{P_1}{P_0}$ while *decreasing* it by the amount of $c_p \ln \frac{T_1}{T_0}$. In other words, the mechanical component of the physical exergy is a direct result of the entropy reduction via isothermal rise in pressure.

The relationship between entropy and pressure deserves a closer look. Consider the piston-cylinder system in Figure 5.5, which depicts an isothermal process with $W = Q$ (i.e., the change in internal energy, $\Delta U = 0$). The entropy change for the process is given by

$$\Delta s = s_a - s_0 = -R \ln \frac{P_a}{P_0} = +R \ln \frac{V_a}{V_0} < 0. \tag{5.30}$$

The reduction in entropy can be explained by the reduction in volume, V, which is accompanied by a rise in pressure, P, because the temperature of the system does not

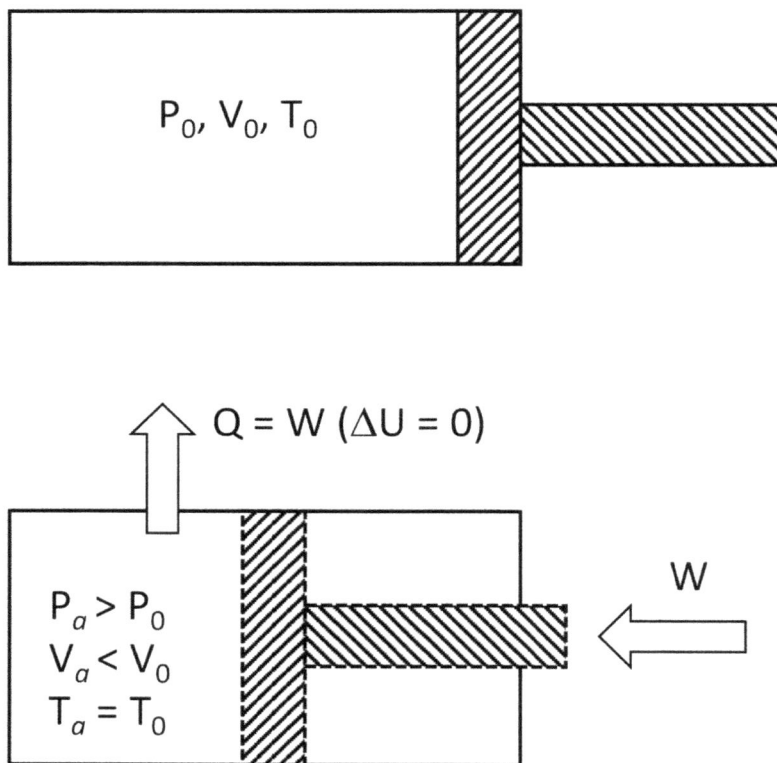

FIGURE 5.5 Isothermal process in a piston-cylinder system.

change. Recall the definition of the *Gibbs's entropy* that was derived in Section 4.2.2, reproduced subsequently for convenience,

$$S = -k\sum_{i} P_i \ln P_i,$$

(5.48)

where P_i is the probability that the system is in energy state ε_i given by

$$P_i = \frac{\exp\left(-\dfrac{\varepsilon_i}{kT}\right)}{Q}.$$

(5.49)

For equal probability of energy microstates (i.e., $P_i = \dfrac{1}{Q}$), Equation 4.48 leads to the *Boltzmann entropy*, $S = k\ln Q$ Note that Q is the *partition function* given by (for a system of N molecules)

$$Q = \frac{q^N}{N!}.$$

(4.22)

where $q = \sum_{j} \exp\left(-\dfrac{\mu_j}{kT}\right)$ is the partition function for energy states ε_j for a molecule (in a system) or a subsystem (in an *ensemble* of systems). Recall that, in Section 4.2.2, it was shown that (with mean-free path of molecules, λ)

$$q = \left(\frac{2\pi mkT}{h^2}\right)^{\frac{3}{2}} V = \frac{V}{\lambda^3}.$$

(4.33)

Equation 4.22 can be expanded to result in

$$\ln Q = -N\left(\ln N - 1 - \ln q\right),$$

(4.38)

so that the Boltzmann entropy becomes

$$S = -kN\left(\ln N - 1 - \ln\frac{V}{\lambda^3}\right) = kN\ln\frac{V}{\lambda^3} + kN\left(1 - \ln N\right), \text{ or}$$

$$s = \frac{S}{N} = k\ln\frac{V}{\lambda^3} + k\left(1 - \ln N\right)$$

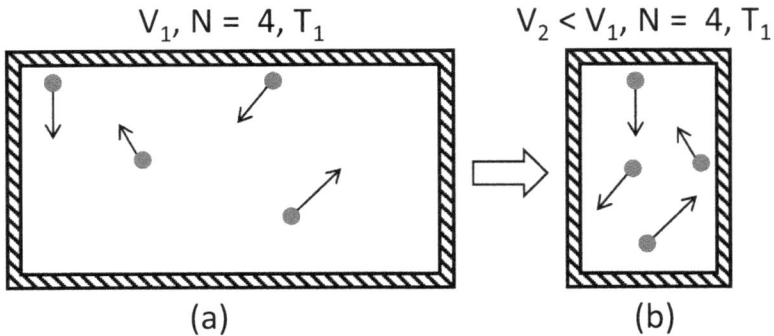

FIGURE 5.6   Isolated box example to demonstrate the relationship between boxes.

Thus, from a microscopic perspective, the relationship between entropy, S, and volume, V, can be traced to the *available number of energy states* for the particles in the system. For a given number of particles, when the volume occupied by them decreases, the available number of energy states for them decreases as well. One way to *visualize* the significance of this statement is to look at a system comprising only four particles in an isolated box as shown in Figure 5.6. In box (a), the space as a proxy for energy states available to the particles is much larger than in box (b). Hence, the entropy of the system is higher in (a) than in (b).

5.2   FUEL EXERGY

5.2.1   GENERAL

Exergy (also known as *availability*) is defined as the maximum theoretical work obtainable as the system in question interacts with the environment to equilibrium. For a given cycle working fluid, *physical* exergy per unit mass is a thermodynamic property and can be calculated from two other (known) properties, say, pressure and temperature, using a suitable equation of state. The ability of producing useful work embedded in a given fuel is quantified by its *chemical* exergy. Chemical exergy is equal to the maximum amount of work obtainable when the substance under consideration (i.e., the fuel) is brought from the environmental state to the dead state by a process involving heat transfer and exchange of substances only with the environment (p. 44 in Kotas, Ref. [1] in Chapter 1).

The *environmental state* is defined as the state of *restricted* equilibrium with the environment (p. 34 ibid.). In other words, the system under consideration and the environment have the same pressure and temperature. The term *restricted* means that the substances of the system are restrained by a physical barrier preventing the exchange of matter between the system and the environment. For fuels, this state is typically defined as 1 atm (101,325 Pa) pressure and 25°C (298.15 K) temperature. For power plant analysis, it is the prevailing site ambient conditions. For generic analysis, ISO conditions, 1 atm and 15°C are adopted.

The term *dead state* refers to an *unrestricted* (mechanical, thermal, and chemical) equilibrium between the system and the environment. In addition to pressure and temperature, chemical potentials of the substances of the system and the environment are equal as well (p. 34 ibid.). For example, CO_2 at pressure P and temperature T in a vessel, where P and T are the same as those of the ambient air outside the vessel, is in an *environmental* state. If, however, pressure of CO_2 in the vessel is equal to its partial pressure in the ambient air outside the vessel, it is in a *dead* state.

Calculating the chemical exergy of a combustible substance (i.e., fuel) is straightforward but tedious and requires access to property data. Full equations and worked-out examples can be found in several textbooks (e.g., chapter 13 in Moran and Shapiro, Ref. [1] in Chapter 2). For a more in-depth coverage, the monograph by Kotas is recommended (ibid.). The difficulty and/or tediousness are even higher for solid and liquid fuels (e.g., coal and fuel oil) because they comprise a large number of constituents, some of unknown nature, which makes it difficult to determine the entropy of reaction. One approach is to assume that the ratio of the chemical exergy of a solid or liquid fuel to its *lower heating value* (LHV) is the same as for pure chemical substances having the same ratios of constituent chemicals; that is,

$$\varphi = \frac{e_f}{LHV}$$

Relationships for φ as a function of the mass fractions of C, H, O, and N can be found in Table C.1 in Appendix C of Kotas (ibid.). For different types of coal, φ is 1.06 to 1.10. For natural gas, φ is $1.04 \pm 0.5\%$ (i.e., between 1.035 and 1.045). In other words, chemical exergy of coal is 6 to 10% higher than its LHV (interestingly, almost equal to its *higher heating value*, HHV). For natural gas, the HHV/LHV ratio is around 1.1.

In Bejan et al. (Ref. [4] in Chapter 2), two values are provided for molar chemical exergy of methane (Table C.2, on p. 522 of the cited work):

- 824,348 kJ/kmol (51,393.3 kJ/kg), reference conditions 1.019 atm and 25°C
- 831,650 kJ/kmol (51,849.5 kJ/kg), reference conditions 1.0 atm and 25°C

(Sources from which the listed values have been obtained can be found in Ref. [4] in Chapter 2.) Higher heating value (HHV) of methane was calculated as 55,507 kJ/kg in Section 2.3.2.2; its LHV was calculated as 50,019 so that HHV/LHV = 1.11. Based on the numbers cited earlier, the ratio of chemical exergy of methane to its LHV is 1.027 to 1.036, which is in concurrence with the range cited by Kotas for natural gas (which is, after all, mostly methane).

5.2.2   A Detailed Look

In order to evaluate the exergy of a fuel, we have to account for two separate contributions to exergy: thermomechanical and chemical. The thermomechanical (physical) part is the same as before; that is, in per unit mole terms

$$\overline{a}_{tm} = \left(\overline{h} - \overline{h}_0\right) - T_0\left(\overline{s} - \overline{s}_0\right). \tag{5.31}$$

For the fuel stream going into the combustor of a gas turbine, the thermomechanical exergy is typically negligible. This can be readily illustrated for the fuel (CH_4) conditions in Figure 2.2, 499.8 K and 30.64 bar, using the equations provided in Section 14.1.1; that is,

$$\overline{a}_{tm} = 8,205\,\text{kJ/kmol} - 298.15\,\text{K}\cdot\left(171.33 - 178.89\right)\text{kJ/kmol-K} = 5,951\,\text{kJ/kmol}$$

which comes to 5,951/16 = 371 kJ/kg, which is 371/50,019 = 0.0074 or less than 1% of the LHV.

The chemical part is applicable to hydrocarbon fuels and pure water at T_0 and P_0. For the fuels, the basis for chemical exergy is the reaction with O_2 given by

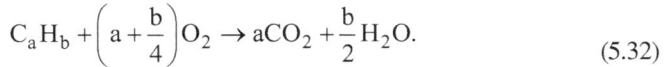

$$C_a H_b + \left(a + \frac{b}{4}\right)O_2 \rightarrow aCO_2 + \frac{b}{2}H_2O. \tag{5.32}$$

For example, for methane (CH_4), $a = 1$ and $b = 4$ so that we have

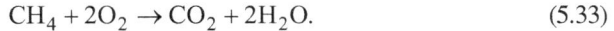

$$CH_4 + 2O_2 \rightarrow CO_2 + 2H_2O. \tag{5.33}$$

For calculation of fuel chemical exergy, a hypothetical device, also known as a *van't Hoff box*, as shown in Figure 5.7, is assumed. What is happening in this device is based on the following assumptions:

- The fuel, $C_a H_b$, enters the device at T_0 and P_0
- The environment is air at $T_0 = 298.15$ K and $P_0 = 1$ atm with the following volume or molar composition: oxygen, 20.738%; nitrogen, 77.292%; water vapor, 1.009%; carbon dioxide, 0.030% (i.e., 300 ppmv, which is not true these days at more than 400 ppmv); and argon, 0.931%
- Heat transfer between the device and the environment takes place at T_0
- The volume of the device is fixed; that is, no work is done to displace the environment
- The oxidant, O_2, enters the device at T_0 and its partial pressure in the environment, which is the product of its mole fraction and P_0
- Inside the device, the fuel and O_2 react completely to produce CO_2 and H_2O (as water vapor, i.e., in a gaseous state)
- CO_2 and H_2O exit the device at their environmental conditions; that is, at T_0 and their respective partial pressures (products of their mole fractions and P_0)
- Kinetic and potential energy contributions are negligible
- There are no irreversibilities

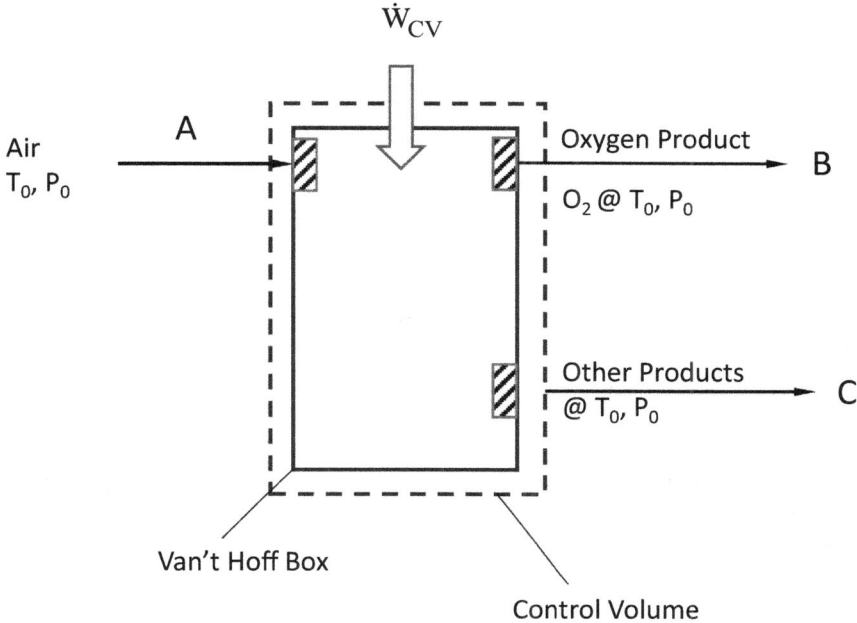

FIGURE 5.7 Hypothetical device (van't Hoff box) as a framework for fuel exergy calculation.

It should be pointed out that, while the device shown in Figure 5.7 looks like a *fuel cell*, it is not. A fuel cell is a device that produces electric power directly from a chemical reaction. The device in Figure 5.7 produces shaft work, but it is not a *heat engine* either because while it receives heat from the environment, it does *not reject* heat into the environment. Also note that each reactant, or product, is shown as an individual stream, which is deliberate. Each species enters the device (a reactant) and/or exits from it (a product) through a *selectively permeable membrane*, which allows the passage of only one species.

In order to calculate the standard chemical exergy of individual gases in gas mixtures and the mixtures themselves, we start with the SSSF exergy balance for a control volume:

$$0 = \dot{Q}\left(1 - \frac{T_0}{T}\right) - \dot{W} + \dot{m}\left(a_1 - a_2\right) - \dot{I}. \qquad (5.9)$$

Next, we imagine a control volume, into which a particular gas enters at T_0 and P_0 and expands isothermally and reversibly (i.e., the last term on the RHS of Equation 5.9 is zero) to its partial pressure in the standard environment. In other words, the gas exits the control volume at T_0 and $P = y_e P_0$, where y_e is the mole fraction of the gas in the standard environment. During this process, heat transfer between the control volume and the environment takes place at T_0 (i.e., the first term on the RHS of Equation 5.9 is zero). Thus, Equation 5.9 becomes,

$$\dot{W}_{rev} = \dot{m}(a_1 - a_2).$$ (5.34)

Equation 5.34, can be rewritten in per unit mole terms as

$$\overline{w}_{rev} = \overline{a}_1 - \overline{a}_2,$$ (5.35)

which is the *chemical exergy per unit mole* of the particular gas. Expanding the RHS of Equation 5.35, we obtain

$$a_{ch} = \left(\overline{h}_1 - \overline{h}_2\right) - T_0\left(\overline{s}_1 - \overline{s}_2\right).$$ (5.36)

Since enthalpy is a function of temperature only, the first term on the RHS of Equation 5.36 is zero. Finally, expanding the entropy terms, we obtain for the chemical exergy of the gas

$$a_{ch} = -T_0\left(\overline{s}_1 - \overline{s}_2\right),$$ (5.37)

$$a_{ch} = -T_0\left(\overline{s}_1^{\,o} - \overline{R}\ln\left(\frac{P_0}{P_0}\right) - \overline{s}_2^{\,o} + \overline{R}\ln\left(\frac{y_e P_0}{P_0}\right)\right),$$ (5.38)

$$a_{ch} = -\overline{R}\,T_0\ln y_e.$$ (5.39)

The physical interpretation of chemical exergy is better understood if Equation 5.39 is rewritten in full; that is,

$$a_{ch} = -\overline{R}\,T_0\ln\left(\frac{y_e P_0}{P_0}\right) = \overline{R}\,T_0\ln\left(\frac{P_0}{y_e P_0}\right),$$ (5.40)

which quantifies the maximum amount of work possible in a process involving (i) reversible, isothermal expansion from P_0 to $y_e P_0$, and (ii) reversible heat transfer with the environment at T_0.

The chemical exergy of an ideal gas mixture with N components is given by

$$a_{ch} = -\overline{R}\,T_0\sum_{i=1}^{N} y_i\ln\frac{y_{e,i}}{y_i},$$ (5.40)

where y_i is the mole fraction of component i in the mixture and $y_{e,i}$ is the mole fraction of component i in the environment. Derivation of Equation 5.40 is left as an exercise for the reader.

Now, we can apply Equation 5.39 to the device in Figure 5.7 to evaluate the chemical exergy of a fuel. Based on the assumptions listed earlier, first and last terms on the RHS of the equation are zero. Converting from mass flow rate basis to molar flow rate basis, we have once again Equation 5.34, which can be expanded as follows:

$$\dot{W}_{rev} = \sum_{R} \dot{n}_i \left(\overline{a}_{i,tm} + \overline{a}_{i,ch} \right) - \sum_{P} \dot{n}_j \left(\overline{a}_{j,tm} + \overline{a}_{j,ch} \right), \tag{5.41}$$

where subscripts R and P denote reactants (i.e., C_aH_b and O_2) and products (i.e., CO_2 and H_2O), respectively. Let us now expand the four terms on the RHS of Equation 5.41 separately. In doing that, we will use the subscript F for the fuel. Also note that, since reactants are products are both at T_0, deltas for temperature-dependent terms (i.e., $\overline{h}$ and $\overline{s}°$) are equal to zero. We start with the individual exergy terms:

$$\overline{a}_{R,tm} = \left(\overline{h} - \overline{h}_0 \right)_F - T_0 \left(\overline{s} - \overline{s}_0 \right)_F + \left(\overline{h} - \overline{h}_0 \right)_{O2} - T_0 \left(\overline{s} - \overline{s}_0 \right)_{O2}, \tag{5.42}$$

$$\overline{a}_{R,tm} = -T_0 \left(\overline{s} - \overline{s}_0 \right)_F - T_0 \left(\overline{s} - \overline{s}_0 \right)_{O2}, \tag{5.43}$$

$$\overline{a}_{R,ch} = \overline{a}_{ch,F} - \overline{R}T_0 \ln y_{e,O2}, \tag{5.44}$$

$$\overline{a}_{P,tm} = -T_0 \left(\overline{s} - \overline{s}_0 \right)_{CO2} - T_0 \left(\overline{s} - \overline{s}_0 \right)_{H2O(g)}, \tag{5.45}$$

$$\overline{a}_{P,ch} = -\overline{R}T_0 \ln y_{e,CO2} - \overline{R}T_0 \ln y_{e,H2O(g)}. \tag{5.46}$$

Substituting Equations 5.43–5.46 into Equation 5.41 and dividing both sides with the molar flow rate of the fuel, we find that

$$\frac{\dot{W}_{rev}}{\dot{n}_F} = \overline{w}_{rev} = \overline{a}_{ch,F} - T_0 \left(\overline{s} - \overline{s}_0 \right)_F - \left(a + \frac{b}{4} \right) T_0 \left(\overline{s} - \overline{s}_0 \right)_{O2} - \left(a + \frac{b}{4} \right) \overline{R}T_0 \ln y_{e,O2}$$
$$-a\overline{R}T_0 \ln y_{e,CO2} - \frac{b}{2}\overline{R}T_0 \ln y_{e,H2O(g)} + aT_0 \left(\overline{s} - \overline{s}_0 \right)_{CO2} + \frac{b}{2}T_0 \left(\overline{s} - \overline{s}_0 \right)_{H2O(g)} \tag{5.47}$$

Solving for $\overline{a}_{ch,F}$ in Equation 5.47, the result is

$$\overline{a}_{ch,F} = \overline{w}_{rev} + T_0 \left(\overline{s} - \overline{s}_0 \right)_F$$
$$-a\overline{R}T_0 \ln y_{e,CO2} - \frac{b}{2}\overline{R}T_0 \ln y_{e,H2O(g)} + \left(a + \frac{b}{4} \right)\overline{R}T_0 \ln y_{e,O2}$$
$$+\left(a + \frac{b}{4} \right) T_0 \left(\overline{s} - \overline{s}_0 \right)_{O2} - aT_0 \left(\overline{s} - \overline{s}_0 \right)_{CO2} - \frac{b}{2}T_0 \left(\overline{s} - \overline{s}_0 \right)_{H2O(g)}. \tag{5.48}$$

Applying the first and second law equations to the control volume and combining the two results in the following expression:

$$\overline{w}_{rev} = -\left\{a\overline{h}^{\circ}_{CO2} + \frac{b}{2}\overline{h}^{\circ}_{H2O(g)} - \overline{h}^{\circ}_F\right\} - T_0\left\{\overline{s}_F + \left(a + \frac{b}{4}\right)\overline{s}_{O2} - a\overline{s}_{CO2} - \frac{b}{2}\overline{s}_{H2O(g)}\right\}. \tag{5.49}$$

The first term (in curly brackets) on the RHS of Equation 5.49 is the *lower heating value* of the fuel, LHV, per unit mole at T_0 and P_0 so that

$$\overline{w}_{rev} = LHV - T_0\left\{\overline{s}_F + \left(a + \frac{b}{4}\right)\overline{s}_{O2} - a\overline{s}_{CO2} - \frac{b}{2}\overline{s}_{H2O(g)}\right\}. \tag{5.50}$$

Substituting Equation 5.50 into Equation 5.48 gives us the following equation for the chemical exergy of the fuel:

$$\overline{a}_{ch,F} = LHV + \left\{a\overline{R}T_0 \ln y_{e,CO2} + \frac{b}{2}\overline{R}T_0 \ln y_{e,H2O(g)} - \left(a + \frac{b}{4}\right)\overline{R}T_0 \ln y_{e,O2}\right\}$$
$$-T_0\left\{\overline{s}_{0,F} + \left(a + \frac{b}{4}\right)\overline{s}_{0,O2} - a\overline{s}_{0,CO2} - \frac{b}{2}\overline{s}_{0,H2O(g)}\right\}, \tag{5.51}$$

For methane (CH_4) with $a = 1$ and $b = 4$, Equation 5.51 becomes

$$\overline{a}_{ch,CH4} = LHV_{CH4} - \left\{\overline{R}T_0 \ln y_{e,CO2} + 2\overline{R}T_0 \ln y_{e,H2O(g)} - 2\overline{R}T_0 \ln y_{e,O2}\right\}$$
$$-T_0\left\{\overline{s}_{0,CH4} + 2\overline{s}_{0,O2} - \overline{s}_{0,CO2} - 2\overline{s}_{0,H2O(g)}\right\}, \tag{5.52}$$

Finally, recalling the definition of Gibbs free energy, Equation 5.52 can be reformulated as

$$\overline{a}_{ch,CH4} = \left\{\overline{g}_{0,CH4} + 2\overline{g}_{0,O2} - \overline{g}_{0,CO2} - 2\overline{g}_{0,H2O(g)}\right\}$$
$$-\left\{\overline{R}T_0 \ln y_{e,CO2} + 2\overline{R}T_0 \ln y_{e,H2O(g)} - 2\overline{R}T_0 \ln y_{e,O2}\right\}. \tag{5.53}$$

The first term in the curly brackets on the RHS of Equation 5.53 is the negative of Gibbs free energy change for the reaction, $-\Delta G$, so that

$$\overline{a}_{ch,CH4} = -\Delta G - \overline{R}T_0\left\{\ln y_{e,CO2} + 2\ln y_{e,H2O(g)} - 2\ln y_{e,O2}\right\}. \tag{5.54}$$

The first term was calculated in Section 3.3.1; that is, using Equation 3.58, $-\Delta G = 817{,}950$ kJ/kmol with H_2O as liquid in the products, which is the *maximum reversible work* associated with combustion of methane. Substituting the value for

H_2O as gas (vapor) into Equation 3.58, we calculate $-\Delta G = 800,770$ kJ/kmol. Using the air composition in the assumptions listed earlier, the term in the brackets on the RHS of Equation 5.54 is evaluated as

$$\ln(0.030/100) + 2 \times \ln(1.009/100) - 2 \times \ln(20.738/100) = -14.1577,$$

so that the second term on the RHS of Equation 5.54 becomes

$$-8.314 \text{kJ} / \text{kmol} - \text{K} \times 298.15\text{K} \times -14.1577 = 35,094.5 \text{kJ} / \text{kmol}.$$

Finally, the chemical exergy of CH_4 is calculated as

$$\bar{a}_{ch,CH4} = 800,770 + 35,095 = 835,864 \text{kJ} / \text{kmol, or}$$

$$a_{ch,CH4} = 835,864 / 16.04 = 52,111 \text{kJ} / \text{kg}.$$

This value is 835,864/831,650 = 1.00507 or 0.51% higher than the value listed in Bejan et al. (ibid.) using model II. It is also 52,111/50,019 = 1.0418 or 4.18% higher than the LHV of CH_4, which was calculated earlier in Section 2.3.2.2. Standard chemical exergy of methane in Kotas (Appendix A, Table A.4 in Ref. [1] in Chapter 1) is 836,510 kJ/kmol and differs from the value calculated earlier by only 646 kJ/kmol (0.08% higher).

In Section 3.3.1, entropy generation during combustion of methane with air in the example gas turbine combustor was calculated as 171,722 kJ/s. This is the combustor irreversibility. In terms of fuel exergy,

$$\dot{E}_{fuel} = \dot{m}_{fuel} e_{fuel} = 13.52 \text{kg} / \text{s} \times 52,111 \text{kJ} / \text{kg} = 704,544 \text{kJ} / \text{s(kW)},$$

combustor irreversibility is 171,722/704,544 = 0.2437 or 24.4 %.

What is the difference between the maximum reversible work calculated in Section 3.3.1 (50,994 kJ/kg from Equation 3.58) and the chemical exergy calculated herein (52,111 kJ/kg from Equation 5.54)? After all, isn't exergy a measure of maximum (reversible) work? The answer to the second question is, of course, affirmative. The difference between the two values, 1,117 kJ/kg or 2.19%, can be traced back to the assumptions used in calculations. The simplifying assumption that each combustion product leaves the reactor control volume at P_0 does not agree with the reality in an actual reactor. In the latter, combustion products comprise a single gas mixture with each species (component) is at a partial pressure, which is the product of the mole fraction of the species in the mixture with the mixture pressure, P_0.

5.3   EXERGY – A HISTORY

In Section 5.1, we presented a derivation of the thermodynamic property exergy by combining the first and second law of thermodynamics within the framework of a *useful work* production process (i.e., a turbine). For a comprehensive review of the development of the concept of exergy and an extended bibliography, the reader is referred to the paper by Sciubba and Wall [2]. In this section, several key points from that paper will be used to introduce the reader to the origins of the concept.

The beginnings of the concept can be traced to the *available energy* defined by Gibbs:

> the greatest amount of mechanical work which can be obtained from a given quantity of a certain substance in a given initial state, without increasing its total volume or allowing heat to pass to or from external bodies, except such as at the close of the processes are left in their initial condition [3].

The term *free energy* was adopted early in the twentieth century with the adjective *free* denoting that the *Gibbs energy* is available for useful work. Another term for the same quantity is Gibbs free *enthalpy*, which fits the flow exergy concept rather nicely. (It has also been proposed to drop the term *free*.)

Modern definition of the exergy can be traced back to early 1950s [2]. The term itself was first proposed in 1953 as *Exergie* (in German) to denote *technische Arbeitsfähigkeit* (German for technical work capability). It was derived from the Greek word *ergon* (work) with the prefix *ex* in contrast to *Energie* (energy) derived from the same word with the prefix *en*. In other words, exergy literally means *external* work (whereas energy means *internal* work). The term eventually replaced other terms used for the same quantity (e.g., available energy, availability, available work, and potential work). Until quite recently, the term *availability* was still used in US textbooks (e.g., Moran and Shapiro, Ref. [1] in Chapter 2).

The modern definition of *exergy* is as follows: Exergy is the maximum theoretical useful work obtained if a system is brought into thermodynamic equilibrium with the environment by means of processes in which the said system interacts only with this environment. There are different forms of energy so that a full-blown equation to define exergy must account for all of them; for example, in specific (i.e., per unit mass) terms:

- Kinetic energy, $V^2/2$, for a system traveling at a speed of V
- Potential energy, $g\Delta z$, for a system at a height of Δz (g is the gravitational constant)
- Heat interaction between system and surroundings, q
- Work interaction between system and surroundings, w
- Chemical energy or potential, $\mu - \mu_0$

In addition, in certain applications, one should also account for other types of energy transfers associated between the system and its surroundings (e.g., electrical and radiation). For the applications of interest herein, they can be ignored. Note that kinetic and potential energies can be fully recovered into any other form. Therefore, the exergy and energy terms for them are identical. In systems of interest herein, both forms of energy are negligible in comparison to the thermal energy represented by *enthalpy*. Chemical exergy is of interest for reacting systems, e.g., combustion of fuel in a gas turbine combustor. Its derivation and formulae are, while straightforward, rather tedious. The reader should consult a thermodynamics textbook for a rigorous coverage (e.g., Moran and Shapiro cited earlier, chapter 13, Section 13.6 of the cited textbook), which is the textbook preferred by the author. An application of the chemical potential concept will be discussed in Chapter 12 (carbon capture).

5.4   EXERGO-ECONOMICS

In academic research, thermodynamics is combined with costing calculations (CAPEX and OPEX) using the exergy concept. It is referred to as *thermo-economics* (first used by Myron Tribus in his MIT lectures [2]) or *exergo-economics*. The basic idea is to apply the concepts of cost engineering, including capital cost of equipment (components) and material stream costs (e.g., fuel price), to material stream exergies and exergy transfers (associated with heat and work). For a comprehensive coverage of exergo-economic concepts and applications, the reader is referred to the book by Bejan et al. (Ref. [3] in Chapter 3). The book is very well written with clearly defined concepts and in-depth examples that one can readily adapt to his or her problems.

To the best of this author's knowledge, based on his professional experience over several decades, including stints in a major OEM and a major EPC firm, exergo-economic concepts are not used in the industry for design and optimization of power generation equipment and/or systems. This is not surprising because the complexity introduced by additional equations (as mentioned earlier, quite tedious for combustion processes and other processes including chemical reactions), with a few exceptions, do not provide additional insight above and beyond what can be gained from straightforward heat and mass balance analysis (i.e., first law of thermodynamics) combined with basic economic concepts (e.g., CAPEX, OPEX, and *levelized cost of electricity*, LCOE). Yet every year, literally thousands of papers are churned out by professors under publish-or-perish pressure in their institutions and their students, who apply the exergo-economic principles superbly outlined in the book cited earlier (published in 1996 and yet to be surpassed) over and over again, almost on autopilot, to the same old problems with minimal variations. The author regularly receives paper review requests from academic journals, and every year, he has to reject at least ten such papers packed with an avalanche of graphs, tables (not to mention simplistic formulae, which have been long discarded in the industry in favor of advanced commercial software, such as Thermoflow Suite and Aspen-HYSYS) but with not a shred of scientific insight.

5.5   EXERGY – HOW (WHEN) TO USE IT

The concept of exergy provides the engineer with a powerful tool to conduct an *autopsy* on a power or process plant. The best example of this application is identifying the huge amount of *energy* lost in the steam turbine condenser to be of little significance in terms of useful work generation ability (i.e., having low *exergy*). Let us demonstrate the validity of this statement by an example, specifically, a thermal power plant operating in Rankine cycle with a fossil-fired boiler and steam turbine generator (STG). Total heat transfer to the working fluid (i.e., feed water and steam) in the boiler is about 290 MWth with 127 MWe of electric power generated by the

STG. Heat rejected to the cooling water in the condenser is 160 MWth (i.e., roughly the balance between cycle heat input and power output). In other words, 55% of cycle heat input is simply thrown out as heat to the environment. This is a significant amount of wasted energy. Assuming that the condenser operates at 60 mbar, turbine exhaust steam condenses at 36.2°C (309.3 K). If one could build a Carnot engine operating with this heat source, with 15°C (288.2 K) ambient (i.e., the *dead* state) temperature, the power that could be generated would be a mere 160 × (1 − 288.2/309.3) = 11 MW, which, of course, is equal to the *exergy* of the exhaust steam from Equation 5.2 and is less than 4% of cycle heat input. This is the theoretical maximum, mind you. The best that one could hope is maybe 2 to 3 MW with an *organic Rankine cycle* (ORC), which, while technically possible on paper, is an utterly uneconomic proposition. (This, after all, is why no one has ever attempted it.)

Academic literature is full of exergy analyses of gas turbines. A recent example is the paper by Salpingidou et al. on a variety of intercooled, recuperative gas turbines [4]. (This paper is strongly recommended as an example of application of the second law concepts using the SSSF exergy balance, Equation 5.9.) Quite unsurprisingly, as identified by hundreds of earlier papers, the largest source of irreversibility (i.e., exergy destruction) by a wide margin is the combustor of the gas turbine. (The reader is referred to Section 3.3 for a quantitative demonstration of the futility of exergy analysis of gas turbines.) Depending on the cycle pressure ratio (CPR), 60 to 80% of the total cycle irreversibility can be attributed to the gas turbine combustor. (Combustor irreversibility is found to increase with CPR.) The other four loss buckets are the compressor (with intercoolers); high-pressure and power turbines, HPT and PT, respectively; and the recuperator (the heat exchanger transferring heat from the turbine exhaust gas to the compressed air). One finding in the paper is that the recuperator exergy loss decreased significantly with increasing CPR (i.e., from nearly 25% at CPR = 5 to less than 3% at CPR = 40). This, of course, is not surprising because higher CPR means turbine exhaust cooling and compressor discharge heating lines are closer to each other. In the paper, the authors also investigate two recuperator variants:

- The recuperator using the HPT exhaust (instead of the PT exhaust as in the conventional recuperative cycle design) – the *alternative recuperator* (AR) cycle in the paper
- Staged recuperation with recuperators using first the HPT exhaust and then the PT exhaust – *staged heat recovery* (SHR) cycle in the paper

The reader is referred to the paper for details on the exergy loss analysis of each variant. Not surprisingly, the AR variant comes out the worst, and its performance deteriorates with improving recuperator effectiveness. This is so because higher effectiveness is requisite for minimal cycle external heat input, but the penalty is lower PT inlet temperature and cycle output. This does not even require a lengthy first law analysis let alone a full-blown second law analysis with exergy loss bookkeeping. The point is that in all variants, unsurprisingly and of little informative value, the key finding is that the combustor is the most irreversible component. The other finding (i.e., that the recuperator effectiveness) plays an important role on the distribution of the exergy losses through the cycle is also of little informative value. Finally, that the pressure ratio also has a strong impact on the exergy losses is also unsurprising and could have been easily gleaned from a cursory look at the *T-s* diagrams of the cycle. In order to demonstrate the validity of this assertion, let us look at the SHR cycle with the following design parameters:

- 90% compressor polytropic efficiency
- 87% turbine polytropic efficiency
- 3% pressure loss in cycle heat exchangers
- Recuperator 1 and 2 effectiveness 90% and 30%, respectively
- HP turbine inlet 1,500°C
- Cycle pressure ratio 12.33:1

Simplified cycle schematic, *T-s* diagram, and state-point data are shown in Figure 5.8. Cycle efficiency is calculated as 56.47%. Note that the two compressors and the HP turbine (HPT) comprise a balanced shaft (e.g., net shaft output is zero). Cycle power output is provided by the power turbine (PT), which has a pressure ratio of 5.45:1. Each compressor section has a pressure ratio of 3.57:1. Key cycle performance parameters are summarized in Table 5.2. Key temperatures and irreversibilities for both recuperators are summarized in Table 5.3. Note that, for each recuperator

$$\text{METH} = \frac{\text{TH1} - \text{TH2}}{\ln\left(\dfrac{\text{TH1}}{\text{TH2}}\right)}, \text{METC} = \frac{\text{TC1} - \text{TC2}}{\ln\left(\dfrac{\text{TC1}}{\text{TC2}}\right)}, \text{and}$$

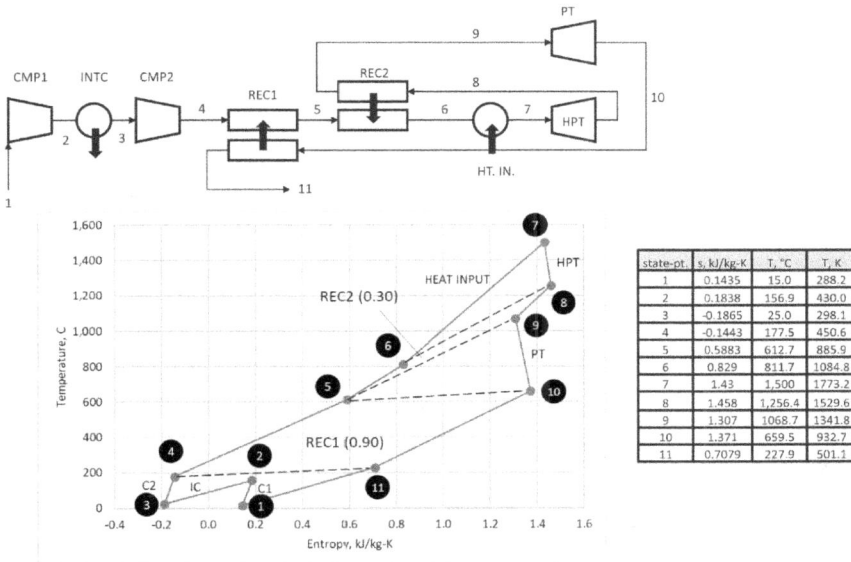

state-pt.	s, kJ/kg-K	T, °C	T, K
1	0.1435	15.0	288.2
2	0.1838	156.9	430.0
3	-0.1865	25.0	298.1
4	-0.1443	177.5	450.6
5	0.5883	612.7	885.9
6	0.829	811.7	1084.8
7	1.43	1,500	1773.2
8	1.458	1,256.4	1529.6
9	1.307	1068.7	1341.8
10	1.371	659.5	932.7
11	0.7079	227.9	501.1

FIGURE 5.8   SHR cycle, CPR = 12.33:1, HPT inlet temperature 1,500°C.

$$\frac{i}{c_p} = \left(1 - \frac{T_0}{METH}\right)DTH + \left(1 - \frac{T_0}{METC}\right)DTC. \tag{5.55}$$

Equation 5.55 is the SSSF exergy balance for each recuperator (both sides divided by c_p). Proving this, by starting from Equations 5.7 and 5.8 and using the perfect gas model for enthalpy and entropy terms (at constant pressure), is left as an exercise for the reader. Note that the first term on the RHS of Equation 5.55 quantifies (when multiplied by c_p) the work of a hypothetical Carnot engine operating between temperature reservoirs at METH and T_0. Similarly, the second term on the RHS of Equation 5.55 represents the work of a hypothetical Carnot engine operating between temperature reservoirs at METC and T_0. In other works, the irreversibility of a counterflow heat exchanger (similar to the recuperators in Figure 5.8) is quantified by the lost work in the heat exchanger, which is the sum of work outputs of the two Carnot engines, operating between the mean-effective temperatures of the hot and cold streams and the ambient (i.e., the dead state) temperature.

Performance parameters of the SHR cycle are summarized in Figure 5.8. Recuperator performances for the SHR cycle are listed in Figure 5.8 (for nomenclature refer to Figure 5.9). A graphical measure of the irreversibility calculated using

TABLE 5.2
Performance Parameters of SHR Cycle in Figure 5.8

$\overline{T}_H$ C (K)	1,128 (1,401)
$\overline{T}_L$ C (K)	101.8 (375)
Carnot efficiency	83.7%
Equivalent Carnot efficiency	73.2%
Air-standard Brayton Cycle efficiency	51.2%
Actual cycle efficiency	56.5%
Carnot factor	0.874
Technology factor	0.771

TABLE 5.3
Recuperator Performances for the SHR Cycle in Figure 5.8 (for Nomenclature, Refer to Figure 5.9)

Parameter	REC1	REC2
TH1, K	932.7	1,529.6
TH2, K	501.1	1,341.8
TC1, K	450.6	885.9
TC2, K	885.9	1,084.8
DTH, K	431.6	187.8
DTC, K	−435.2	−199.0
METH, K	694.7	1,433.6
METC, K	643.9	982.0
T_0, K	288.2	288.2
i/c_p, K	12.1	9.4

$$DTC = TC1 - TC2$$

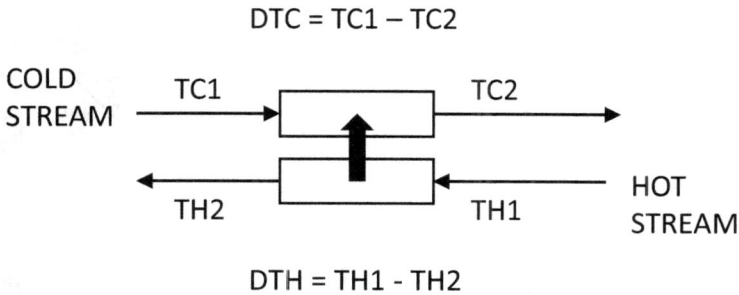

$$DTH = TH1 - TH2$$

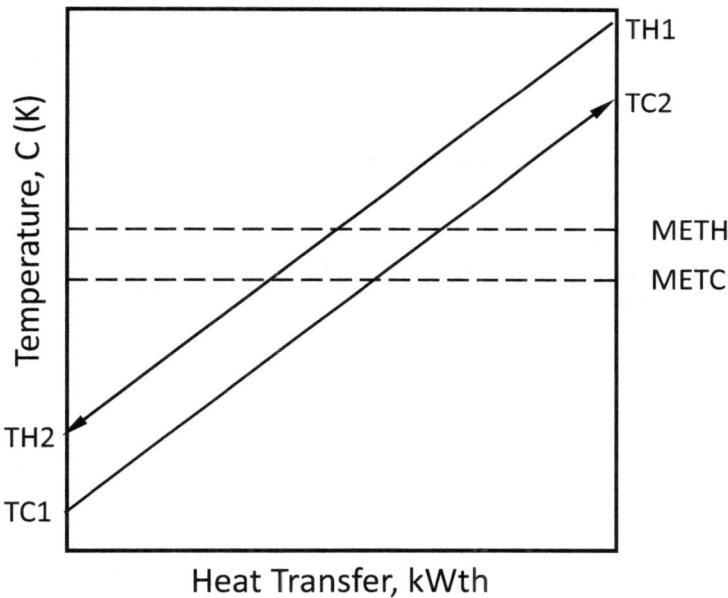

FIGURE 5.9 Generic recuperator (counterflow heat exchange) schematic and heat release diagram.

Equation 5.55 can be obtained from the area between the heat release lines of hot and cold streams. (In the practically impossible upper limit of an infinitely large heat exchange surface, those two lines would coincide, i.e., $i/c_p = 0$.)

What happens when the second recuperator's effectiveness is reduced to 10%? What happens when CPR is increased to, say, 24.7:1 (up from 12.3:1)? The answers to these questions can easily be obtained from the cycle *T-s* diagrams in Figure 5.10 and Figure 5.11, respectively. Note that each new cycle is superimposed on the base cycle shown in Figure 5.8. Performance parameters and recuperator performance for the new SHR cycle in Figure 5.10 are summarized in Table 5.4 and Table 5.5, respectively. Performance parameters and recuperator performance for the new SHR cycle in Figure 5.11 are summarized in Table 5.6 and Table 5.7, respectively. A comparison of all three cases is presented in Table 5.8.

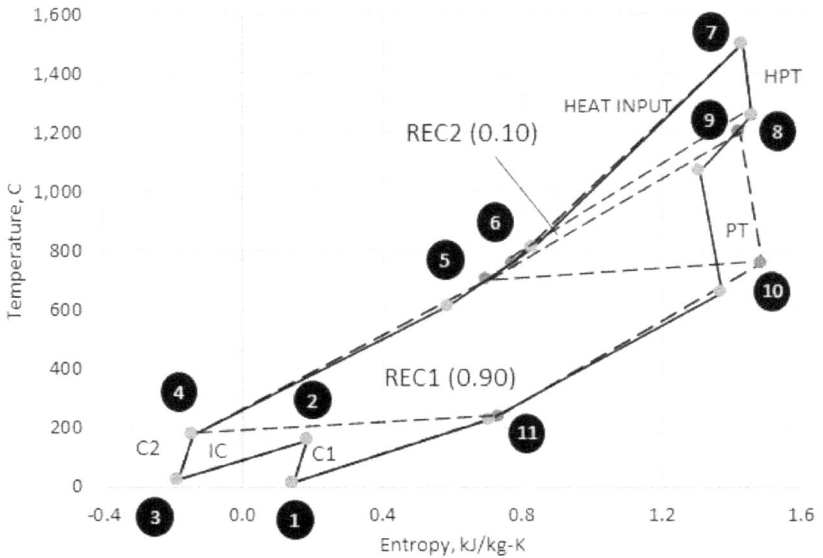

FIGURE 5.10 SHR cycle, CPR = 12.33:1, HPT inlet temperature 1,500°C, REC2 effectiveness 10% (dashed lines); base cycle (in Figure 5.8) with solid lines.

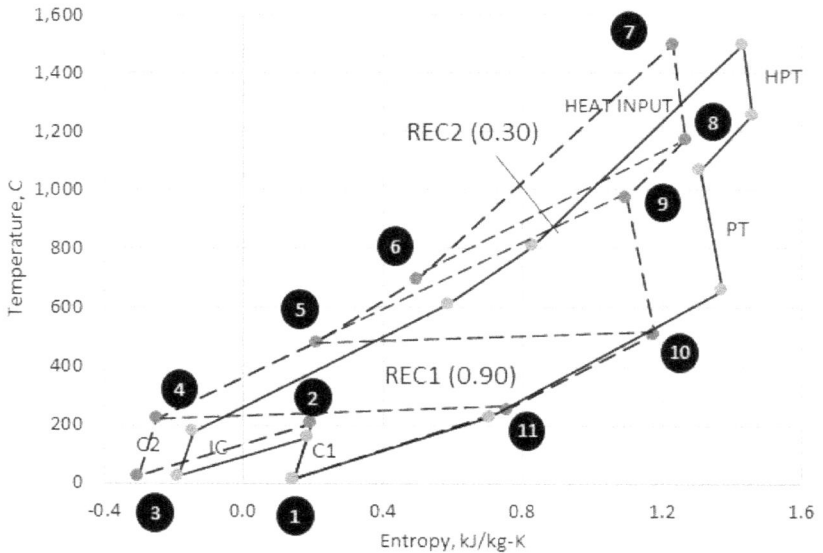

FIGURE 5.11 SHR cycle, CPR = 24:7:1, HPT inlet temperature 1,500°C (dashed lines); base cycle (in Figure 5.8) with solid lines.

TABLE 5.4
Performance Parameters of SHR Cycle in Figure 5.10

$\overline{T}_H$ C (K)	1,097 (1,370)
$\overline{T}_L$ C (K)	104.8 (378)
Carnot efficiency	83.7%
Equivalent Carnot efficiency	72.4%
Air-standard Brayton cycle efficiency	51.2%
Actual cycle efficiency	58.1%
Carnot factor	0.865
Technology factor	0.802

TABLE 5.5
Recuperator Performances for the SHR Cycle in Figure 5.10 (for Nomenclature, Refer to Figure 5.9).

Parameter	REC1	REC2
TH1, K	1,031.4	1,529.6
TH2, K	512.0	1,475.8
TC1, K	450.6	975.5
TC2, K	975.5	1,032.7
DTH, K	519.4	53.8
DTC, K	−524.8	−57.3
METH, K	741.6	1,502.5
METC, K	679.6	1,003.8
T_0, K	288.2	288.2
i/c_p, K	15.3	2.7

TABLE 5.6
Performance Parameters of SHR Cycle in Figure 5.11

$\overline{T}_H$ C (K)	1,059 (1,332)
$\overline{T}_L$ C (K)	115.8 (389)
Carnot efficiency	83.7%
Equivalent Carnot efficiency	70.8%
Air-standard Brayton cycle efficiency	60.0%
Actual cycle efficiency	54.1%
Carnot factor	0.845
Technology factor	0.765

Even a cursory examination of the *T-s* diagrams Figures 5.8, 5.10, and 5.11 and the data in Table 5.8 should suffice why *staged heat recovery* (SHR) is not a good idea. (In passing, 10% effectiveness for the second recuperator is the minimum allowed by the simulation software THERMOFLEX. It is a stand-in for the removal

TABLE 5.7
Recuperator Performances for the SHR Cycle in Figure 5.11 (for Nomenclature, Refer to Figure 5.9)

Parameter	REC1	REC2
TH1, K	932.7	1,529.6
TH2, K	501.1	1,341.8
TC1, K	450.6	885.9
TC2, K	885.9	1,084.8
DTH, K	431.6	187.8
DTC, K	−435.2	−199.0
METH, K	694.7	1,433.6
METC, K	643.9	982.0
T_0, K	288.2	288.2
i/c_p, K	12.1	2.7

TABLE 5.8
Comparison of SHR Cycles in Figures 5.8, 5.10, and 5.11, Cases A, B, and C, Respectively

CASE	A	B	C
Recuperator 1 effectiveness, %	90	90	90
Recuperator 2 effectiveness, %	30	10	30
Cycle *PR*	12.33	12.33	24.67
Power turbine *PR*	5.45	5.45	8.34
$\overline{T}_H$ C (K)	1,128 (1,401)	1,097 (1,370)	1,059 (1,332)
$\overline{T}_L$ C (K)	101.8 (375)	104.8 (378)	115.8 (389)
Recuperator 1 i/c_p	12.1	15.3	5.2
Recuperator 2 i/c_p	9.4	2.7	14.8
Equivalent Carnot efficiency	73.2%	72.4%	70.8%
Air-standard Brayton cycle efficiency	51.2%	51.2%	60.0%
Actual cycle efficiency	56.5%	58.1%	54.1%
Specific work, kJ/kg	470.4	518.9	522.6
Carnot factor	0.874	0.865	0.845
Technology factor	0.771	0.802	0.765

of that recuperator between the two turbines.) Elimination of the second recuperator increases PT exhaust temperature (via higher PT inlet temperature) and compensates for some of the reduction in combustor inlet temperature (higher heat input to the cycle). The remaining inefficiency is reflected in lower $\overline{T}_H$. However, this default is more than compensated by the elimination of pressure losses in both legs of the recuperator with the net impact of increased cycle efficiency.

The more interesting case is the detrimental impact of increased CPR on cycle efficiency. The explanation of this is left as an exercise to the reader. It should be noted that the elimination of recuperator 2 in the high CPR case would have the same favorable impact on cycle efficiency as explained earlier.

REFERENCES

1. ASME Steam Tables, Compact Edition, *ASME Research and Technology, Committee on Water and Steam in Thermal Systems, Subcommittee on Properties of Steam*, New York: ASME, 2006.
2. Sciubba, E., and Wall, G., A Brief Commented History of Exergy from the Beginnings to 2004, *International Journal of Thermodynamics*, 10 (1), pp. 1–26, 2007.
3. Gibbs, J.W., A Method of Geometrical Representation of the Thermodynamic Properties of Substances by Means of Surfaces, *Transactions of the Connecticut Academy*, II, pp. 382–404, 1873.
4. Salpingidou, C., Misirlis, D., Vlahostergios, Z., Donnerhack, H., Flouros, M., Goulas, A., and Yakinthos, K., Exergy Analysis and Performance Assessment for Different Recuperative Thermodynamic Cycles for Gas Turbine Applications, *Journal of Engineering for Gas Turbines Power*, 140 (7), p. 071701, 2018.

6 Waste Heat Recovery

For want is nexte to waste, and shame doeth synne ensue.
—Richard Edwardes (1576), *The Paradise of Dainty Devices*

The subject matter of waste heat recovery is probably the best place to start the application of second law principles including entropy and exergy to practical problems. At the risk of being repetitive, the subject is covered here in two takes. Each take uses a simple cycle gas turbine as the framework. In the first take, the focus is on defining the reference or *dead* state to come up with a numerical value for the energy and exergy content of the waste heat *stream* (i.e., the exhaust gas of a gas turbine). In this take, the emphasis is on exergy. In the second take, a closer look is taken at the enthalpy (energy) of the waste heat stream.

Before delving into calculations, let us start with the *Merriam-Webster* (i.e., dictionary) definition of the term *waste heat*: It is heat rejected or escaping from furnaces of various types (as coke ovens, cement kilns, or steel furnaces) after it has served its primary purpose. This definition is, of course, correct but incomplete because it does not mention heat content of the exhaust gas of an internal combustion engine. Another correct but incomplete definition leaves out furnaces: Waste heat is the unused heat given to the surrounding environment (in the form of thermal energy) by a *heat engine* in a thermodynamic process in which it converts heat to useful work. A better definition is from *Wikipedia*: Waste heat is heat that is produced by a machine, or other process that uses energy, as a byproduct of doing work.

Enough with definitions. As far as this monograph is concerned, the use of the term *waste heat* is strictly limited to a process or processes with the following characteristics:

- Thermal power is generated by combustion of a fuel in the form of hot gas (combustion products)
- Thermal power is converted to mechanical (shaft) power
  - In an internal combustion engine (e.g., gas turbine)
  - In an external combustion engine (e.g., boiler-turbine combination)
  - At the end of the conversion, hot gas is rejected through a duct (e.g., the stack, flue, exhaust pipe, or whatever else one may call it) to the atmosphere without putting it to any other use

Note that heat rejected to the cooling water in a steam condenser (or to ambient air in an air-cooled condenser) is also wasted energy (i.e., waste heat). In fact, in terms of thermal power (i.e., measured in kWth or MWth), it is a huge quantity. In terms of work-producing capability, however, it is readily negligible (i.e., it has low exergy). This was numerically demonstrated in Section 5.5.

DOI: 10.1201/9781003247418-8

In a vast majority of cases, the waste heat (i.e., *thermal* energy) of primary interest is the energy of the exhaust gas stream of a fossil-fired prime mover; for example, an internal combustion engine such as a gas turbine or recip engine, or a boiler (fired heater). The amount of energy contained in the engine exhaust gas (also referred to as *stack* or *flue* gas) is a function of gas composition, temperature, and mass flow rate. (Gas pressure has a negligible effect and can be safely ignored.) The thermodynamic property for measuring the thermal energy is *enthalpy* (see Section 14.2), which can be rigorously calculated using an *equation of state* as a function of temperature and composition. Gas composition is determined from combustion stoichiometry as a function of fuel-air flow ratio and fuel heating value. Requisite thermochemical calculations are tedious but straightforward. While they can be easily programmed in an Excel spreadsheet, at present, a wide variety of commercial software programs (even smartphone apps) that can easily take care of them are readily available.

6.1   WASTE HEAT RECOVERY – TAKE ONE

The exergy concept will come in quite handy in practical calculations involving *waste heat recovery*. Specifically, the useful work potential of the exhaust gas of an internal combustion engine (e.g., a gas turbine or a diesel engine) is exactly equal to the exergy of the exhaust gas stream, which is a function gas temperature, pressure, and composition. From a purely academic perspective, this does not introduce a meaningful difficulty. From a practical engineering analysis perspective, however, it presents a *standardization* difficulty due to the following reasons:

- Exhaust gas composition of each gas turbine or engine varies slightly with efficiency (fuel impact is eliminated by assuming 100% CH_4 (methane) fuel for rating and design calculations). Resulting variation in reference (dead state) enthalpy and entropy, although small in magnitude relative to actual system properties, creates numerical problems.
- Conventional dead state conditions (e.g., $T_0 = 15°C$ or $25°C$ and atmospheric P_0) implies that part of the water vapor (H_2O) in the exhaust gas (typically, around 10%(v), i.e., by volume, of the exhaust gas stream) would condense. In other words, at those temperatures and pressure, the exhaust gas would be a mixture of gas (with still some water vapor in it) and water.

The implication of the second practical problem listed earlier is increasing the exhaust gas exergy artificially by implicitly including the latent heat of condensation of water vapor knocked out of the gas mixture. This, however, represents very low-grade energy and is difficult (i.e., a costly and complex process) to utilize for useful work production. Thus, several simplifying assumptions will be made for calculating exhaust gas exergy. These points are illustrated by a numerical example below.

Let us consider an advanced-class gas turbine, GE's 7HA.02, which is a 60 Hz machine rated at nearly 400 MWe (ISO, base load). The gas turbine is modeled in THERMOFLEX using built-in engine library model (ID number 681) with natural gas fuel (see Figure 6.1). Gas turbine exhaust gas data is as shown in the figure. At the end of heat recovery, stack gas is assumed to be at 87.78°C (190°F). Total exhaust gas pressure loss is assumed to be 13 in. H_2O (32.4 mbar).

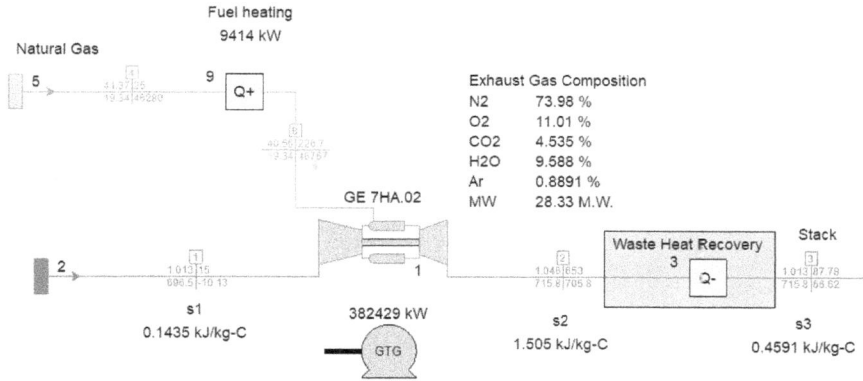

FIGURE 6.1 Gas turbine exhaust gas heat recovery example.

TABLE 6.1

Exhaust Gas Dew Point Calculations

Water vapor in exhaust gas, $\%(v)$	9.588
Exhaust gas flow rate	715.8 kg/s
	(25.3 kmol/s)
Water vapor flow rate, kg/s	43.61
Partial pressure of H_2O @ 1 atm, psia	1.41
Saturation temperature @ partial pressure	45.3°C
	(113.5°F)
Latent heat of condensation, kJ/kg	2,394.3
Liquid water knocked out, kg/s	36.61
Water vapor remaining in gas flow, kg/s	7.00
Total gas flow @ 15°C and 1 atm, kg/s	679.19

If we specify the dead state as $T_0 = 15°C$ and $P_0 = 1$ atm (1.01324 bara), exhaust gas stream data will be as shown in Table 6.1. Based on the conditions in the table, enthalpy and entropy of the gas-liquid mixture at the dead state are $h_0 = -137.1$ kJ/kg and $s_0 = -0.1974$ kJ/kg-K, respectively.

The hypothetical process between the stack and dead state is summarized in Table 6.2. The amount of heat energy available is considerable at almost 145 MWth. However, at an average temperature of about 313 K (~40°C), it is of little value from useful work generation perspective. This is represented by the Carnot efficiency of the hypothetical reversible heat engine, which is less than 10%, corresponding to reversible work (theoretical maximum) of 11.5 MW. In practice, possibly with an *organic Rankine cycle* (ORC) engine, barely 1 MW would be achievable at significant cost (mainly, the requisite heat exchanger).

The other possibility for specifying the dead state properties is to assume a *hypothetical* gas state; that is, H_2O is still in gas phase in the gas mixture. Enthalpy and entropy values at this hypothetical state can be determined by extrapolation of gas properties to 15°C (see Table 6.3) as $h_0 = -10.66$ kJ/kg and $s_0 = 0.2286$ kJ/kg-K,

TABLE 6.2
Work Potential by Cooling Stack Gas to the Dead State

	Stack	Dead State
Flow rate, kg/s	715.8	715.8
Enthalpy, kJ/kg	66.62	−136.69
Entropy, kJ/kg-K	0.4591	−0.1909
Exergy, kJ/kg	16.01	0
Temperature	87.78	15
Heat release, kWth	145,529	
$T_H = \Delta h/\Delta s$, K	312.78	
T_0, K	288.15	
Carnot efficiency	7.88%	
Reversible work, kW	11,462	

TABLE 6.3
Hypothetical (Gaseous) Dead State Property Calculation

T, °C	h, kJ/kg	s, kJ/kg-K
60	37.09	0.3739
55	31.78	0.3579
50	26.48	0.3416
45.3 (dew point)	21.46	0.3264
15 (dead state)	−10.66	0.2286

respectively. As summarized in Table 6.4, maximum theoretical (reversible) work is reduced by one-third (about 3.7 MW).

Finally, each property package has its own assumption for setting the zero enthalpy and entropy. (For example, the zero-enthalpy temperature for gases and saturated vapor H_2O in THERMOFLEX, which is the software used for the calculations herein, is 77°F (25°C).) For second law analysis of heat recovery systems using exergy, it is convenient to set reference (dead state) enthalpy and entropy to 0 each. This simply requires redefining (or shifting) the properties as $h'(T) = h(T) - h(T_{ref})$ and $s'(P,T) = s(P,T) - s(P_{ref}, T_{ref})$. Going with this convention, second law calculations for exhaust gas (water) heat recovery analysis for the example gas turbine are summarized in Table 6.5.

The following observations are made from the data presented in Table 6.5:

- Gas turbine exhaust gas exergy is 348.7 kJ/kg. This is the maximum theoretical work that can be done by a bottoming Carnot cycle.
- Heat recovery in a waste heat recovery boiler amounts to 639.2 kJ/kg (bringing the stack gas temperature to about 88°C), corresponding to *heat recovery effectiveness* of 639.2/716.5 = 89.2%.
- Of that amount, 13.2 kJ/kg (about 2%) is utilized to heat the natural gas to 226.7°C (440°F) for improved gas turbine efficiency.

TABLE 6.4
Work Potential by Cooling Stack Gas to the Hypothetical (Gaseous) Dead State

	Stack	Dead State
Flow rate, kg/s	715.8	715.8
Enthalpy, kJ/kg	66.62	−10.66
Entropy, kJ/kg-K	0.4591	0.2286
Exergy, kJ/kg	10.86	0
Temperature	87.78	15
Heat release, kWth	55,315	
$T_H = \Delta h/\Delta s$, K	335.26	
T_0, K	288.15	
Carnot efficiency	14.05%	
Reversible work, kW	7,772	

TABLE 6.5
Exhaust Gas Heat Recovery Second Law Analysis

Reference (dead state) temperature, °C	15.0
Reference pressure, bara	1.01324
Reference enthalpy, kJ/kg	0.00
Reference entropy, kJ/kg-K	0.0000
Exhaust gas enthalpy, kJ/kg	716.5
Heat recovered, kJ/kg	639.2
Fuel heating, kJ/kg	13.2
Steam cycle efficiency (assumed)	40.0%
Actual work, kJ/kg	250.4
Max. possible work, kJ/kg (Btu/lb)	348.7
(Exhaust gas exergy)	(149.9)
Lost work, kJ/kg	98.3
Exergetic efficiency	71.8%
a/h	48.7%
$T_H = Dh/Ds$, K	561.3
Carnot efficiency	48.66%

- The remainder is converted to useful work in a Rankine (steam) bottoming cycle with 40% efficiency to 250.4 kJ/kg of useful work, corresponding to *exergetic efficiency* of 250.4/348.7 = 71.8%.

As mentioned earlier, maximum theoretical work that can be done by a bottoming Carnot cycle, whose efficiency is determined as

$$\eta = 1 - \frac{T_0}{T_H}, \quad T_H = \frac{h_{exh} - h_0}{s_{exh} - s_0} = \frac{h_{exh}}{s_{exh}}.$$

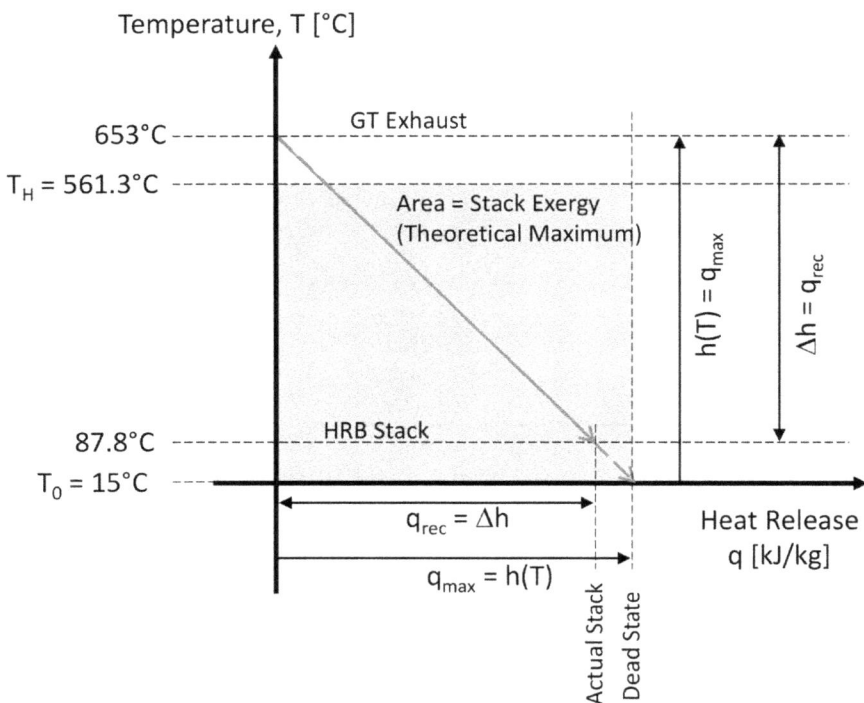

FIGURE 6.2 Graphical illustration of theoretical maximum work via gas turbine exhaust gas heat recovery.

Thus, $T_H = 716.5/1.2764 = 561.3$ K, $T_0 = 288.15$ K, and $\eta = 1 - 288.15/561.3 = 48.66\%$ and $w = 48.66\% \times 716.5 = 348.7$ kJ/kg, which is equal to exhaust gas exergy. This is illustrated graphically in Figure 6.2.

The next question stems from the fact that we set the stack gas temperature to a practical value; that is, $87.78°C$ ($190°F$). What is the upper limit to useful work production with this condition? This would be another (hypothetical, of course) Carnot cycle operating between

$$T_H = \frac{h_{exh} - h_{stck}}{s_{exh} - s_{stck}} = \frac{716.5 - 77.3}{1.2764 - 0.2305} = 611.1\,\mathrm{K}$$

and $T_L = 87.78 + 273.15 = 360.9$ K. The efficiency of that Carnot cycle would be $\eta = 1 - 360.9/611.1 = 40.94\%$ and $w = 40.94\% \times (716.5 - 77.3) = 261.2$ kJ/kg, which is the *practical* maximum and $261.2/348.7 = 0.75$ (i.e., 75% of exhaust gas exergy). This is illustrated graphically in Figure 6.3.

As it turns out, the arbitrarily specified 40% bottoming Rankine (steam) cycle efficiency in Table 6.5 was close to the entitlement (about 41%). As a what-if exercise, let us set the stack temperature to $71.11°C$ (344.3 K). In that case, $T_H = 599.3$

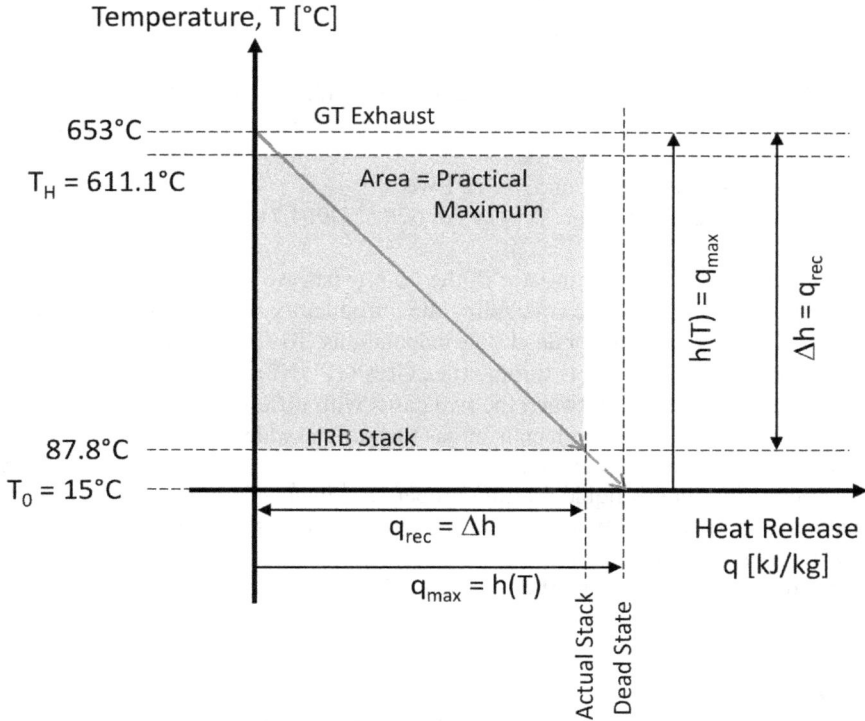

FIGURE 6.3 Graphical illustration of practical maximum work (via specified stack gas temperature) via gas turbine exhaust gas heat recovery.

TABLE 6.6
Exhaust Gas Heat Recovery with 71.1°C Stack Gas Temperature

Exhaust gas enthalpy, kJ/kg	716.5
Stack gas enthalpy, kJ/kg	59.5
Heat recovered, kJ/kg	656.9
Heat recovery effectiveness	91.7%
Steam cycle efficiency (assumed)	42.6%
Actual work, kJ/kg	273.9
Lost work, kJ/kg	74.7
Exergetic efficiency	78.6%

K and $\eta = 1 - 344.3/599.3 = 42.55\%$. Improved performance is summarized in Table 6.6.

This simple exercise illustrates the power of second law (exergy) analysis for realistic assessment of combined cycle potential of a gas turbine with simple yet unassailable calculations drawing upon first principles. For the given gas turbine and performance fuel heating, maximum possible combined cycle gross output with a heat recovery boiler (HRB) with 71.1°C (160°F) stack temperature is found as (with 904,716 kWth gas turbine fuel consumption)

- Gas turbine generator output of 382,429 kWe (42.72% gross LHV efficiency)
- Steam turbine output of 715.8 kg/s × 273.9 kJ/kg = 196,083 kWe (roughly half of gas turbine output)
- Total combined output of 578,512 kWe
- Combined efficiency (gross LHV) of 578,512/904,716 = 63.94%

These numbers will pop again later in the book when we look at gas turbine combined cycles in-depth. Before concluding the introductory second law discussion, let us bring clarity to reference or dead state calculations. To do that, the example earlier is repeated with an E class gas turbine (i.e., GE's GT11N2) with a lower exhaust gas temperature. Differences between the two cases with different gas composition and temperature (essentially at both ends of gas turbine class hierarchy) are summarized in Table 6.7. Real enthalpy and entropy values, as explained earlier, are exact in the sense they represent gas-liquid mixture properties calculated by the built-in ideal gas equation of state of the software program. Hypothetical enthalpy and entropy values are extrapolation to of gas mixture properties to dead state temperatures. Exergy values are calculated based on the hypothetical gas reference/dead state.

TABLE 6.7
Reference (Dead) State Properties for Two Different Exhaust Gas Conditions

	7HA.02	GT11N2
Nominal TIT, °C	1,600 (H Class)	1,300 (E Class)
Exhaust gas temperature, °C	653	532
Exhaust gas composition		
N_2, %(v)	73.98	75.03
O_2, %(v)	11.01	14.11
CO_2, %(v)	4.535	3.101
H_2O, %(v)	9.588	6.857
Ar, %(v)	0.8891	0.9024
H_2O dew point, °C	45.3	38.9
Dead state temp., °C	15	
Real h_0, kJ/kg	−137.1	−92.5
Real s_0, kJ/kg-K	−0.1974	−0.0661
Hypothetical h_0, kJ/kg	−10.66	−10.50
Hypothetical s_0, kJ/kg-K	0.2286	0.2135
Exhaust gas exergy, kJ/kg	348.7	248.9
Dead state temp., °C	25	
Real h_0, kJ/kg	−103.7	−59.5
Real s_0, kJ/kg-K	−0.0782	0.0464
Hypothetical h_0, kJ/kg	−0.05	−0.04
Hypothetical s_0, kJ/kg-K	0.2609	0.2454
Exhaust gas exergy, kJ/kg	334.9	237.0

Clearly, running a simulation software every time when exhaust gas property information is required is neither possible nor practical. The most readily available piece of information for a gas turbine or gas engine is the exhaust gas temperature. The author developed several simple relationships for exhaust gas exergy as a function of gas temperature. They are listed in Section 14.1.2 and repeated subsequently for convenience:

$$a[\text{in Btu/lb}] = 0.001628 \cdot T_{exh}^{1.60877}[\text{in }°F], \text{and} \tag{6.1}$$

$$a[\text{in Btu/lb}] = 0.1961 \cdot T_{exh}[\text{in }°F] - 86.918. \tag{6.2}$$

Both equations are for 0 enthalpy and entropy at 1 atm and 15°C dead state. The formulae are developed from a curve-fit to actual gas turbine exhaust data with 100% CH_4 methane fuel between 900°F (480°C) and 1,250°F (680°C). However, they can be extrapolated to higher and lower temperatures with little loss in fidelity. Exhaust gas exergy calculations using Equations 6.1 and 6.2 are graphically summarized in Figure 6.4. Also shown in the chart are exergy values of exhaust gas from two different gas turbines compared in Table 6.7. Clearly, for front-end bottoming cycle calculations, predictions of the two equations are adequate for accurate estimates for a broad range of gas turbine classes.

6.2   WASTE HEAT RECOVERY – TAKE TWO

For many front-end calculations and illustration of key principles, simple formulae with constant specific heat assumption are sufficient (e.g., Equations 6.1 and 6.2). In the preceding section, we looked at the waste heat *exergy*. Let us now look at waste heat energy (*enthalpy*) calculation. Assume that we have a prime mover operating in the field with exhaust gas flow, composition of y_i (mole fraction of component *i*, e.g.,

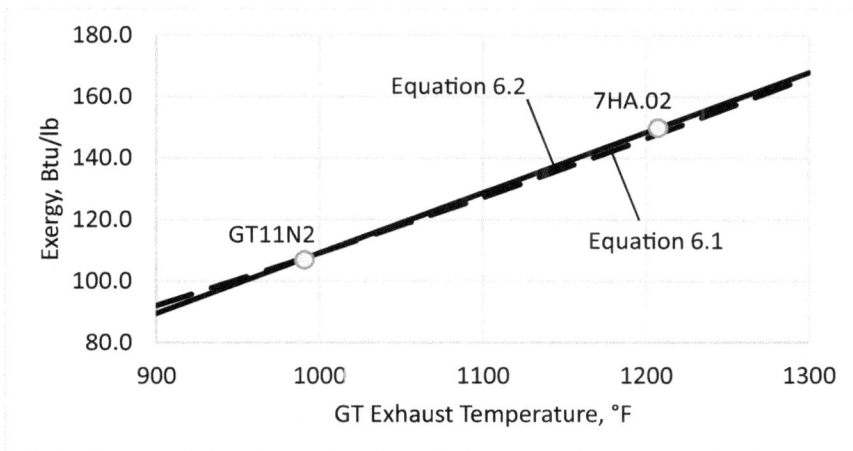

FIGURE 6.4   Gas turbine exhaust gas exergy.

O_2, N_2, etc.), and temperature, $\dot{m}_x$ and T_x. The energy (i.e., *enthalpy*) content of the said gas flow is given by

$$\dot{Q}_x = \dot{m}_x h(T_x, y_i).$$ (6.3)

where enthalpy, h, can be calculated using a suitable equation of state (EOS), such as JANAF tables. Equation 6.3, while strictly speaking correct, is problematic without specifying a zero-enthalpy reference temperature. Typical choices are 15°C (ISO temperature) or 25°C. Without the knowledge of a specific EOS with a known zero-enthalpy reference temperature (also known as the *datum*), the foolproof version of Equation 6.3 is

$$\dot{Q}_x = \dot{m}_x \left[h(T_x, y_i) - h(T_{ref}, y_i) \right],$$ (6.4)

with an arbitrary choice of T_{ref}, say, 15°C. Equation 6.4 can also be written as a function of specific heat of the gas in question as

$$\dot{Q}_x = \dot{m}_x \int_{T_{ref}}^{T_x} c_p(T, y_i) \, dT,$$ (6.5)

which can be approximated as

$$\dot{Q}_x = \dot{m}_x c_p (T_x - T_{ref}).$$ (6.6)

As an example, let us consider the exhaust gas of a vintage Frame 6 gas turbine (firing 100% CH_4 fuel at 25°C) with the following properties:

- 139.1 kg/s exhaust flow rate
- 537.7°C temperature
- Volumetric (or *molar*) composition, 13.955%(*v*) O_2, 3.102%(*v*) CO_2, 7.124%(*v*) H_2O, 74.917%(*v*) N_2, and 0.902% Ar

The common datum condition for gas property functions in THERMOFLEX is dry gases and saturated vapor H_2O at 77°F (25°C). Based on that datum, gas enthalpy at 537.7°C is 559.9 kJ/kg, and average gas specific heat is 1.0945 kJ/kg-K (from $c_p = dh/dT$, see later). Here is a small (or *not-so-small*, it depends on the specific process under investigation) problem: 25°C is *below* the dew point of water vapor in the gas mixture at nearly atmospheric pressure. Thus, some of the water vapor in the exhaust gas, 3.635 kg/s to be precise, condenses out of the *dry* mixture. In other words, at 25°C, we end up with a mixture of dry gas and liquid water. The enthalpy of the mixture is −63.84 kJ/kg (i.e., it is *not* 0 kJ/kg). Using Equation 6.4, we find that

$$\dot{Q}_X = 139.1[559.9--63.84] = 86,762 \text{kWth},$$

that is, when the exhaust gas is cooled to 25°C, heat extracted from the cooling process, *including* the *latent heat of condensation* of water vapor, is about 87 MWth. This is the *exact* finding from a rigorous calculation using a suitable EOS and software. Using Equation 6.6, the proverbial quick and dirty answer is

$$\dot{Q}_X = 139.1 \cdot 1.0945(537.7 - 25) = 78,056 \text{kWth}.$$

The error is about −9 MWth or −10%. At 25°C, latent heat of condensation of water vapor is 2,442.5 kJ/kg. Heat released via condensation of 3.635 kg/s water vapor is 8,879 kWth, which, when added to the answer found from Equation 6.6 earlier, results in 86,935 kWth and brings the heat release calculation error down to about +173 kWth or +0.2%. A linear curve-fit to the gas enthalpy with the specified composition is given by

$$h[\text{in kJ} / \text{kg}] = 1.0945 \cdot T[\text{in}°C] - 33.168 \,(\text{no } H_2O \text{ condensation}),$$

which implies a constant c_p of 1.0945 kJ/kg-K. (This equation is essentially the SI unit's version of the simple transfer functions provided in Section 14.1.2.)

Specific heat of the gas as a function of temperature can be expressed as a second-order polynomial; that is,

$$c_p[\text{in kJ} / \text{kg} - K] = 0.0014 \cdot t^2 + 0.0151 \cdot t + 1.0351,$$

where $t = T$ [in °C]/100. It should be noted that this example is for a given combination of prime mover and fuel. Substituting the second-order specific heat polynomial earlier into Equation 6.5 and integrating, exhaust gas heat content is found as 77,859 kWth, which is 197 kWth lower than that found from Equation 6.6 with the average value of 1.0945 kJ/kg-K. Correction via inclusion of latent heat of H_2O condensation ends up with 86,738 kWth, which, as expected, is very close to the value that was obtained with Equation 6.4.

There is a wide variety of prime movers and fuels in addition to fired furnaces, boilers, and others. Obviously, to go through the same exercise to develop new curve-fits and suitable averages for each individual case is not a practical proposition. Nevertheless, the situation is not as dire as it seems at the first glance. This assertion is based on taking a close look at a selected range of data summarized in Tables 6.8 and 6.9. Data in the tables is plotted in Figure 6.5.

As shown in Figure 6.5, across a wide variety of equipment and fuels, exhaust/stack/flue gas enthalpy exhibits a consistent, linear trend. For gas- or liquid-fuel-fired prime movers, average c_p is 1.1579 kJ/kg-K. For furnaces (or fossil boiler burners) using solid or liquid fuels, average c_p is 1.0581 kJ/kg-K. The entire series can be represented by the curve-fit formula ($R^2 = 0.9991$)

TABLE 6.8

Frame 6B Exhaust Properties with Different Fuels (Gas Composition Data in Percent by Volume or mol)

	100% CH_4	Natural Gas	NG 180°C Fuel	#2 Distillate
Temperature, °C	537.7	537.5	537.5	538.4
O_2	13.955	13.963	14.016	14.086
CO_2	3.102	3.167	3.143	4.178
H_2O	7.124	6.983	6.936	5.136
N_2	74.917	74.986	75.004	75.689
Ar	0.902	0.902	0.902	0.912
Enthalpy, kJ/kg	559.9	559.23	559.1	554.36

TABLE 6.9

Different Flue Gas Properties – Prime Movers with 100% CH_4 Fuel at 25°C (Furnace A: 93% Boiler Efficiency, Illinois #6 Coal; B: 90% Boiler Efficiency, Texas Lignite; C: 90% Boiler Efficiency, Heavy Fuel Oil, All with 10% Excess Air; WI: Water Injection, 2 kg/s; RICE: Reciprocating Internal Combustion Engine)

	Frame 6B	Frame 7H	Aero	Aero + WI	RICE		Furnace, A	Furnace, B	Furnace, C
Temperature, °C	537.7	664.4	534.0	493.8	374.1	133.5	158.4	212.0	
O_2	13.955	10.745	13.547	12.687	9.71	1.768	1.532	1.794	
CO_2	3.102	4.556	3.287	3.343	5.025	14.867	13.026	13.089	
H_2O	7.124	10.018	7.491	11.108	10.951	9.476	21.651	10.493	
N_2	74.917	73.793	74.774	71.995	73.43	72.614	62.89	73.584	
Ar	0.902	0.889	0.901	0.867	0.884	0.873	0.757	0.886	
SO_2	0	0	0	0	0	0.40173	0.1441	0.15414	
Enthalpy	559.9	721.44	556.85	521.12	382.89	113.77	150.11	199.0	

$$h[in kJ/kg] = 1.1227 \cdot T\left[in °C\right] - 35.529 \,(no\, H_2O \,condensation),$$

which implies a globally applicable, constant c_p value of 1.1227 kJ/kg-K.

In conclusion, there is enough quantitative evidence that, for practical purposes, we can use the following simple formulae with a judicious choice of average (constant) specific heat for the exhaust gas:

$$h_2 - h_1 = c_p \left(T_2 - T_1\right). \tag{6.7}$$

$$s_2 - s_1 = c_p \ln\left(\frac{T_2}{T_1}\right) - R \ln\left(\frac{P_2}{P_1}\right) \cong c_p \ln\left(\frac{T_2}{T_1}\right). \tag{6.8}$$

$$a_2 - a_1 = c_p \left(T_2 - T_1\right) - T_0 c_p \ln\left(\frac{T_2}{T_1}\right). \tag{6.9}$$

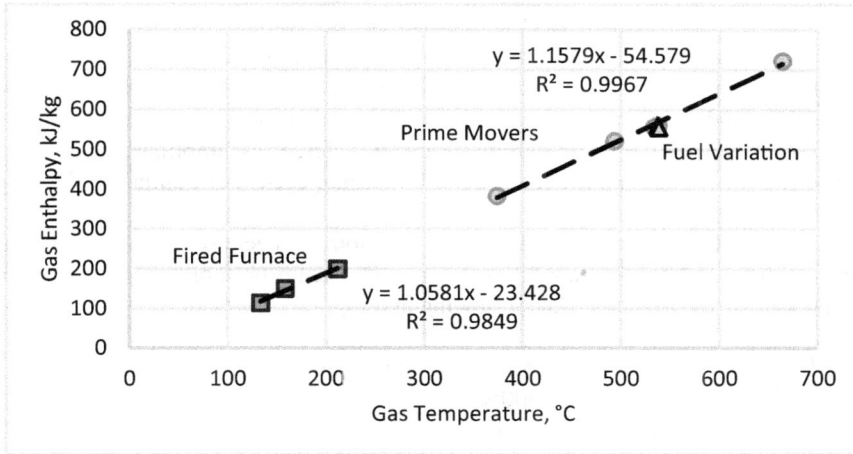

FIGURE 6.5   Exhaust (or stack/flue) gas enthalpy for various equipment and fuels.

$$\bar{T} = \frac{(T_2 - T_1)}{\ln\left(\dfrac{T_2}{T_1}\right)}.$$

(6.10)

6.3   PUTTING WASTE HEAT TO GOOD USE

Going back to the Frame 6 gas turbine example earlier, let us apply the second law concepts to the estimation of steam turbine power output in a bottoming cycle utilizing the exhaust gas of the said turbine. The energy content of the exhaust gas was found to be 78 MWth using Equation 6.6 with c_p = 1.0945 kJ/kg-K (ignoring the latent heat of condensation of water vapor in the exhaust gas). According to the Kelvin-Planck statement of the second law of thermodynamics, maximum work that can be generated using this rate of thermal energy transfer is from a Carnot engine operating between two temperature *reservoirs*. The hot temperature reservoir is at the logarithmic average of the exhaust gas and zero-enthalpy reference temperatures (known as the *mean-effective* value). Using Equation 6.10

$$\bar{T}_H = \frac{(537.7 - 25)}{\ln\left(\dfrac{537.7 + 273.15}{25 + 273.15}\right)} = 512.45\,K = 239.3\,^{\circ}C.$$

With the low temperature reservoir at 25°C, the efficiency of the Carnot engine is found as

$$\eta_c = 1 - \frac{\bar{T}_L}{\bar{T}_H} = 1 - \frac{25 + 273.15}{512.45} = 0.4182,$$

so that the output of the Carnot engine is calculated to be $0.4182 \times 78 = 32.6$ MW. Including the latent heat of H_2O condensation, the available exhaust gas thermal power is 86.8 MWth, and the Carnot engine output becomes 36.3 MW. If the datum is pushed down to 15°C, it can be shown that the available thermal power becomes 91.4 MWth, and the Carnot efficiency is 0.4377 so that the Carnot engine output is $0.4377 \times 91.4 = 40.0$ MW. In summary, using different assumptions, we arrive at theoretically possible maximum values of bottoming cycle work output in the range of 32.6 MW to 40.0 MW.

What would be a realistic value for the bottoming cycle output? For this type of gas turbine and exhaust conditions, a suitable steam bottoming cycle would be two-pressure with no reheat. The following assumptions are made:

- ISO ambient conditions (15°C, 60% relative humidity, 1 atm)
- Main steam at 55 bar and 500°C
- LP admission steam at 5 bar and 240°C
- Once-through (open loop) water-cooled condenser at 50 mbar pressure
- HRSG pinch and approach temperature deltas 10 and 4°C, respectively

Using Thermoflow Inc.'s GT PRO software, the bottoming cycle performance is found as follows:

- 93.3°C HRSG stack exit temperature
- 21,755 kWe steam turbine generator output
- 1,000 kWe plant auxiliary power consumption (excluding gas turbine, including transformers)
- Net bottoming cycle output of 20.8 MWe

Note that at stack temperature of 93.3°C, there is no H_2O condensing out of the flue gas. The appropriate second law benchmark in this case is 32.6 MW calculated earlier. Consequently, the *exergetic* efficiency of the bottoming cycle is 20.8/32.6 = 0.638. Furthermore, heat recovery from the gas turbine exhaust gas is estimated as (noting the three degrees higher exhaust gas temperature due to extra exhaust loss imposed by the HRSG),

$$\dot{Q}_r = 139.1 \cdot 1.0945 \left(540.3 - 93.3 \right) = 68,053 \, \text{kWth},$$

which corresponds to a *heat recovery effectiveness* of 68/78 = 0.872. Net steam cycle (first law) efficiency is found to be 20.8/68 = 0.306. The overall *bottoming cycle efficiency* is $0.872 \times 0.306 = 0.267$.

Before moving on, let us estimate the error introduced by ignoring the effect of pressure in the entropy formula, Equation 6.8. The molecular weight of the gas is 28.5 kg/kmol so that the gas constant is $R = R_{unv}/MW = 8.3145/28.5 = 0.292$ kJ/kg-K. For a total pressure drop of 22.5 mbar, including the transition duct and the HRSG, the pressure ratio, $P_2/P_1 = 1.023$, so that the pressure-related term is evaluated to be $0.292 \times ln(1.023) = 0.0066$ kJ/kg-K. In comparison, the temperature related term is

$c_p \times ln(T_2/T_1) = 1.0945 \times ln(813.5/366.5) = 0.8727$ kJ/kg-K. Consequently, the error involved is about 0.8%.

Another way of expressing the Kelvin-Planck statement in conjunction with the exhaust gas stream of a prime mover (or the stack/flue gas of a boiler, furnace, etc.) is as follows: Maximum possible theoretical work that can be extracted from the thermal power of a waste gas stream is exactly equal to the (flow) exergy cf the said stream. Exergy is a property of the gas stream in question and can be calculated using an EOS for known gas composition and temperature (neglecting the small contribution of gas pressure). Specifically, with the assumption of perfect gas (i.e., ideal gas with constant specific heat), said exergy can be calculated using the following equation:

$$a_x = c_p\left(T_x - T_0\right) - T_0 c_p \ln\left(\frac{T_x}{T_0}\right).$$

(6.11)

Equation 6.10 can be reformulated so that

$$a_x = c_p\left(T_x - T_0\right)\left[1 - T_0 \frac{\ln\left(\dfrac{T_x}{T_0}\right)}{\left(T_x - T_0\right)}\right], \text{or}$$

(6.12)

$$a_x = q_x\left[1 - \frac{T_0}{T_x}\right] = q_x \eta_c.$$

(6.13)

Equation 6.13 shows that the flow exergy of the gas is equal to the work produced by a Carnot engine operating between two temperature reservoirs, one at the mean-effective temperature of the gas in question (*high* temperature) and the other at the reference temperature (*low* temperature). Another way to look at the Carnot efficiency, based on Equation 6.13, is that it is the *ratio* of gas exergy to gas enthalpy because $q_x = h_x$ with the datum, $h(T_0) = 0$.

What can actually be done with a realistic cycle design is only a fraction of the theoretical maximum quantified by the exergy of the gas stream in question; that is,

$$w_x = \eta_{ex} \eta_c q_x.$$

(6.14)

It should be emphasized that we do not have to specify a thermodynamic cycle or a cycle working fluid. Obviously, in the vast majority of applications, the cycle in question the *Rankine* cycle with H_2O as the working fluid (going through liquid, gas-liquid, and gaseous phases). Other possibilities, especially for waste gas streams with low temperatures, include the *organic* Rankine cycle (ORC) with working fluids such as retrograde hydrocarbons (e.g., pentane). Recently, supercritical CO_2 as a working

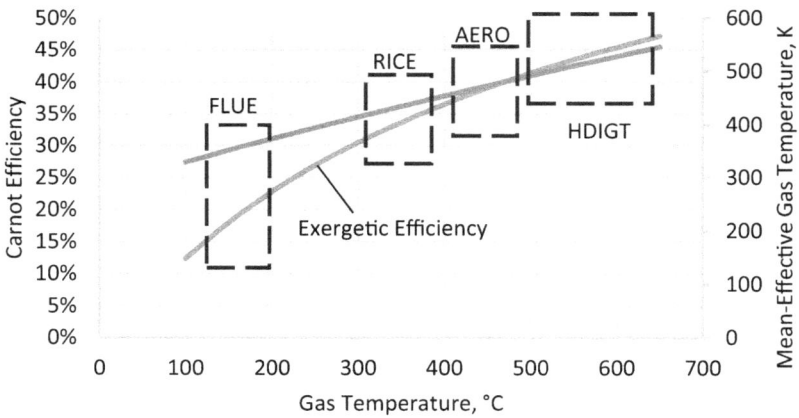

FIGURE 6.6 Exhaust gas heat recovery potential of various prime movers and boiler flue gas.

fluid in a Rankine cycle has attracted wide interest for low- to medium-temperature waste heat recovery applications.

For a range of waste energy (i.e., exhaust, flue, or stack gas) temperatures, T_x; Carnot efficiency, η_c; and the mean-effective gas temperature, $\overline{T}_X$ are plotted in Figure 6.6 for $T_0 = 15°C$. At the high end, temperatures are representative of heavy-duty industrial gas turbines, such as F, H, or J class units used in combined cycle power plant applications. The next group is typical of aeroderivative gas turbines with high cycle pressure ratios (i.e., 30:1 or higher). Gas-fired reciprocating internal combustion engines (RICE) are the prime movers with lowest exhaust gas temperatures. The low end of temperature range covers the flue gas temperatures of fossil-fired utility boilers.

6.4   HEAT RECOVERY STEAM GENERATOR (HRSG)

In the examples earlier, using the concept of exergy, it was shown how to reliably estimate the useful work generation potential of the gas turbine exhaust gas stream. In the preceding section, the exergy of the exhaust gas stream (i.e., the second law benchmark) was calculated to be 32.6 MW. A realistic heat balance calculation showed that that the net bottoming cycle output would be 20.8 MWe. Consequently, the *exergetic* efficiency of the bottoming cycle was calculated as 20.8/32.6 = 0.638. What happened to the missing 11.8 MW or about 36% of the exhaust gas exergy? If you think of the bottoming cycle as a *control volume* encompassing the HRSG, steam turbine generator (STG), steam condenser, and the *balance of plant* (BOP) encompassing the feed and condensate pumps, pipes, and valves, several things contribute to the lost work, 11.8 MW:

- Exergy destruction (irreversibility) in the HRSG
- Exergy destruction in the STG
- Exergy destruction in the condenser

- Exergy transfer out of the system (i.e., exergy loss) via radiation from the HRSG casing (heat loss, typically less than 1% of exhaust gas heat), heat rejected to condenser cooling water, and stack gas
- Miscellaneous (small) irreversibilities in the BOP equipment

Note the distinction made between exergy *transfer* and exergy *destruction* as exergy *loss* mechanisms. Furthermore, exergy transfer that is bookkept as exergy loss is exergy transfer *via heat transfer*. Exergy transfer out of the system *via work* is not a loss; it is the raison d'être of the system in the first place.

In heat recovery from the exhaust gas of a gas turbine, the main source of irreversibility is the constant-pressure–temperature evaporation characteristic of water. The best way to illustrate the fundamental thermodynamics in an HRSG is to look at the heat release diagram. In this diagram, the horizontal axis is the amount of heat transfer; the vertical axis is steam and gas temperature. The conceptual heat release diagram for a *single-pressure, no-reheat* (1PNR) HRSG is shown in Figure 6.7. The diagram depicts hot exhaust gas flowing from left to right across the superheater, the evaporator, and the economizer, and exiting through the stack. Its temperature drops from *TEXH* to *TSTACK*. Heat transfer increases from zero at *TEXH* to its maximum value at *TSTACK*. Feedwater enters from the right and heated in the economizer to a temperature slightly below the saturation temperature at the evaporator pressure. Feedwater boils in the evaporator at constant pressure and temperature to become steam and further heated in the superheater before exiting the HRSG on the left.

The irreversibility in the HRSG is quantified by the mean-effective gas and steam/water temperatures in the HRSG. The former is given by

$$\bar{T}_{gas} = \frac{h_{exh} - h_{stck}}{s_{exh} - s_{stck}} \approx \frac{T_{exh} - T_{stck}}{\ln\left(\frac{T_{exh}}{T_{stck}}\right)}. \qquad (6.15)$$

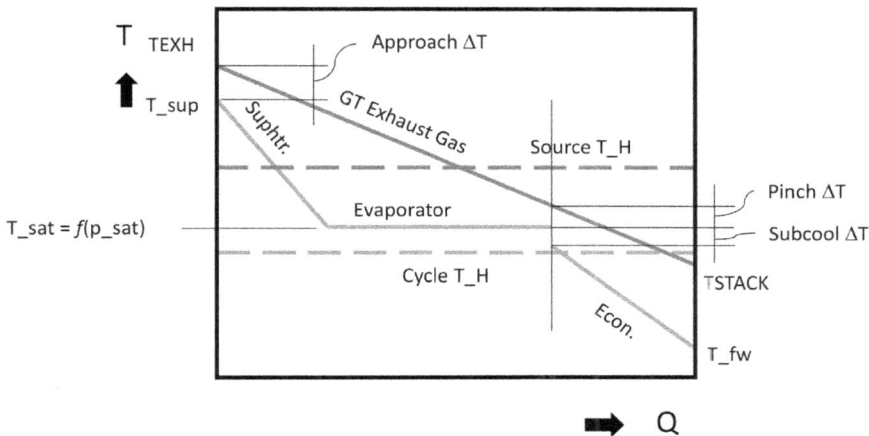

FIGURE 6.7 Conceptual HRSG heat release diagram (1PNRH).

For the steam/water, we have

$$\overline{T}_{stm} = \frac{h_{sup} - h_{fw}}{s_{sup} - s_{fw}}. \tag{6.16}$$

If one could build a Carnot engine operating between two temperature reservoirs at $\overline{T}_{gas}$ and T_0, the work output would be

$$\dot{W}_{id,gas} = \left(1 - \frac{T_0}{\overline{T}_{gas}}\right)\dot{Q}_{rec}. \tag{6.17}$$

Similarly, if one could build a Carnot engine operating between two temperature reservoirs at $\overline{T}_{stm}$ and T_0, the work output would be

$$\dot{W}_{id,stm} = \left(1 - \frac{T_0}{\overline{T}_{stm}}\right)\dot{Q}_{rec}. \tag{6.18}$$

Even without a rigorous quantitative evaluation, just by looking at the heating and cooling lines in Figure 6.7, one could deduce that

$$\overline{T}_{stm} < \overline{T}_{gas} \text{ so that}$$

$$\dot{W}_{id,stm} < \dot{W}_{id,gas}.$$

Therefore, the irreversibility in the simple HRSG depicted in Figure 6.7 is

$$\dot{I}_{HRSG} = \dot{W}_{id,gas} - \dot{W}_{id,stm},$$

$$\dot{I}_{HRSG} = \left(1 - \frac{T_0}{\overline{T}_{gas}}\right)\dot{Q}_{rec} - \left(1 - \frac{T_0}{\overline{T}_{stm}}\right)\dot{Q}_{rec} = \left(\frac{T_0}{\overline{T}_{stm}} - \frac{T_0}{\overline{T}_{gas}}\right)\dot{Q}_{rec},$$

$$\dot{I}_{HRSG} = \frac{T_0}{\overline{T}_{stm}}\left(1 - \frac{\overline{T}_{stm}}{\overline{T}_{gas}}\right)\dot{Q}_{rec}. \tag{6.19}$$

In order to demonstrate the application of these formulae, Frame 6B gas turbine used earlier is used again with a 1PNR bottoming cycle in GT PRO. The bottoming cycle design is as follows:

- 34 bar, 510°C main steam
- 58 mbar condenser pressure
- 10°C/5°C HRSG evaporator pinch/subcool temperature deltas

Resulting HRSG heat release diagram is shown in Figure 6.8. Calculated values are as follows:

- From Equation 6.15, $\overline{T}_{gas}$ is found as 603.8 K (exact, via $\Delta h/\Delta s$) or 595.4 K (approximate, using temperatures only)
- From Equation 6.16, $\overline{T}_{stm}$ is found as 498.2 K
- Q_{rec} is 60,464 kW (the difference between GT exhaust and HRSG stack energy, see Figure 6.8)
- From Equation 6.19, using $T_0 = 25°C$ (298.15 K), $\dot{I}_{HRSG}$ is found as follows:
  - 6,332 kW with $\overline{T}_{gas} = 603.8$ K (exact)
  - 5,910 kW with $\overline{T}_{gas} = 595.4$ K (approximate)

The exergy balance of the GTCC power plant calculated by GT PRO is shown in Figure 6.9. HRSG exergy loss calculated by the program via rigorous exergy balance around the HRSG control volume is 6,479 kW, which is 2.3% higher than calculated using Equation 6.19 with the exact value of $\overline{T}_{gas}$. The difference can be traced to 0.75% radiant heat loss bookkept by GT PRO. Setting it to 0%, HRSG exergy loss is 6,299 kW, which is practically the same as calculated by Equation 6.19 (which does not account for any casing heat loss from the HRSG).

A visual measure of HRSG irreversibility is the space between hot (i.e., exhaust gas) and cold (i.e., water and steam) streams in the heat release diagram, which is significantly large for a 1PNR unit. Indeed, HRSG irreversibility is 6,479/33,501 = 19.3% of GT exhaust exergy. The next two largest contributors are exergy losses in the STG and the steam condenser, 8.2% and 6.1% of GT exhaust energy, respectively.

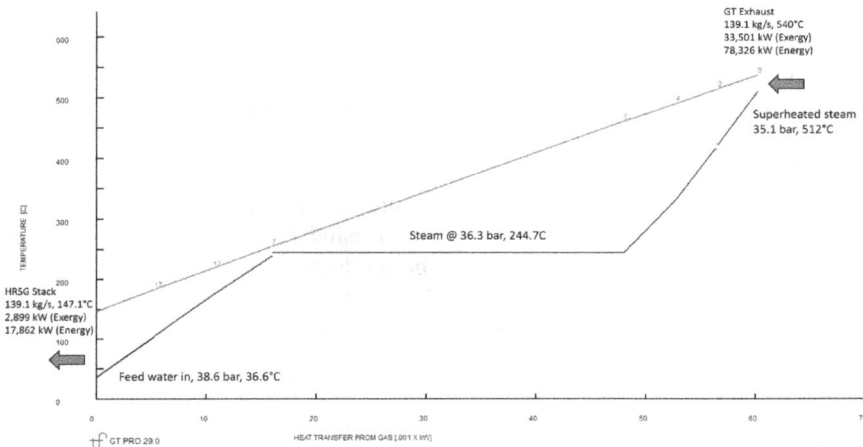

FIGURE 6.8 HRSG heat release diagram (one-pressure, no-reheat steam cycle).

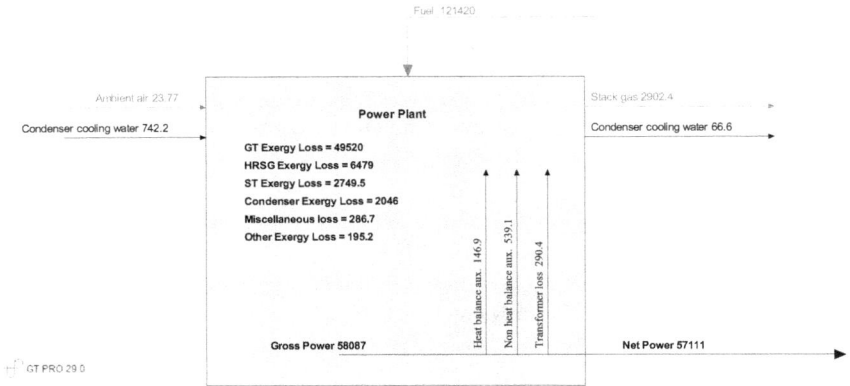

FIGURE 6.9 Exergy balance of a GTCC with a Frame 6B gas turbine and one-pressure, reheat HRSG.

In order to reduce HRSG irreversibility (i.e., minimize the space between hot and cold streams), additional evaporators are introduced, each with steam generation at different pressure levels. Presently, modern HRSG practice is *three-pressure design with reheat* (3PRH). For such designs, calculation of $\overline{T}_{stm}$ using a simple formula like Equation 6.19 is impossible. An empirical formula for estimation of $\overline{T}_{stm}$ in a 3PRH HRSG can be found in the paper by Gülen and Smith [1].

6.5 RECAP

Waste heat recovery is arguably the proverbial poster child process for the application of the second law principles to practical problems in thermal engineering. The second law enters the scene in two forms, which are equivalent to each other:

- *Exergy* of a waste heat stream (e.g., flue gas from a furnace, boiler, or internal combustion engine)
- *Mean-effective temperature* of the heat transfer process from the waste heat stream

Exergy is a stream property and can be calculated directly from stream properties (e.g., pressure, temperature, and composition) using a suitable equation of state. It quantifies the maximum work that can be produced in a process that brings the stream to an equilibrium with the surroundings. That maximum work is the same as the work output of a (hypothetical, of course) Carnot engine that is receiving the heat released by the said process and rejects a portion of it to the surroundings.

Carnot engine is an ideal construction that is based on the Kelvin-Planck statement of the second law. Heat transfer into and out of the Carnot engine takes place at constant temperature (i.e., it is an isothermal process). For the isothermal heat rejection process, the characteristic temperature is the temperature of the ambient or the surroundings of the Carnot engine. For the isothermal heat addition process, the characteristic temperature is the mean-effective temperature of the process, which, in most cases, can be approximated as the logarithmic average of the waste stream temperature and the temperature of the surroundings.

From a practical perspective, the most important application of waste heat recovery from prime mover exhaust in thermal power applications is the HRSG. For practical aspects of HRSG operation and reference, the reader is referred to the handbook issued by the HRSG Users Group [2]. For different HRSG design configurations and their comparison, Refs. [3–5] can be consulted. For more references and calculated examples, including using the exergy method, refer to chapter 6 in the author's monograph on gas turbine combined cycle power plants (Ref. [6]).

REFERENCES

1. Gülen, S.C., Smith, R.W., Second Law Efficiency of the Rankine Bottoming Cycle of a Combined Cycle Power Plant, *ASME Journal of Engineering for Gas Turbines Power*, 132, p. #011801, 2008.
2. Robert, S. (Editor), *Guidelines for the Operation and Maintenance of HRSGs*, The HRSG Users Group, 2003.
3. Bolland, O., A Comparative Evaluation of Advanced Combined Cycle Alternatives, *Journal of Engineering for GTs & Power*, 113, pp. 191–197, 1991.
4. Galopin, J.F., Going Supercritical: Once-through is the Key, *Modern Power Systems*, December, 1998.
5. Zhang, W., Magee, J., Singh, H., Ruchti, C., and Selby, G., HRSG Development for the Future, *PowerGen Europe 2012*, June 12–14, 2012, Köln, Germany.
6. Gülen, S.C., *Gas Turbine Combined Cycle Power Plants*, Boca Raton, FL: CRC Press, pp. 127–147, 2019.

7 Heat Rejection

The production of motion in steam-engines is always accompanied by . . . the re-establishing of equilibrium in the caloric; that is, its passage from a body in which the temperature is more or less elevated, to another in which it is lower.
—Sadi Carnot (1824), *Reflections on the Motive Power of Heat*

According to the Kelvin-Planck statement of the 2LT, there cannot be a heat engine operating in a thermal cycle with no rejection of part of its heat input to the surroundings. This would imply a heat engine with 100% cycle efficiency (ignoring other, smaller losses). It is the ratio of heat rejected from the cycle to the heat injected into the cycle that determines the efficiency of the cycle. In actual heat engines, said heat rejection takes place in a heat exchanger, wherein a hot stream (i.e., the working fluid of the engine cycle in question) transfers heat to a cold stream, i.e., the coolant, which is typically air or water (or water mixed with another liquid to prevent freezing, etc.). Examples are the radiators in car engines and the steam condensers, air- or water-cooled. For an in-depth discussion of the latter equipment, the reader is referred to a recent monograph by the author (Ref. [6] in Chapter 6). Herein, application of key 2LT concepts to cycle heat rejection will be explained within the framework of a water-cooled steam condenser, which is also known as a *surface condenser*.

A simple schematic diagram of the surface condenser is provided in Figure 7.1. There are three important design/sizing parameters:

- Terminal temperature difference (TTD)
- Cooling water temperature rise (TRISE)
- Condensate subcool (TSUBC)

The heat balance of the system is simple and can be written from three perspectives:

$$\text{Condensing steam} : \dot{Q}_C = \dot{m}_{stm} \left(1 - y\right) h_{fg} \tag{7.1}$$

$$\text{Cooling water} : \dot{Q}_C = \dot{m}_{cw} c_{p,cw} \text{TRISE} \approx \dot{m}_{cw} \text{TRISE} \tag{7.2}$$

$$\text{Condenser design} : \dot{Q}_C = UA \times \Delta T_{LM} \tag{7.3}$$

where y is the moisture fraction of steam, h_{fg} is the latent heat of evaporation at TSTM/PSTM, U is the condenser heat transfer coefficient (Btu/ft^2-h-F), and A is the condenser heat transfer surface area (ft^2). (The approximation in Equation 7.2 is valid when using the British units.) The overall heat transfer coefficient is UA in Btu/h-F. The log-mean temperature difference is defined as

DOI: 10.1201/9781003247418-9

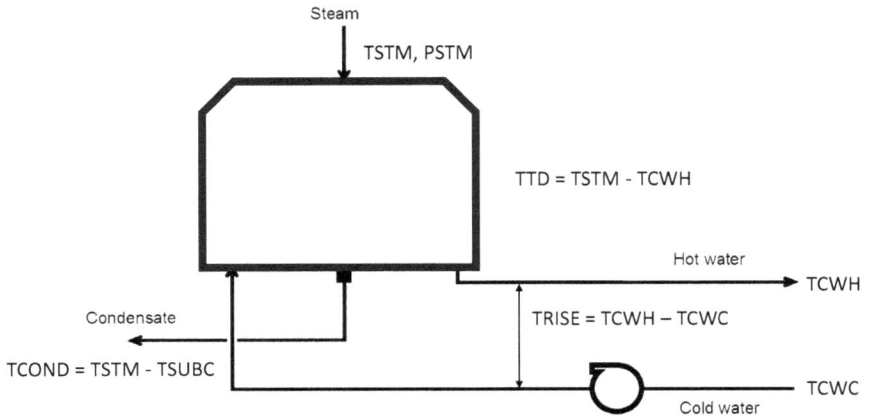

FIGURE 7.1   Schematic diagram of water-cooled steam surface condenser.

$$\Delta T_{LM} = \frac{TRISE}{\ln\left(\dfrac{TSTM - TCWC}{TTD}\right)}. \tag{7.4}$$

From Equation 7.2, it is easy to see that, for a given bottoming cycle, parasitic power consumption of the cooling water circ pump is inversely proportional to the cooling water temperature rise in the condenser. From Equations 7.3 and 7.4,

$$UA = \frac{\dot{Q}_C}{\Delta T_{LM}} = \frac{\dot{Q}_C}{TRISE} \ln\left(\frac{TSTM - TCWC}{TTD}\right) \tag{7.5}$$

which, recognizing the connection between TTD and TRISE for given TSTM and TCWC, can be rewritten as

$$\frac{UA}{\dot{Q}_C} = \frac{-1}{TRISE} \ln\left(1 - \frac{TRISE}{TSTM - TCWC}\right). \tag{7.6}$$

A typical steam condenser *heat release* diagram is shown in Figure 7.2 (condenser pressure is 1 psia or 145 mbar). Cooling water enters the condenser at 59°F (15°C) and exits at 77°F (25°C) for a temperature rise of 18°F (10°C). On the diagram, TCWM is mean-effective cooling water temperature, which is the logarithmic average of TCWC and TCWH. It can be shown that, however, for typical values of TCWC and TCWH,

$$TCWM = \frac{TCWH - TCWC}{\ln\left(\dfrac{TCWH}{TCWC}\right)} \cong \frac{1}{2}(TCWH + TCWC);$$

that is, TCWM is equal to the arithmetic average of TCWH and TCWC (because cooling water specific heat is practically constant).

In Figure 7.2, two second law loss mechanisms in the condenser are identified as follows:

- *Exergy destruction* (irreversibility) due to temperature mismatch between hot and cold streams at TSTM and TCWM, respectively
- *Exergy transfer* out of the Rankine steam cycle via heat transfer to the cooling water

Using the same approach in Section 6.4, exergy destruction in the condenser is quantified as

$$\dot{I}_C = \dot{Q}_C \frac{TAMB}{TCWM}\left(1 - \frac{TCWM}{TSTM}\right). \tag{7.7}$$

The exergy transfer via heat transfer to the cooling water is simply

$$\dot{E}_C = \dot{Q}_C \left(1 - \frac{TAMB}{TCWM}\right), \tag{7.8}$$

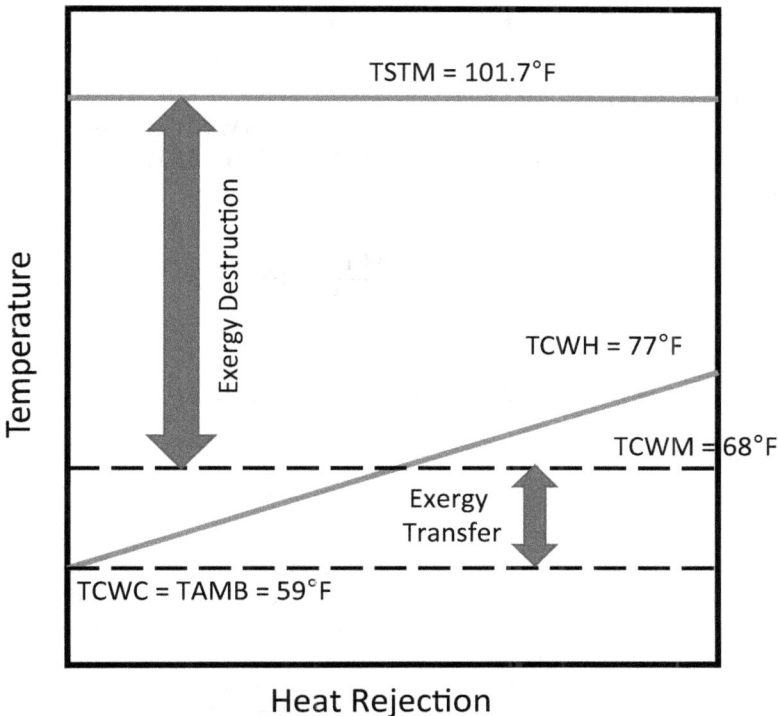

FIGURE 7.2   Heat release diagram of water-cooled steam surface condenser.

which, of course, is the work output of a hypothetical Carnot engine operating between hot and cold temperature reservoirs at TCWM and TAMB, respectively. Total lost work in the condenser is, thus,

$$\dot{W}_{LOST} = \dot{I}_C + \dot{E}_C \text{ or} \tag{7.9}$$

$$\dot{W}_{LOST} = \dot{Q}_C \left(1 - \frac{TAMB}{TSTM}\right), \tag{7.10}$$

which quantifies the work output of a hypothetical Carnot engine operating between hot and cold temperature reservoirs at TSTM and TAMB, respectively. Using the numbers in Figure 7.2, from Equations 7.7, 7.8, and 7.9

$$\dot{E}_C = \dot{Q}_C \left(1 - \frac{519}{528}\right) = 0.017 \dot{Q}_C,$$

$$\dot{W}_{LOST} = \dot{Q}_C \left(1 - \frac{519}{562}\right) = 0.076 \dot{Q}_C \text{ so that}$$

$$\dot{I}_C = \dot{W}_{LOST} - \dot{E}_C = (0.076 - 0.017) \dot{Q}_C = 0.059 \dot{Q}_C.$$

In Figure 6.9 in the preceding chapter, condenser exergy loss was shown as 2,046 kW. Heat rejected to cooling water was 742.2 − 66.6 = 675.6 kWth. Condenser parameters for that case are as follows: TCWC = 15°C, TCWH = 25°C, TSTM = 35.76°C (308.9 K), TAMB = 15°C, and $\dot{Q}_C$ = 39,388 kWth. Thus,

$$\dot{E}_C = 39,388 \left(1 - \frac{288.15}{293.15}\right) = 672 \text{ kW},$$

$$\dot{W}_{LOST} = 39,388 \left(1 - \frac{288.15}{308.9}\right) = 2,646 \text{ kW}$$

$$\dot{I}_C = \dot{W}_{LOST} - \dot{E}_C = 2,646 - 672 = 1,974 \text{ kW}.$$

The difference between the two numbers is a result of the difference between reference temperatures. In GT PRO, the reference or *dead state* temperature is specified as 25°C, which is 10°C higher than the ambient temperature. In that case, exergy destruction

$$\dot{I}_C = \dot{Q}_C \left(1 - \frac{TREF}{TSTM}\right) = 39,388 \left(1 - \frac{298.15}{308.9}\right) = 1,371 \text{ kW},$$

so that

$$\dot{W}_{LOST} = 1,371 + 672 = 2,043 \, kW.$$

From the heat release diagram in Figure 7.2 and the second law loss correlations, the most obvious conclusion is that the condenser losses can be brought down by reducing PSTM (i.e., TSTM). This will work in two ways:

- It will close the temperature gap and reduce the irreversibility.
- It will reduce the amount of heat rejection.

Limitations imposed by increasingly larger (i.e., costly) equipment and parasitic losses (via circ pump power consumption) put a brake on excessively aggressive designs. Another factor to consider is the increase in exhaust loss if a suitable LSB is not available.

8 Cogeneration

Kill two birds with one stone.

—Popular idiom

Cogeneration, in most cases a natural extension of waste heat recovery, describes fossil-fired power plants that generate multiple product streams, usually thermal energy, and electricity. Outside the USA, the term commonly used for cogeneration is *combined heat and power* (CHP). In most cogeneration (*cogen* for short in the industry jargon) applications, the thermal energy product comprises one or more streams of steam at different pressures and temperatures that are used by an end-use customer for district heating and/or cooling, manufacturing process needs, or similar industrial and/or residential uses. The most common material stream that is a product of a cogen power plant is steam or hot water at a known pressure and temperature. Using steam tables, the exergy of steam or hot water can be exactly calculated. The numerical value that is obtained is exactly equal to the power that can be generated in a *Carnot engine*, which utilizes that particular product stream as its heat source.

Before moving on, let us clarify one persistent deficiency in the terminology. By definition, *all* heat engines, including the gas turbine, are cogeneration (i.e., *combined* heat and power generation) devices. This can be easily deduced from the simple gas turbine heat balance given by

$$\dot{Q}_{exh} + \dot{W}_{shft} \approx HC$$

where heat consumption (*HC*) quantifies the fuel energy consumption rate by the unit (i.e., the product of fuel mass flow rate and its heating value). In short, a gas turbine burns fuel and generates two products: shaft output (work) and exhaust energy (heat). In advanced-class heavy-duty machines, the former is roughly equal to 40% of heat consumption (thermal efficiency) and the latter to about 60% (i.e., the remainder). Consequently, what is really meant by the terms of *cogeneration* and *CHP* is that the second product (i.e., exhaust gas energy/heat) is utilized for some useful purpose (instead of being dumped into the atmosphere through the exhaust stack).

A simple numerical example illustrates the calculation process using data taken from the steam tables. Suppose that a GTCC power plant supplies an industrial customer with 25,000 pph of saturated steam at 125 psia. The enthalpy of the steam is 1,191.1 Btu/lb for a total thermal energy supply of 29.8 million Btu/h (or 8,725 kWth). The exergy of 125 psia (saturated) steam is 369.9 Btu/lb for a total thermal exergy supply of 9.25 million Btu/h or 2,710 kWth. In other words, a hypothetical Carnot engine utilizing 29.8 million Btu/h of saturated steam at 125 psia as its sole heat source and rejecting heat to the ambient at 59°F would generate 2,710 kW of power, which implies a maximum theoretical thermal efficiency of 31.1%. Figure 8.1 shows

DOI: 10.1201/9781003247418-10

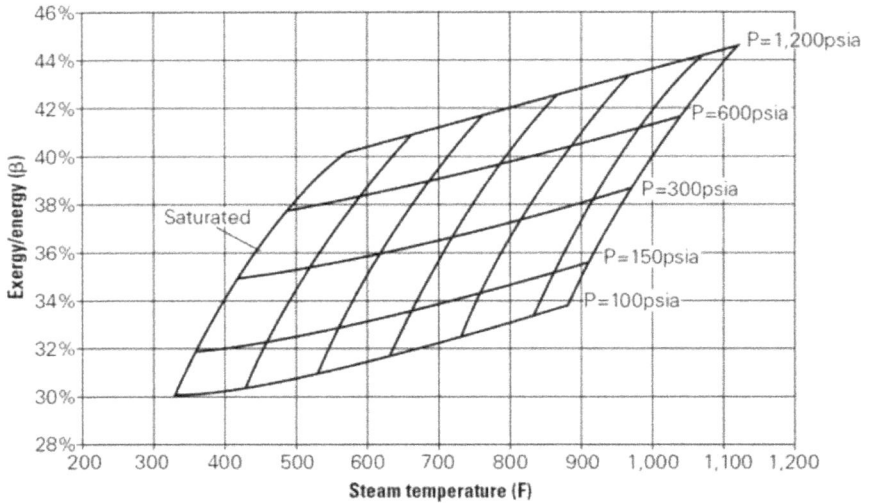

FIGURE 8.1 Exergy-to-energy ratio for steam.

the calculated results for a range of steam pressures and temperatures in terms of β, which is the ratio of steam exergy to steam energy (that is, enthalpy).

Obviously, it is impossible to design a Carnot engine. What is possible is, say, a steam turbine with 85% isentropic efficiency discharging to a condenser at a pressure of 2.5 inches of mercury. Using 29.8 million Btu/h (~9 MWth) of saturated steam at 125 psia (about 9 bar), this steam turbine would generate 1,840 kW of shaft power for a thermal efficiency of only 21.1%. Though this seems to be a paltry number, consider that, when compared to the theoretically possible maximum of 2,710 kW (only from a hypothetical Carnot engine), the second law or rational efficiency of the example steam turbine is 21.1%/31.1% = 67.9%. The term *rational* conveys the underlying concept of using a reference point that is theoretically possible instead of using a reference point that is (even theoretically) impossible.

At this point, it should be obvious that the rational efficiency of a cogen power plant with $\dot{W}$ MWe electric power output and $\dot{Q}$ MWth process heat supply is exactly equal to the efficiency of the same power plant operating in pure power mode. In that mode, it is easy to show that the electric power output of the power plant is

$$\dot{W}_{max} = \dot{W} + \eta_{rat}\beta \times \dot{Q} \text{ or} \tag{8.1}$$

$$\dot{W}_{max} = \dot{W} + \eta_{mar} \times \dot{Q}, \tag{8.2}$$

where η_{rat} is the rational efficiency and η_{mar} is the marginal efficiency of $\dot{Q}$ Furthermore,

$$\dot{W}_{Carnot} = \beta \times \dot{Q} \text{ with} \tag{8.3}$$

$$\beta = \frac{\alpha}{h} \text{ and} \qquad (8.4)$$

$$\alpha = (h - h_0) - T_0(s - s_0). \qquad (8.5)$$

In Equations 8.4 and 8.5, the symbol for exergy is a (from the term *availability* common in US textbooks). The subscript 0 denotes the zero-exergy reference (typically 15°C/59°F and 14.7 psia/1 atm). Enthalpy, h, and entropy, s, can be evaluated from the ASME steam tables as a function of pressure and temperature.

At the risk of being too pedantic, it should be pointed out that Equations 8.4 and 8.5 implicitly assume that steam delivered to the process (i.e., mass flow out of the bottoming cycle control volume) is compensated by makeup water at p_0 and T_0. This is illustrated schematically in Figure 8.2a. In most cogen applications, however, there usually is a process return stream (e.g., warm water) at p_r and T_r, which is illustrated schematically in Figure 8.2b. The exact location of process return entry into the bottoming cycle is a function of p_r and T_r. (It cannot be simply added to the condenser because of deaeration requirements. The same goes for the makeup water as well.) It should also be pointed out that there is a slight error in Equation 8.4, which should have been

$$\beta = \frac{\alpha}{h - h_0}. \qquad (8.6)$$

This is so because the zero-enthalpy reference for the ASME steam tables is the triple point of the water.[1] Consequently, at the zero-exergy reference of 59°F and 14.7 psia, h_0 is not 0; it is 27.1 Btu/lb (63.0346 kJ/kg). Obviously, choosing the triple point as

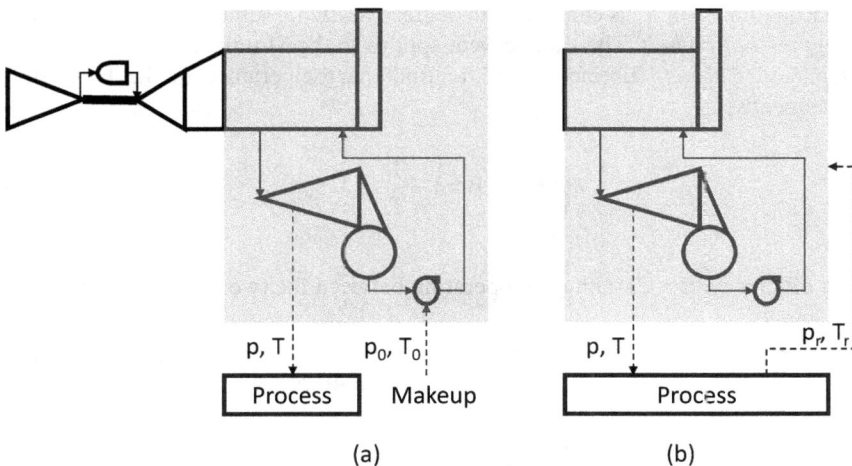

p, T   p₀, T₀   p, T   pᵣ, Tᵣ

Process   Makeup   Process

(a)   (b)

FIGURE 8.2   Process return and makeup water options.

the zero-exergy reference would have prevented the introduction of this error. But then it would be silly to assume makeup water at 32°F (0°C) and less than 0.1 psia. (This, however, is really not an issue if one goes the thermodynamically rigorous route; please read on.) As far as the error introduced, the enthalpy of 150 psig saturated steam is 1,195.5 Btu/lb, and its exergy is 385.7 Btu/lb. From Equation, 8.4, β is calculated as 0.3226 vis-à-vis 0.3301 from Equation 8.6. As long as one is fully aware of the underlying assumption or simplification, this is an error that one can live with.

For the more generic case of process return at p_r and T_r, the applicable equations are

$$\beta = \frac{\Delta\alpha}{h - h_r} \text{ and} \tag{8.7}$$

$$\Delta\alpha = \left(h - h_r\right) - T_0\left(s - s_r\right). \tag{8.8}$$

Combining Equations 8.7 and 8.8,

$$\beta = 1 - T_0\left(\frac{s - s_r}{h - h_r}\right). \tag{8.9}$$

From the discussion of mean-effective Brayton cycle heat addition and rejection temperatures in Chapter 4, it is hoped that the reader will easily recognize that

$$\overline{T}_p = \frac{h - h_r}{s - s_r} \tag{8.10}$$

is the mean-effective temperature of process supply and return streams. (Note that from Equation 8.10, $\overline{T}_p$ is calculated in degrees *Rankine*; subtract 459.67 to convert it to degrees *Fahrenheit*. The same caveat applies to the SI units as well with degrees *Kelvin* and *Celsius*; subtract 273.15 to find the temperature in the latter units.) Consequently,

$$\beta = 1 - \frac{T_0}{\overline{T}_p} \tag{8.11}$$

is the efficiency of a Carnot cycle operating between the two temperature reservoirs at $\overline{T}_p$ and T_0.

Using the same example as earlier – that is, process steam supply at 150 psig (saturated), let us assume that the process return is at 40 psia and 90°F:

- From Equation 8.10, is 324.6°F (note that saturated steam temperature at 150 psig is 365.9°F)
- From Equation 8.11, β is found as 0.3383 or 33.8%

Note that this rigorous approach can be applied to the makeup water case in Figure 8.2a as well. One simply assumes 14.7 psia and 59°F for p_r and T_r respectively:

- From Equation 8.10, $\overline{T}_p$ is 315.1°F
- From Equation 8.11, β is found as 0.3301 or 33% (which, of course, is what we calculated earlier)

Note that this would also be the answer one would get if one chose the zero-exergy reference point as the triple point. All this effort to pinpoint a number, which, depending on one's assumptions (or the boundary conditions of the problem), varies between 0.3226 and 0.3383, can be deemed to be a waste of resources. This is indeed a fair assessment if one is only interested in back-of-the-envelope type estimates. Nevertheless, for a thermodynamically rigorous analysis, it is best to dot the proverbial i's and cross the proverbial t's – especially since the days of pen and paper with a $10 drugstore calculator are long gone.

If one is designing the cogen system using a heat and mass balance simulation software, these types of calculations are not really necessary. Simply by turning the process export stream on and off, one can calculate the marginal efficiency of the net heat to process. As demonstrated earlier, the bona fide efficiency of the cogen power plant, with or without the process steam export, is exactly equal to the net electric efficiency of the facility in pure power mode. (As stated in the beginning of the chapter, the GTCC is the ultimate or perfect cogeneration system.) Trying to inflate the cogeneration's importance (it is, by the way, very important) by adding exergetic apples and oranges and quoting efficiencies of 80%, 90%, and so on, is thermodynamic cheating. Fuel utilization effectiveness is acceptable but not really informative – one can even get a number greater than 100% if use is made of the lowest grade energy is system (e.g., utilizing the condenser heat rejection for water heating to take showers with it).

NOTE

1 The triple point of pure water is at 0.01°C (273.16K, 32.01°F) and 4.58 mm of mercury (611.2Pa, 0.089 psia).

9 Technology Factor

How often have I said to you that when you have eliminated the impossible, whatever remains, however improbable, must be the truth?
—Sherlock Holmes; Sir Arthur Conan Doyle (1890), *The Sign of the Four*

It is nearly impossible to prove or disprove cycle efficiency claims without delving into the minute details of the heat balance calculations, including myriad hardware design assumptions, thermodynamic property packages, system boundary conditions, and so on. However, by using the fundamental principles developed in the preceding chapters, it is eminently possible to separate proverbial wheat from the chaff quite easily but also rigorously and, most importantly, in an unassailable manner, with absolutely zero room for ifs, ands, or buts by just using two parameters and a few very simple formulae.

The fundamental thermodynamic analysis in Section 3.2.1 showed us that the performance of real gas turbines can be related to the efficiency of a Carnot cycle operating between the two thermal reservoirs at T_3 (set equal to the *TIT* of the gas turbine in question) and T_1 (set to the ambient, e.g., ISO conditions), respectively, via two parameters:

1. Cycle factor, which is the ratio of ideal Brayton cycle efficiency at cycle pressure *PR* and cycle maximum temperature T_3 to the efficiency of the said Carnot cycle.
2. Technology factor, which is the ratio of the actual gas turbine efficiency to the ideal Brayton cycle efficiency.

Since the Carnot efficiency derived from T_3 has no practical value, from now on, this technology factor, which references the actual gas turbine performance to the *Carnot equivalent* performance of the ideal Brayton cycle, will be the metric representing the *true Carnot factor* (TCF). Note that, in the literature, quite frequently in fact, the term *Carnot factor* is used to quantify the overall deviation between a *law of physics* (i.e., the second law thermodynamics as personified by the Carnot cycle) and a *man-made construction* (i.e., the actual gas turbine). This, however, is a fundamentally flawed approach because there is *absolutely nothing whatsoever* within the technological arsenal of mankind that can close the gap between the two cycles, which is fixed once and for all once the heat engine cycle (in this case, the Brayton cycle) is chosen. Furthermore, as rigorously demonstrated in Chapter 3, the ideal Brayton cycle is in fact a Carnot equivalent when expressed in terms of mean-effective heat addition and rejection temperatures. Thus, the *true* Carnot factor, as controlled by technology, must be referred to the ideal, to which we *can* approach, and *not* to the ideal, to which we *cannot*. A summary of the discussion in this paragraph can be

DOI: 10.1201/9781003247418-11

found in Table 9.1. An illustration of the concepts with applicable equations will be based on gas turbine Brayton cycle (see Figure 9.1). Since this is not an introductory textbook, the reader is expected to be comfortably conversant in gas turbine cycle calculations. The best reference that the author can recommend as a companion to have under one's hand while reading this chapter is one of the several editions of the gas turbine book by Saravanamuttoo et al. (Ref [2] in Chapter 1).

TABLE 9.1
Cycle Factors

	Symbol	Numerator	Denominator	Significance
Carnot factor	CAF	Actual cycle efficiency	Carnot efficiency	Measure of goodness vis-à-vis an impossible theoretical ideal
Cycle factor	CYF	Ideal Brayton cycle efficiency	Carnot efficiency	Possible theoretical ideal
Technology factor (true Carnot factor)	TF	Actual cycle efficiency	Ideal Brayton cycle efficiency	Measure of goodness vis-à-vis a possible theoretical ideal

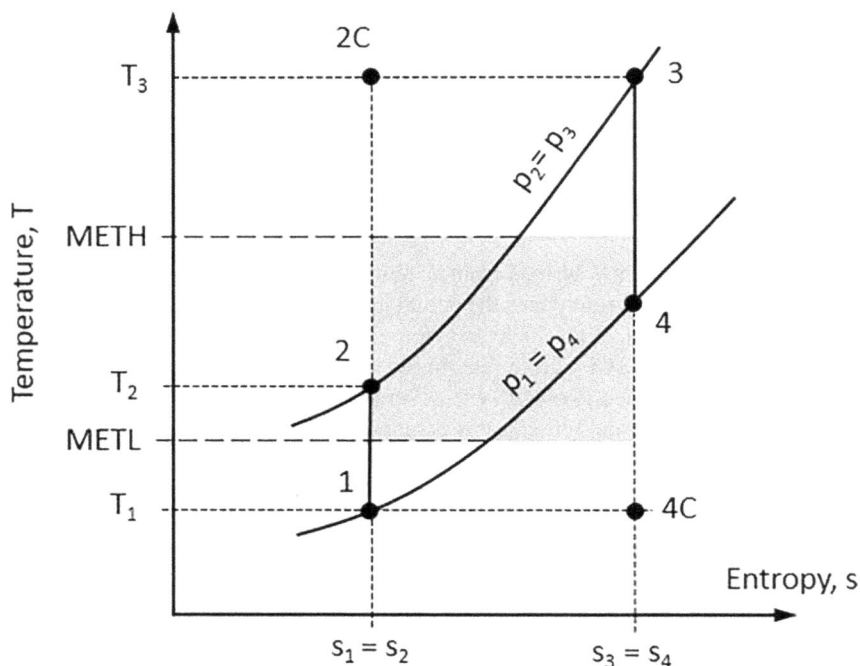

FIGURE 9.1 Gas turbine Brayton cycle {1-2-3-4-1}, Carnot cycle {1-2C-3-4C-1}, and equivalent Carnot cycle (the shaded rectangle).

Cycle factors listed in Table 9.1 are calculated for the gas turbine Brayton cycle in Figure 9.1 as follows:

$$CAF = \frac{\eta_{act}}{1 - \dfrac{T_1}{T_3}},$$ (9.1)

$$METH = \frac{T_3 - T_2}{\ln\left(\dfrac{T_3}{T_2}\right)},$$ (9.2)

$$METL = \frac{T_4 - T_1}{\ln\left(\dfrac{T_4}{T_1}\right)},$$ (9.3)

$$CYF = \frac{1 - \dfrac{METL}{METH}}{1 - \dfrac{T_1}{T_3}},$$ (9.4)

$$TF = \frac{\eta_{act}}{1 - \dfrac{METL}{METH}}$$ (9.5)

$$TF = \frac{CAF}{CYF} \text{ or } CAF = CYF \times TF$$ (9.6)

A numerical example was already provided in Section 3.2.1 using an actual heavy-duty industrial gas turbine (an early example of GE's 60 Hz HA class product lineup) as the basis. The gas turbine in question was defined by a cycle pressure ratio (*PR*) of 21.6:1 and turbine inlet temperature (*TIT*) of 1,600°C (42% net LHV thermal efficiency). Cycle factor was calculated as *CYF* = 0.69. Technology factor was evaluated as *TF* = 0.719. Thus, the Carnot factor for this gas turbine is *CAF* = 0.69 × 0.719 = 0.50.

There is a rigorous approach to deriving the technology factor for any given thermodynamic cycle. Such a derivation for the Brayton cycle technology factor can be found in the chapter written by Desideri, who named it *internal efficiency* (see chapter 3, in Ref. [1]).

Earlier in this chapter, the technology factor for the gas turbine Brayton cycle was defined as (Equation 9.6)

$$TF = \frac{\eta_{act}}{1 - \dfrac{METL}{METH}},$$

which can also be written as

$$TF = \frac{\eta_{act}}{1 - PR^{-k}} \text{ with } k = 1 - \frac{1}{\gamma}. \tag{9.7}$$

An approximate formulation of the actual Brayton cycle efficiency was developed in Chapter 8 of the author's monograph on gas turbines (Ref. [4] in Chapter 1). Herein, it is presented in its final form

$$\eta = \frac{\left(1 - PR_c^{\frac{k_a}{\eta_c}}\right) + (1+f)\chi_g \tau_3 \left(1 - \frac{1}{PR_t^{\eta_t \cdot k_g}}\right)}{f \cdot \ell}. \tag{9.8}$$

Ignoring the difference between compressor and turbine pressure ratios, specific heats and specific heat ratios of air and gas, and using *isentropic* component efficiencies instead of their *polytropic* counterparts, we end up with

$$\eta_{act} = \frac{\left(1 - PR^k\right) + (1+f)\tau_3\left(1 - PR^{-k}\right)\eta_t\eta_c}{\eta_c f \cdot \ell}. \tag{9.9}$$

In Equation 9.9, f is the fuel-air mass ratio with

$$\ell = \frac{LHV}{c_{p,a}T_1} \text{ and}$$

$$\tau_3 = \frac{T_3}{T_1}.$$

Using the heat balance around the combustor and ignoring the contribution of the fuel-air ratio, f, in the numerator (about 0.025 for state-of-the-art heavy-duty industrial gas turbines), Equation 9.9 can be reformulated as

$$\eta_{act} = \frac{\left(1 - PR^k\right) + \tau_3\left(1 - PR^{-k}\right)\eta_t\eta_c}{\eta_c\left(\tau_3 - 1\right) - \left(PR^k - 1\right)}. \tag{9.10}$$

Substituting Equation 9.10 into the definition of the *TF*, Equation 9.7, we obtain

$$TF = \frac{\tau_3\eta_t\eta_c - PR^k}{\eta_c\left(\tau_3 - 1\right) - \left(PR^k - 1\right)}. \tag{9.11}$$

Equation 9.11 is the rigorous definition of the Brayton cycle technology factor. For latest F, H, and J class gas turbines, consider the following values:

- $\eta_t = 0.84$, $\eta_c = 0.89$ (isentropic efficiencies)
- $k = 0.2857$ (i.e., $\gamma = 1.4$)

Equation 9.11 returns about 0.70, which is the average of the state-of-the-art technology (see Chapter 10). For vintage pre-E, E, and F class gas turbines, with $\eta_\tau = 0.83$, $\eta_c = 0.88$, on average, Equation 9.11 returns 0.62, which is also in good agreement with vintage gas turbine data (e.g., see table 4.4 in chapter 4 of Ref. [4] in Chapter 1).

Similarly, with component isentropic efficiencies of 79.5% and 78% for the turbine and the compressor, respectively, Equation 9.11 returns 0.52 for 1944-vintage Jumo-004 engine with cycle *PR* of 3.14 and *TIT* of 1,427°F. This is essentially the same as the technology factor found from the engine data; that is, 0.53 (see table 4.1 in Ref. [4] in Chapter 1).

Clearly, Equation 9.11 is a very handy tool to estimate the technology factor of the Brayton cycle with the two cardinal cycle parameters (i.e., *PR* and *TIT*) and component isentropic efficiencies. Historical development trend is summarized in Table 9.2. If the values in the table seem too low for the later vintage technology, please be aware that the isentropic efficiency for the compressor is a function of the cycle pressure ratio, whereas the turbine isentropic efficiency is a function of the cycle *PR and* the amount of chargeable cooling air. Consequently, the progress in the aerodynamic design of the components, which is reflected by the *polytropic* efficiency, is more difficult to discern in the *overall* performance of the component as reflected by the *isentropic* efficiency.

Note that the technology factor in this discussion is predicated on the air-standard Brayton cycle, whose efficiency is based on the cycle *PR* only. The underlying assumption is that the working fluid (i.e., air) specific heat is constant (i.e., air is treated as a calorically perfect gas). Another ideal cycle definition relaxes this condition; that is,

$$c_p = f(T).$$

Yet another ideal cycle definition can be made with cycle heat input via combustion so that compressor and turbine working fluid compositions are different. These three types of ideal cycle definitions are summarized in Table 9.3. (They are also covered in detail in Chapter 5.) Consequently, depending on which cycle one uses as a

TABLE 9.2
Component Isentropic Efficiency Progress

	1940s	1970s–1990s	1990s–2000s
Turbine	79.5%	83%	84%
Compressor	78%	88%	89%
Technology factor	0.53	0.62 ± 0.02	0.70 ± 0.01

TABLE 9.3
Ideal Brayton Cycles

	Perfect Gas		Ideal Gas
Working fluid	Generic		Atmospheric air
Constant c_p	Yes	No	
Combustion	No	No	Yes – 100% CH_4
TIT	2,912	2,912	2,912
PR	23	23	23
h	59.2%	55.4%	54.0%
METH	1,694	1,701	NA
METL	419	503	NA

reference, the technology factor can vary between 0.70 and 0.77. If one is consistent in his or her assumptions and methods, any approach is equally valid. Since using rigorous property calculations does not add to the predictive power of technology factor methodology, due to its inherent simplicity, ideal cycle approach is perfectly adequate. In this book, the choice of reference is the cold air-standard Brayton cycle (i.e., constant c_p) because of its fundamental importance and theoretical simplicity.[1]

NOTE

1 For instance, if the combustion variant is repeated with 100% H_2 fuel, the cycle efficiency is 55.1%.

REFERENCES

1. Jansohn, P. (Editor), *Modern Gas Turbine Systems: High Efficiency, Low Emission, Fuel Flexible Power Generation (Woodhead Publishing Series in Energy)*, 1st Edition, Cambridge: Woodhead Publishing Ltd, 2013.

10 Internal Combustion Engines

Fire that's closest kept burns most of all.
—William Shakespeare, *The Two Gentlemen of Verona*

Heat engines convert one form of energy, *heat*, into another, *work* In order to have a heat engine, two reservoirs must be present: a *heat source* (the *high temperature reservoir*) and a *heat sink* (the *low temperature reservoir*). On paper, the heat engine is represented by a *thermodynamic cycle*, which is a *collection of thermodynamic states* through which a working fluid goes through while the following happens:

- Receiving heat from the high temperature reservoir
- Rejecting heat to the low temperature reservoir
- Generating useful work in between

The said collection of thermodynamic states constitute a cycle because the starting and ending states are one and the same.

The ultimate heat engine cycle is the Carnot cycle. Practical heat engine cycles are attempts to approach the Carnot ideal. Pretty much all practical heat engines, if one has to be really hair-splitting about it, are not heat engines per se. They are *energy conversion* devices. To be precise, they convert chemical energy of fuel to mechanical energy. The theoretically neat cycle heat addition process is replaced by the quite messy combustion process.

Nevertheless, this author is not that fussy about precise terminology and academic hair-splitting. The bottom line, from a practical engineering perspective, is that as long as the underlying *ideal cycle* can be represented on a thermodynamic surface – for example, pressure-volume (*P-v*) or temperature-entropy (*T-s*) – with a working fluid that (i) obeys the ideal gas equation of state (i.e., $Pv = RT$) and (ii) has constant specific heat, the work-producing device in question deserves the title of heat engine. This means that the device in question can be described precisely by two cycle parameters, for example:

- Cycle *maximum temperature* at the end of the heat addition process
- Cycle *pressure ratio* for the gas turbine
- Cycle *compression ratio* for the piston-cylinder (reciprocating) engine

The maximum temperature in question (for the ideal cycle) is the *proxy* for the temperature of combustion products (hot gas) at the end of the combustion process; that is, the turbine inlet temperature (for a gas turbine) or the temperature at the beginning

DOI: 10.1201/9781003247418-12

of the expansion (work) stroke for a reciprocating engine. That temperature and the pressure or compression ratio set the Carnot limit for those devices via mean-effective heat addition and heat rejection temperatures. How those two temperatures can be calculated was demonstrated in great detail in Section 3.2.1 for the gas turbine Brayton cycle. The same can be done for the Otto cycle describing a spark-ignition reciprocating engine. This is the maximum theoretical efficiency for the cycle/engine in question. It is substantially lower than the *ultimate* Carnot efficiency defined by the cycle maximum temperature. The ratio of the two is the *cycle factor*. The ratio of the actual engine efficiency to the *equivalent* Carnot cycle efficiency sets the *technology factor*.

10.1 GAS TURBINE – BRAYTON CYCLE

Basic second law analysis of ideal (air-standard) Brayton cycle is provided in Section 3.2.1. When one discusses the operation of a gas turbine, one does not make a distinction between the heat engine and the (ideal) thermodynamic cycle it is operating in (i.e., the *Brayton* cycle). They are one and the same. Admittedly, this is a problematic statement. The term *cycle* implies that the working fluid completes a closed loop during the process. This is indeed the case for the ideal, *air-standard* Brayton cycle. However, (i) the actual gas turbine does not operate in a *closed loop*, and (ii) the working fluid changes during the process (i.e., from air to combustion products). The combustor, where cycle heat addition takes place, is an integral part of the gas turbine (i.e., the heat engine). This is why the gas turbine is classified as an internal combustion engine. Nevertheless, the Brayton cycle provides a very reliable albeit idealized framework to analyze and quantify the operation of the gas turbine (i.e., the heat engine itself). (As a side note, closed-loop or *closed-cycle* gas turbines were actively investigated in the past and even made their way from the drawing board to the field installation. Ultimately, however, they ended up as commercial and (arguably) technical flops. For an in-depth discussion, refer to chapter 22 of Gülen (Ref. [4] in Chapter 2) and the monograph by Frutschi [1].)

Gas turbines on their own are referred to be operating in *simple cycle*. The term comes from the fact that in electric power generation, gas turbines operate as a combined cycle, which means that their exhaust energy is used to make steam in a heat recovery steam generator (HRSG) and produce additional electric power by using that steam in a steam turbine generator. The simple cycle in question is, of course, the Brayton cycle. The second cycle implicit in the term *combined cycle* is the Rankine cycle, which is the thermodynamic cycle describing the operation of a steam turbine (see Chapter 11).

Best-in-class gas turbine ISO base load simple cycle performances (50 and 60 Hz) from three OEMs are summarized in Table 10.1 [2]. Six gas turbines from three major OEMs in the table are still designated by the letters of *H* (nominally, 1,500°C *TIT*) and *J* (nominally, 1,600°C *TIT*), but they are clearly 1,700°C class machines (see later). Class designation confusion started with the introduction of steam-cooled *H System*™ of one OEM (General Electric) in 2003 [3] and continues to date long after the commercial demise of the steam-cooled technology (except in a restricted use by another OEM [4]). These details and more can be found in author's monograph on gas turbines (Ref. [4] in Chapter 1). Starting from fundamental cycle thermodynamics, *turbine inlet temperature* (*TIT*) or *firing temperature*, which is a colloquial term for the hot gas temperature at the inlet of the first turbine stage (also known as *rotor inlet temperature*, RIT), can be estimated rather accurately from known cycle pressure ratio (*PR*) and exhaust temperature (*TEXH*) [5]. Using a simple empirical formula (based on the well-known isentropic temperature and pressure ratio relationship) introduced in Gülen (ibid.), estimated *TIT* (in °F) is given by

$$TIT = (TEXH + 460) \cdot PR^{0.2345} - 460 \qquad (10.1)$$

and listed in Table 10.1. Estimated values confirm that, within the error of such a simple estimation, these gas turbines are indeed 1,700°C *TIT* class machines. Their high *TIT* and *TEXH* are ideally suitable to advanced, three-pressure, reheat (3PRH) steam cycles with up to 2,500–2,600 psig (170–180 barg) main steam pressure, and 1,112°F (600°C) main and hot reheat steam temperatures.

The current status of gas turbine technology did not happen overnight. It started with Hans von Ohain's hydrogen-fired radial turbojet HeS-1 in his laboratory (1936–1937) and BBC's 4-MW gas turbine in Neuchâtel, Switzerland, for electric power generation (1939). The first gas turbine as we know it today (i.e., the core of Jumo-004 turbojet engine) went into serial production and active war service (in Messerschmitt Me 262 interceptor) in 1944, near the end of WWII. In the ensuing eight decades, thermal efficiency of gas turbines progressed from barely 15% (net LHV) to well above 40%, as shown in Figure 10.1. While the basic engine configuration stayed unchanged, the technology development took place primarily in the

TABLE 10.1

ISO Base Load Simple and Combined Cycle Ratings of 1,700°C Class Advanced-Class Gas Turbines [2]

	OEM A		OEM B		OEM C	
Gas Turbine	A1	A2	B1	B2	C1	C2
Frequency, Hz	50	60	50	60	50	60
Exhaust flow, kg/s	1,040	779	989	738	1,050	725
Exhaust temperature, °C	640	658	649	649	670	670
TIT class, °C	1,700	1,700	1,700	1,700	1,700	1,700
Estimated *TIT*, °C	1,647	1,684	1,690	1,690	1,714	1,714
Cycle pressure ratio (*PR*)	23.8	23.7	25	25	24	24
SC output, MWe	571	430	563	425	593	405
SC efficiency, %	44	43.3	43.6	44	42.8	42.6
Fuel consumption, MWth	1,297.7	993.1	1,291.3	965.9	1,385.5	950.7
Fuel flow, kg/s	25.9	19.8	25.8	19.3	27.7	19.0
Airflow, kg/s	1,014	759	963	718	1,022	706
Weight, tonne	432	293	560	347	497	305
Power density, kJ/kg of airflow	563	566	584	592	580	574
Power density, kW/kg of weight	1.32	1.47	1.01	1.22	1.19	1.33

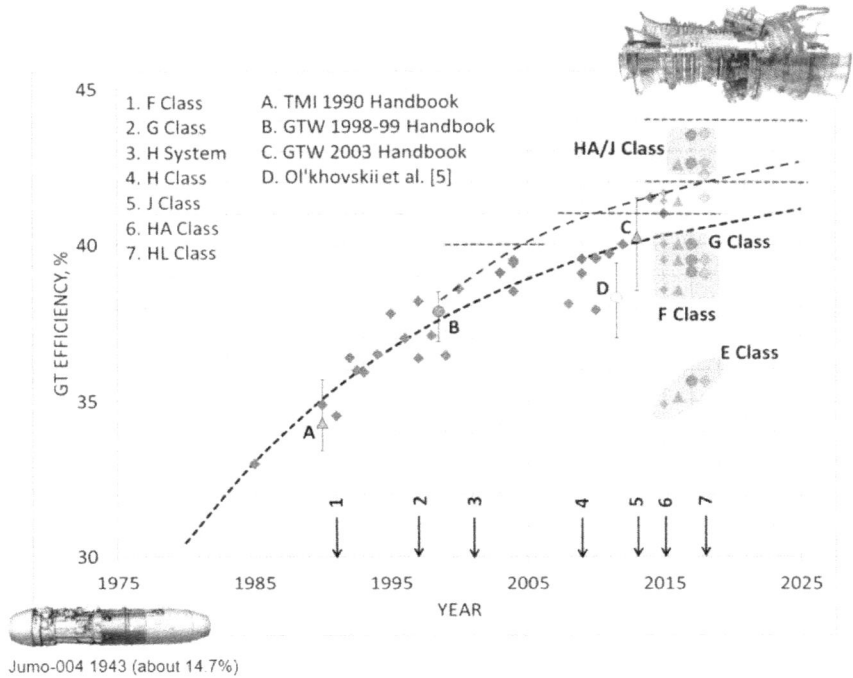

FIGURE 10.1   Gas turbine history, 1985–2018 (an updated version of the chart first published in a 2015 paper by Gülen [6]).

engine thermodynamic (Brayton) cycle; that is, *cycle pressure ratio (PR)* and *turbine inlet temperature (TIT)*. This is illustrated by the class hierarchy of heavy-duty industrial gas turbines (commonly known as frame machines to distinguish them from the aero-derivatives):

- E Class – 1,300°C *TIT* (*PR* = 12:1)
- F Class – 1,400°C *TIT* (*PR* = 15:1)
- G Class – 1,500°C *TIT* (*PR* = 20:1)
- H System (discontinued) – 1,500°C *TIT* (*PR* = 23:1)
- H Class – 1,500°C *TIT* (*PR* = 21:1)
- J/JAC Class – 1,600°C *TIT* (*PR* = 23:1)
- HA Class – 1,600°C *TIT* (*PR* = 23:1)
- HL Class – 1,600°C *TIT* (*PR* = 24:1)

Note that G, J, and H systems have steam-cooled hot gas path (HGP, i.e., the turbine) parts. Furthermore, listed *TIT* values are introductory; HA, J/JAC, and HL class *TIT*s are now most likely 1,650+°C (*PR* of 25:1 for J/JAC). For the direct relationship between cycle *PR* and *TIT* and the underlying thermodynamics, the reader is referred to the monograph by Gülen (Ref. [4] in Chapter 1). Early developers of gas turbines and jet engines already knew that higher cycle *PR* and *TIT* were requisite for higher thermal efficiency. The missing ingredient was HGP component (blades, vanes, disks) materials that could withstand such high temperatures without melting away in seconds. Eventually, development of nickel-based superalloys, thermal barrier coatings (TBC), directionally solidified (DS) and single crystal (SX) casting technologies, and increasingly intricate film cooling techniques brought the technology to 1,700°C *TIT* and 25:1 pressure ratio [1].

Not surprisingly, the advance in cycle thermodynamics came with its own baggage; that is, excessive CO and NOx (a criteria pollutant) production in the combustor, secondary (cooling), and combustion airflow management problems due to elevated compressed air temperatures (result of higher cycle pressure ratio), increasing HGP cooling requirements, and larger compressors for high airflows requisite for gas turbine outputs pushing beyond 500 MW (50 Hz units) – see Gülen (ibid.). Solving all these problems required significant advances in the following areas, in addition to those mentioned earlier, especially in the last three decades (starting with the US Department of Energy's (DOE) *Advanced Turbine Systems* program in the 1990s): premix or Dry-Low NOx (DLN) combustion, 3D aerodynamics, advanced seals, tight clearances, and model-based adaptive control. Recently, two emerging areas of technology can be added to the mix as well; that is, *additive manufacturing* (colloquially known as 3D printing) and *digital twins.*

In specific hardware terms, modern gas turbines share the following characteristics:

- Single-shaft, two-bearing design
- Can-annular DLN combustor
- Four-stage turbine (hot gas path)
- 12–14 stage compressor with 3D airfoils and variable guide vanes (VGVs)

In addition to these common features, OEMs have their own proprietary technologies, such as hydraulic (active) clearance control, single tie-bolt rotor design (same as in 1940s vintage Jumo-004, combining individual discs with Hirth serrations), axial fuel staging, rotor cooling air cooling, enhanced air cooling, and steam cooling, to name a few.

Apart from the conventional gas turbine architecture described earlier, one should also mention two noteworthy variants: reheat (sequential combustion) gas turbine and steam-cooled H system (now defunct). The former is the practical implementation of a fundamental Brayton cycle enhancement (i.e., reheat), whereas the latter comprises an externally cooled hot gas path (mostly). Reheat gas turbines, GT24 (60 Hz) and GT26 (50 Hz), former ABB/Alstom, now owned by GE (the 60 Hz version) and Ansaldo (the 50 Hz version) are still around, whereas the H system was eventually dropped by GE due to its complexity and cost even though it operated quite successfully in the field.

Gas turbine performance can be easily estimated (or verified) with known cycle *PR* and *TIT*, as illustrated in Figure 10.2. Ideal (air-standard) Brayton cycle efficiency is a function of *PR* only. As shown in the figure, existing technology (at a given point in time) is characterized by *TIT* and can be expressed as a technology factor (*TF*).

Note that turbine inlet temperature is a technology classification parameter, and as such, it is a public number. The problem is that it is a nominal number. In other

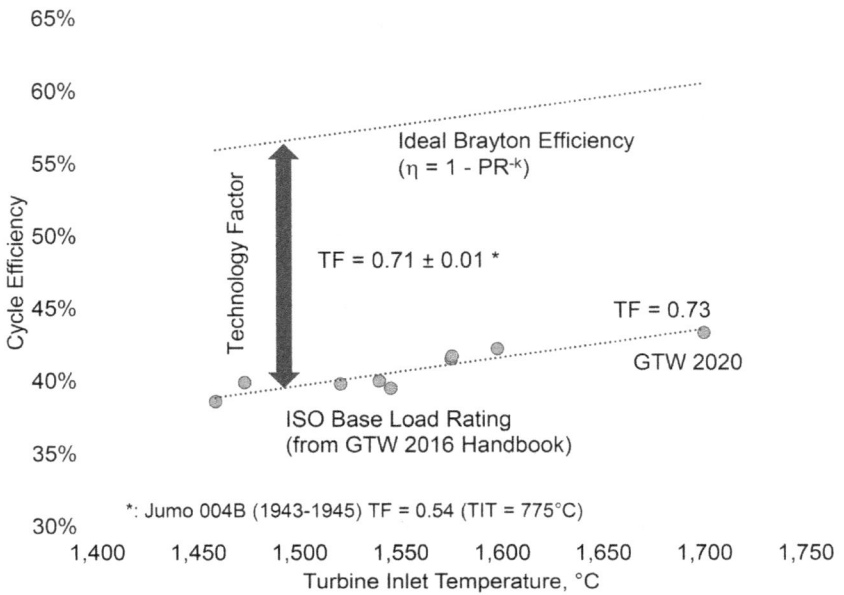

FIGURE 10.2   State-of-the-art gas turbine Brayton cycle and technology factors.

words, nominal J class (or HA and HL by some OEMs) *TIT* is 1,600°C, which was
the introductory number back in 2011. At the time of writing, in 2022, it is a certainty
that HA/HL or J class gas turbines have *TIT*s well above 1,600°C. Turbine inlet
temperature can be estimated from cycle *PR* and exhaust temperature (*TEXH*) using
Equation 10.1. (Those two parameters are publicly available in OEM brochures and
trade publications.)

Actual gas turbine *firing temperatures*, however, are closely guarded trade secrets.
As a rough check, if the turbine inlet temperature (*TIT*) is known, the firing temperature should be in the following proverbial ballpark:

- Subtract 200°F from the *TIT* (advanced F, H, and J class)
- Subtract up to 250°F from the *TIT* (early F class)
- Subtract 125°F from the *TIT* (for E class with *PR* ≤ 14)

For rough estimates of the firing temperature from *TIT* for a given turbine, the following linear transfer function can be used (both temperatures are in °F)

$$\text{TFIRE} = 0.7744 \times \text{TIT} + 438.32. \tag{10.2}$$

Application of Equations 10.1 and 10.2 to published rating data for 60 Hz and 50 Hz
advanced-class gas turbines from major OEMs resulted in the numbers summarized
in Tables 10.2 and 10.3. Average technology factor (*TF*) for both 60 and 50 Hz units
is 0.721 ± 0.12. Average *TIT* is 1,675°C ± 34°C.

TABLE 10.2

Gas Turbine World (GTW) 2021 Handbook 60 Hz Simple Cycle Performance Data (ISO Base Load) – OEM: Original Equipment Manufacturer

OEM	Output, MW	Efficiency, %	PR	TEXH, °F	Ideal Efficiency, %	Technology Factor	TIT, °C	TFIRE, °C
A	369	42.3	24	1,166	59.7	0.709	1,630	1,502
B	430	43.3	23.7	1,217	59.5	0.727	1,684	1,544
C	435	44	25	1,193	60.1	0.732	1,680	1,541
D	405	42.6	24	1,238	59.7	0.714	1,714	1,567

TABLE 10.3

Gas Turbine World (GTW) 2021 Handbook 50 Hz Simple Cycle Performance Data (ISO Base Load) – OEM: Original Equipment Manufacturer

OEM	Output, MW	Efficiency, %	PR	TEXH, °F	Ideal Efficiency, %	Technology Factor	TIT, °C	TFIRE, °C
A	538	42.8	26	1.150	60.6	0.707	1,647	1,515
B	571	44	23.8	1.184	59.6	0.739	1,647	1,515
C	574	43.4	25	1.196	60.1	0.722	1,684	1,543
D	593	42.8	24	1.238	59.7	0.717	1,714	1,567

10.1.1 OPTIMUM BRAYTON CYCLE PRESSURE RATIO

Let us derive a non-dimensional version of the ideal, air-standard cycle net output (per unit mass):

$$\frac{\dot{W}}{\dot{m}c_p} = -(T_2 - T_1) + (T_3 - T_4) \tag{10.3}$$

$$\frac{\dot{W}}{\dot{m}c_p T_1} = w = -\left(\frac{T_2}{T_1} - 1\right) + \frac{T_3}{T_1}\left(1 - \frac{T_4}{T_3}\right) \tag{10.4}$$

$$k = \frac{\gamma - 1}{\gamma}, \quad \tau_3 = \frac{T_3}{T_1} \tag{10.5}$$

$$w = \left(PR^k - 1\right) \times \left(\frac{\tau_3}{PR^k} - 1\right) \tag{10.6}$$

Setting the first derivative of *w* with respect to *PR*, we find that the *optimum PR* for *maximum* net specific output is given by

$$PR_{opt} = \sqrt[2k]{\tau_3} \tag{10.7}$$

Thus, for $T_3 = 1,675°C$ (average *TIT* for the advanced-class gas turbines from earlier), $T_1 = 15°C$, and $\gamma = 1.4$ (which gives $k = 0.2857$), we have

$$PR_{opt} = \left(\frac{1,675 + 273.15}{15 + 273.15} \right)^{\frac{1}{2 \times 0.2857}} = 28.4,$$

which is right in the ballpark of the actual cycle *PR* values.

The efficiency of the ideal Brayton cycle with maximum power output is given by

$$\eta_{max} = 1 - \frac{1}{PR_{opt}^k} = 1 - \frac{1}{\left(\sqrt[2k]{\tau_3} \right)^k},$$ (10.8)

$$\eta_{max} = 1 - \frac{1}{\left(\sqrt[2k]{\tau_3} \right)^k} = 1 - \frac{1}{\sqrt{\tau_3}} = 1 - \sqrt{\frac{T_1}{T_3}},$$ (10.9)

which is sometimes referred to as the *Curzon and Ahlborn* efficiency [7].

There is little doubt that the most cost-effective gas turbine design intuitively leads to maximizing the specific output per unit mass flow rate. However, there is another consideration at play here, which is also implied by the ideal *T-s* diagram in Figure 3.4. Consider the triangular area {1-4-4C-1} in Figure 3.4, which quantifies the lost work associated with cycle heat rejection. What if we could place another power cycle in there, which has a mean-effective heat addition temperature of $\overline{T}_L$ and mean-effective heat rejection temperature of T_1? This is the beginning of the combined cycle concept, which will be covered in Chapter 12.

10.1.2 Exergy Analysis

No in-depth exergy analysis of the gas turbine is provided. As explained earlier in Section 5.5 and numerically demonstrated in Section 3.3, combustor irreversibility dominates everything. This assertion is also validated by the analysis of a recent study in Section 5.5. Interested reader can check out the paper by Facchini et al. [8] and references listed therein for detailed gas turbine exergy analysis findings (including the reheat or sequential combustion gas turbine). For the earliest application of exergy analysis to air-cooled gas turbines, refer to the paper by El-Masri [9].

10.2   RECIPROCATING ENGINES

The term *internal combustion engine* (ICE) has been historically applied to piston-cylinder engines operating in Otto or Diesel cycles, which have formed the backbone of land, marine, and air propulsion for more than a century. Also known as reciprocating ICE (RICE), these engines burning liquid fuels, such as gasoline and diesel, still constitute the majority in land- and marine-based transportation in the form of car, truck, and ship engines. The other ICE (i.e., the gas turbine) has made significant inroads in aircraft propulsion. Gas turbines are also used in ship propulsion, especially in advanced warships and large vessels, such as cruise ships. The reason that "recips" are covered herein is based on the recent emergence of large, natural gas–fired RICE technology offered by OEMs, such as Wärtsila, MAN, and Jenbacher, which are used in power and cogeneration applications. Their ratings range from a few megawatts up to about 20 MW. Due to the nature of their underlying thermodynamic cycle, these engines have high thermal efficiencies (i.e., 45% net LHV or even higher). Just like the aeroderivative gas turbines, their low exhaust energy, both in terms of mass flow and temperature, renders them unsuitable to high-efficiency combined cycle applications. Nevertheless, due to their increasingly significant role in electric power generation, they merit a brief look in this book.

The difference between RICE and gas turbine can be summarized in two aspects:

- Steady-state, steady-flow (SSSF) versus unsteady-state, unsteady-flow (USUF) processes
- Constant-pressure versus constant-volume combustion

To define the SSSF and USUF processes, we have to first define the terms *control volume* (CV) and *control surface* (CS). Let us say that you have a system that you want to analyze thermodynamically (e.g., a pump). The CV is the imaginary space (let us just say a *bubble*) completely surrounding the system in question (e.g., the pump). The surface of the CV (i.e., the bubble) is referred to as the CS.

The SSSF process for a CV is defined by the following assumptions:

- The CV is stationary relative to some coordinate frame.
- The state of the mass at every point within the CV does not vary with time.
- The mass flow rate into and out of the CV does not vary with time.
- The rates at which heat and work cross the CS remain constant.

Thus, the SSSF concept is readily applicable to flow devices, such as pumps, compressors, and turbines, and heat exchange devices, such as cross-flow heat exchangers.

The USUF process for a CV is defined by the following assumptions:

- The CV is stationary relative to some coordinate frame (same as the SSSF).
- The state of the mass may change at every point within the CV, but at any instant of time, the state is uniform throughout the CV.
- The state of the mass crossing all areas of flow is constant with respect to the CS, but the mass flow rates into and out of the CV may vary with time.

Thus, the USUF concept is useful for the analysis of processes that involve unsteady (continuous or intermittent) flows with rapid mixing within the CV (e.g., the cylinder of a four-stroke ICE).

Consequently, the following is true:

- Gas turbine processes are amenable to thermodynamic analysis under the assumption of SSSF. (Recall that we have used the SSSF exergy conservation formulation in Chapter 5.)
- Piston-cylinder (or *reciprocating*) ICE processes are amenable to thermodynamic analysis under the assumption of USUF.

Furthermore, the following is true:

- The heat addition process of a gas turbine (taking place in the combustor, which is a steady flow device) can be idealized as a constant-pressure process (e.g., the air-standard Brayton cycle).
- The heat addition process of a piston-cylinder (or *reciprocating*) ICE (taking place in the cylinder, which is an unsteady flow device) can be idealized as a constant-volume process (e.g., the air-standard *Atkinson* cycle).

While the ideal gas turbine operation can uniquely be described by the air-standard Brayton cycle, for RICE, we have several possibilities; that is, the air-standard *Otto* cycle, *Diesel* cycle, the dual cycle, and the Atkinson cycle. All RICE cycles, except the dual cycle, which combines Otto and Diesel cycle heat addition processes, are cycles with *four* processes, specifically the following:

- The Otto cycle, on a *T-s* surface, includes (1) isentropic compression, (2) constant-volume (*isochoric*) heat addition, (3) isentropic expansion, and (4) isochoric heat rejection.
- The Diesel cycle, on a *T-s* surface, includes (1) isentropic compression, (2) constant-pressure (*isobaric*) heat addition, (3) isentropic expansion, and (4) isochoric heat rejection.
- The Atkinson cycle, on a *T-s* surface, includes (1) isentropic compression, (2) isochoric heat addition, (3) isentropic expansion, and (4) isobaric heat rejection.

Modern natural gas–fired RICE for electric power generation are spark-ignition engines, for which the ideal engine proxy is the air-standard Otto cycle. It can be shown that the efficiency of the Otto cycle, in a way similar to the gas turbine Brayton cycle, is a function of the cycle *compression ratio*:

$$\eta_{\text{RICE}} = 1 - \frac{1}{\text{VR}^{\gamma-1}}, \tag{10.10}$$

where the compression ratio, *VR*, is the ratio of specific volumes at the beginning and at the end of the compression process (i.e., the *compression stroke* in engine parlance). In reciprocating ICE terminology, it is the ratio of the *swept volume* to the *clearance volume*. Strictly speaking, the compression ratio of a given engine can be calculated from the cylinder geometry; that is, cylinder bore (diameter), stroke, and clearance height at the piston *top dead center* (TDC). The effective *VR* of an actual engine can be lower than the geometrical *VR* mainly due to the finite time required for heat release and ignition retarding (depending on engine design and controls).

In actual engines, high *VR* values can lead to *knocking*; that is, self-ignition of the fuel-air mixture due to the high temperature caused by high compression ratio before the spark ignition. This is one reason why automotive engines with low compression ratios have typically low efficiencies (i.e., below 40%) vis-à-vis their gas-fired power generation counterparts, whose *VR*s can go as high as 11 or 12.

Unfortunately, there is no single source for RICE performance ratings with pertinent engine parameters analogous to the *Gas Turbine World* magazine's annual handbook for industrial gas turbines. Furthermore, engine OEMs do not publish their products' performances in a standardized manner similar to the gas turbine OEMs. This makes a rigorous analysis of technology factors quite difficult. Nevertheless, a good idea can be obtained by using the computational data provided in a paper by Caton [10]. In the cited paper, the author starts with an automotive engine (2,000 RPM, 8 cylinders, 5.7 liters of displaced volume) with *VR* = 8.0 and *equivalence ratio* of λ = 1.0 (i.e., air-fuel ratio is the same as the stoichiometric value). The high-efficiency version of this engine has a *VR* of 16, which is facilitated by two modifications: lean burn with λ = 0.7 and *exhaust gas recirculation* (EGR) to the tune of 45% of the engine exhaust gas flow. (These two engine features are necessary to prevent knocking.) Furthermore, for achieving a *brake mean-effective pressure* (BMEP) of 9 bar, which is the basis of comparison for conventional and high-efficiency engines, inlet pressure is increased to 1.7 bar (vis-à-vis 0.92 bar for the reference engine). From Figure 4 in Caton (ibid.), a transfer function for engine efficiency as a function of *VR* is derived as follows:

$$\eta_b = 1 - \frac{0.933}{VR^{0.216}},$$

$$(10.11)$$

which, when compared to Equation 10.10, suggests a specific heat ratio of γ = 1.216 (instead of 1.4 for the air-standard cycle). In another paper, Klimstra and Hattar [11] from Wärtsila investigate the performance improvement opportunities for large natural gas–fired RICE. Figure 7 in the cited paper provides measured engine efficiencies for two different compression ratios of two different engines (34 cm and 50 cm bore). One data point for Wärtsila 18V50SG (a 19 MW spark ignition, gas-fired engine) is taken from Sutkowski [12]. Two additional data points are from the technical specification documents for Jenbacher J620 and J920 engines (*VR* = 13.5, 48.2% electrical efficiency for J920). Ideal and actual engine efficiencies are plotted in Figure 10.3. Note that Equation 10.11 and Wärtsila data represent *brake thermal efficiencies* (BTE), whereas Jenbacher and Sutkowski data represent engine *electrical efficiencies* (i.e., measured at generator terminals). In RICE, BTE is the ratio of *brake thermal power* to fuel energy input. Brake thermal power is the counterpart of the shaft power output in gas turbines. The difference between BTE and electrical efficiency is the losses incurred in the generator (about 1 to 1.5%). No correction is made when plotting the data in Figure 10.3.

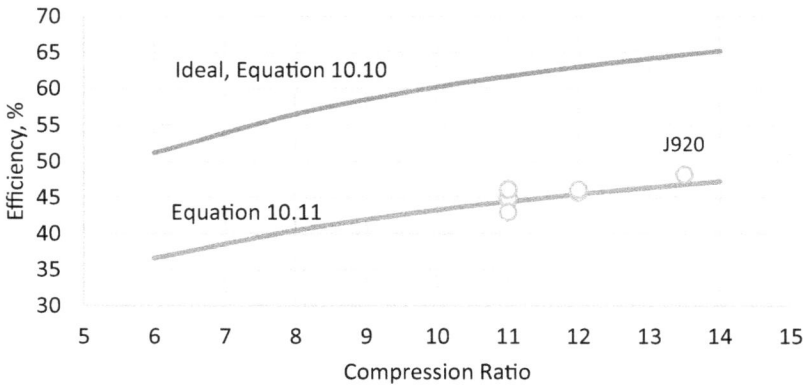

FIGURE 10.3   RICE efficiency as a function of compression ratio.

Technology factors implied by the data in Figure 10.3 is plotted in Figure 10.4. Not surprisingly, RICE technology factors are quite similar to the gas turbine technology factors in Figure 10.2. Both internal combustion engine technologies have gone through a century of intense R&D toward higher thermal efficiencies with concomitantly lower emissions. On the gas turbine side, the progress in performance has been driven by increasing turbine inlet temperatures and cycle pressure ratios. On the RICE side, the key driver of thermal efficiency, as noted by Klimstra and Hattar (ibid.), has been the increase in BMEP from 12 bar to 20 bar or higher mainly via increasing compression ratio. As a second contributor, the authors cite the widening of the cylinder bore from less than 250 mm to values around 340 and 500 cm. To better understand the logic behind these statements, we have to look at the engine heat balance and understand the significance of BMEP. While having the units of pressure, BMEP does not correspond to a measured value inside the engine. It is the average (constant) pressure which, if it were to be imposed on the pistons uniformly from the top to the bottom of each power stroke, would produce the measured brake power output. In other words, BMEP is the measure of the efficiency of a given engine at producing torque from a given displacement.

The heat balance of an advanced 10 MWe (nominal) engine ($N = 1,000$ RPM) is summarized in Table 10.4. The brake thermal efficiency of this RICE is 49%. The electric output is 9,521 kWe with a generator efficiency of 98.3% for net electric efficiency of 48.2%. The engine bore (D) and stroke (L) are 310 and 350 mm, respectively, for a total displacement of 528.3 liter (20 cylinders). The output per cylinder is $9,686/20 = 484.3$ kW, which translates to

$$\text{BMEP} = \frac{1.2 \times 10^9}{V_{\text{disp}} N} = \frac{1.2 \times 10^9}{\left(\frac{\pi}{4} D^2 L\right) N} = \frac{1.2 \times 10^9}{\left(\frac{\pi}{4} 310^2 350\right) 1,000} = 22 \, \text{bar}.$$

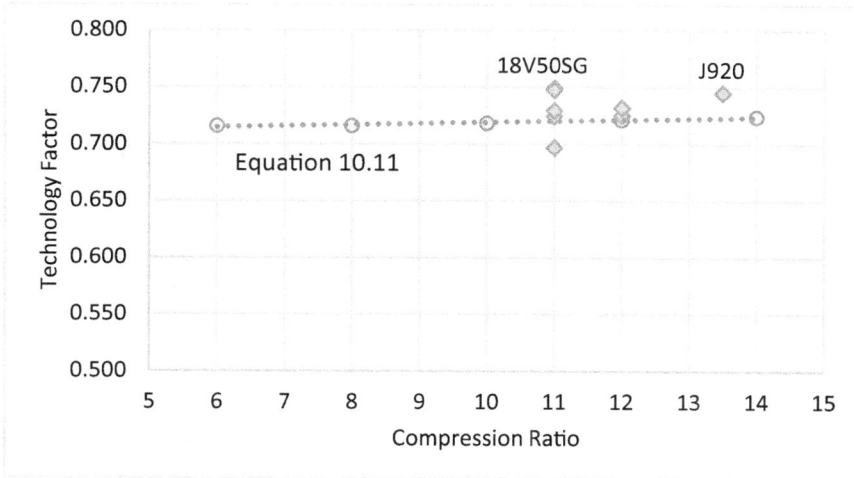

FIGURE 10.4 Technology factor of RICE as a function of compression ratio.

TABLE 10.4
Heat Balance of 10 MW RICE (BMEP = 22 bar, 1,000 RPM)

Heat consumption, kWth	19,760	100.0%
Charge air cooler heat rejection, kWth	3,213	16.3%
Heat transfer to lube oil, kWth	1,052	5.3%
Heat transfer to jacket water, kWth	1,012	5.1%
Surface heat loss, kWth	337	1.7%
Exhaust energy, kWth	4,460	22.6%
Mechanical losses, kW		0.0%
Brake output, kW	9,686	49.0%

The energy balance in Table 10.4 shows that for this advanced RICE with 310 mm bore, total heat transfer to the jacket cooling water and the lube oil constitutes 10.4% of the fuel energy input (engine *heat consumption*). Total engine heat loss, including heat radiation from the cylinder surface, is 12.1%. Heat transfer to the lube oil mostly represent the heat generated by friction between the engine parts and the rest is the heat transferred from the charge air and combustion gas to the cooling water and the surroundings. The exhaust energy is after the turbocharger turbine, which is represented by the charge air cooler heat duty, 3.2 MW, so that about 39% of fuel energy input is rejected as heat into the exhaust manifold. In summary, 39% + 49% = 88% of fuel energy input is accounted for by engine control volume work and exhaust so that the process is quite close to being adiabatic.

Consequently, the benefit of higher BMEP can be traced to the relatively diminished weight of friction and heat losses in the engine heat balance as a fraction of engine heat consumption. A larger bore lowers the area-volume ratio of the combustion chamber and that of the cylinder, which has a positive impact in terms of reduced heat transfer from the working fluid to the cylinder walls.

10.2.1 EXERGY ANALYSIS

As it turns out, a detailed exergy analysis of the RICE does not provide much insight. Just like in a gas turbine engine, the largest source of exergy destruction (irreversibility) is the combustion process. Caton [10] reports that, for what he calls the "conventional engines," the exergy destruction during combustion ranged between 20.5% and 21.5% (of the fuel exergy). For the high-efficiency engine in the cited paper, with lean burn and EGR, exergy destruction during combustion is higher (i.e., 24% of fuel exergy). The higher irreversibility is traced back mainly to the diluted charge and the associated lower combustion temperatures. Nevertheless, the overall cycle results in higher thermal efficiencies due to improved cycle parameters.

For a rigorous exergy analysis of a reciprocating engine, the reader is referred to the paper by Van Gerpen and Shapiro [13]. In this paper, the authors develop the exergy (availability) balance for a diesel engine in detail from the first and second low equations for the USUF process inside the cylinder. They transform the time derivatives (i.e., dX/dt) to crank angle derivatives (i.e., $dX/d\theta$). Furthermore, they also divide the exergy into its two components: *thermomechanical* and *chemical* (e.g., see Section 5.2.2). This enables them to separate the irreversibility due to combustion from that associated with heat transfer to the walls. Their final equation for the exergy balance for a control volume comprising the charge air inside the engine cylinder is as follows:

$$\frac{dA}{d\theta} = \int q'\left(1 - \frac{T_0}{T}\right)dS - \left(W' - P_0\frac{dV}{d\theta}\right) + n'_f\bar{a}_{f,\text{flow}} - I' \tag{10.12}$$

The following physical quantities/parameters can be identified in Equation 10.12:

- A is the exergy, kJ
- θ is the crank angle in degrees
- q' is the heat flux, kJ/m^2 per degree
- S is the cylinder wall area, m^2
- W' is the work done as a function of the crank angle, kJ per degree (i.e., it is the counterpart of $\dot{W}$ in kW)
- V is the volume in m^3
- T is temperature in K
- n'_f is the molar flow rate of fuel per unit degree
- $\bar{a}_{f,\text{flow}}$ is the flow exergy per unit mole
- I' is the irreversibility in kJ per degree
- P_0 is the reference pressure in Pa (N/m^2)
- T_0 is the reference temperature in K

Consequently, the term on the LHS of Equation 10.12 is the rate of change of exergy in the control volume, including thermomechanical (ATM) and chemical components ACH).

$$\frac{dA_{tm}}{d\theta} = \frac{dU}{d\theta} + P_0 \frac{dV}{d\theta} - T_0 \frac{dS}{d\theta} \tag{10.13}$$

$$\frac{dA_{ch}}{d\theta} = -\sum_{i=1}^{n} \frac{dn_i}{d\theta} \mu_i^{\circ} - \sum_{j=1}^{n} \frac{dn_j}{d\theta} \mu_j^{\circ} \tag{10.14}$$

In Equation 10.13, U, V, and S represent the total internal energy, volume, and entropy of the control volume, respectively, comprising the charge air (constant during compression and expansion strokes) and fuel injected into the cylinder. In Equation 10.14, subscripts i and j denote the species present in the reference environment (i.e., N_2, O_2, H_2O, CO_2, and Ar in the air) and not present in the reference environment, respectively. Thus, n_i and n_j are number of moles of species i and j, respectively, whereas μ_i° and μ_j° denote chemical potentials of the species.

The following is on the RHS of Equation 10.12, from left to right:

- The first term is the rate of exergy transfer associated with heat transfer from the gases to the cylinder wall (AQ).
- The second term is the net rate of exergy transfer to the piston associated with work (AW).
- The third term is the rate of exergy input to the control volume with the fuel, including the availability input associated with flow work, and the thermomechanical and chemical exergy of the fuel ($AFUEL$).
- The fourth term is the rate of exergy destruction due to irreversibilities within the control volume (I), dominated by combustion irreversibility.

Variation of the individual terms in the exergy balance equation are plotted in Figure 10.5 as a function of the crank angle, from the *bottom dead center* (BDC) back to the BDC (i.e., from −180 degrees to +180 degrees). (In other words, the process is shown for two turns of the crankshaft or two piston strokes for a four-stroke engine.)

The following observations can be made from the curves in Figure 10.5:

- During the compression stroke:
  - Up to a crank angle of −10 degrees, the value of AW is negative (i.e., work done on the control volume) and the thermomechanical availability (ATM) increases.
  - AQ and I are zero (reversible and adiabatic).
  - The pressure and temperature changes are not enough to cause appreciable dissociation, so the chemical exergy (ACH) is virtually constant.

- Combustion starts at a crank angle of −10 degrees, and thereafter, the following happens:
  - There is a corresponding rise in the exergy because of fuel addition ($AFUEL$).
  - The thermomechanical exergy (ATM) increases rapidly as fuel enters the control volume and burns.
  - The chemical reactions associated with combustion generate irreversibility (I).

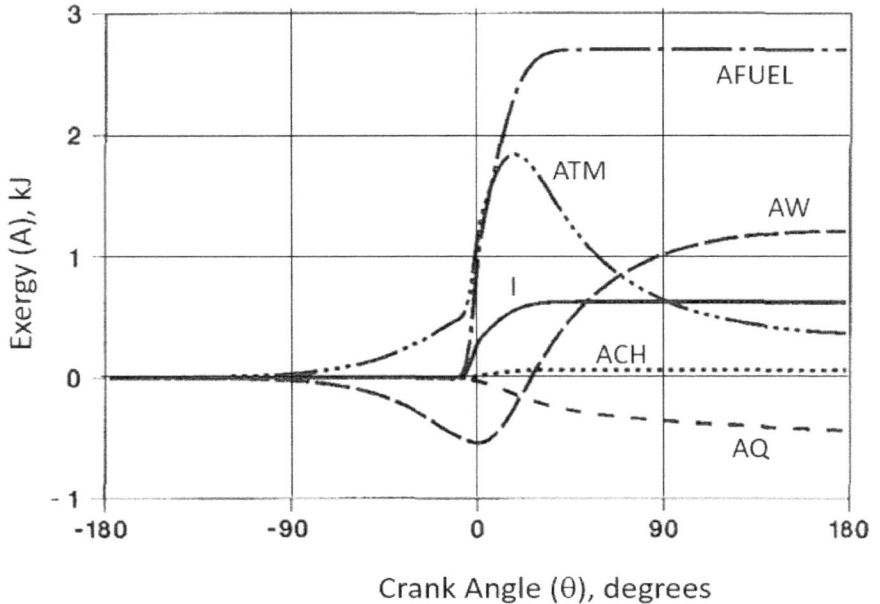

FIGURE 10.5 Cumulative change of exergy balance equation terms [13].

- Just after the TDC (crank angle 0), the heat transfer term (AQ), negative because heat is transferred out of the control volume, increases with increasing system temperature.
- The fuel exergy $(AFUEL)$ and the irreversibility (I) do not change for crank angles greater than about 90 degrees because combustion has essentially ceased.
- ATM decreases as the exergy continues to leave the system with heat (AQ) and work (W).

Also, from the curves in Figure 10.5, we can determine the following, at the end of the expansion stroke:

- Cumulative irreversibility (I) is about 22% of $AFUEL$
- Work transfer to piston (AW) is about 44% of $AFUEL$
- Close to 40% of $AFUEL$ is left over (to be exhausted out of the cylinder by the next piston stroke)

A similar analysis is presented by Mattson et al. for a compression-ignition engine fueled with biodiesel [14]. Their findings were essentially identical to those reported earlier from the paper by Van Gerpen and Shapiro (ibid.). In other words, in terms of the total exergy breakdown, combustion irreversibility and exergy transfer with exhaust gas are the dominating terms. The reader is strongly advised to obtain copies of the two cited papers and study them for further learning and guidance for application to his or her work. Both are rich in detail, basic principles, and step-by-step description of the fundamental equations.

10.3 THE MOST EFFICIENT HEAT ENGINE CYCLE

Both Brayton and Otto air-standard (ideal) cycles are four-process cycles combining basic (fundamental) reversible thermodynamic processes:

- Isentropic compression or expansion (S)
- Isochoric heat transfer (V)
- Isothermal heat transfer (T)
- Isobaric heat transfer (P)

For example, Brayton cycle can be concisely described as an SPSP cycle, whereas the Otto cycle is an SVSV cycle. Amann has investigated 256 such combinations of reversible thermodynamic processes to identify engine cycles with constant energy input per unit mass of working gas that offer the greatest thermal-efficiency potential [15]. He only considered *nonregenerative* cycles (e.g., no heating of compressed working fluid by hot expanded working fluid to reduce cycle heat addition) with no more than four processes. (Since the author's interest was mainly in automotive applications, he also omitted turbocharging.) His findings are as follows:

- For a fixed compression ratio, the SVSP (Atkinson) cycle promises the highest efficiency.
- For a fixed peak cycle pressure, the SPSP (Brayton, or Joule) cycle promises the highest efficiency.
- At fixed compression ratio, the SVSV (Otto) cycle is the preferred cycle over the Atkinson cycle because of its high BMEP (remember that the author is focused on automotive applications).
- Similarly, at fixed peak cycle pressure, the SPSV (Diesel) cycle is the preferred cycle over the Brayton cycle because of its high BMEP.

The Atkinson cycle is an air-standard cycle that describes a piston engine similar to the Otto engine. As such, the working fluid (air in the piston-cylinder assembly) constitutes a *closed* system. However, under the name of *Humphrey*, the similar (but *not* the same) collection of four processes is used for a gas turbine with *constant-volume heat addition* (i.e., a modified Brayton cycle) as a surrogate for a *pulse detonation engine* (PDE). While the PDE so far exists on paper only, the Humphrey cycle can be considered as the air-standard cycle for Holzwarth's explosion turbine (see Chapter 4 in Ref. [4] in Chapter 1), which combined a centrifugal compressor and velocity-compounded turbine (each an *open* system) with an *explosion chamber* (i.e., a closed system) in between. Atkinson and Brayton cycles are shown on a *T-s* diagram in Figure 10.6.

Using the first law for a closed system, the heat addition process in Atkinson cycle is given as follows:

$$du = dq_{in} = c_v \cdot dT \qquad (10.15)$$

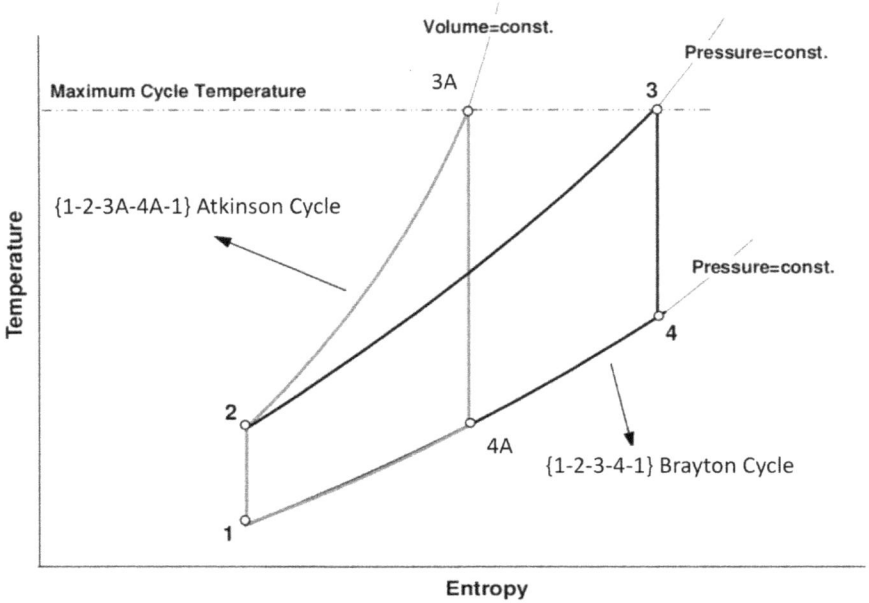

FIGURE 10.6 Comparison of Atkinson and Brayton cycles on *T-s* diagram.

All other processes can be described in a manner identical to the Brayton cycle. Using the ideal gas equation of state for the beginning and the end of the heat addition process, $2 \rightarrow 3A$ and noting that $v_{3A} = v_2$, one can verify that

$$\frac{P_{3A}}{P_2} = \frac{T_{3A}}{T_2} \tag{10.16}$$

Since the temperature rises as a result of heat addition, during a constant-volume process, the pressure will also rise in proportion to the rise in temperature. Combining the pressure ratios for the isentropic compression and expansion processes, and noting that the heat rejection is at constant pressure (i.e., $P_{4A} = P_1$), the following relationships are obtained:

$$\frac{P_{3A}}{P_2} = \frac{\pi}{\pi'} = \pi'' \tag{10.17}$$

$$\pi = \pi' \cdot \pi'' \tag{10.18}$$

where the details are as follows:

- $\pi' = P_2/P_1$ is the pressure ratio of isentropic *precompression*

- $\pi'' = P_{3A}/P_2$ is the pressure ratio of *constant-volume heat addition*
- $\pi = P_{3A}/P_1$, the product of the two, is the *overall* cycle pressure ratio

Using isentropic p–T relationships and going through some algebra, it can be shown that

$$\eta_A = 1 - \frac{1}{\theta_2} \cdot \left\{ \gamma \cdot \frac{\tau^{1/\gamma} - 1}{\tau - 1} \right\} \tag{10.19}$$

where θ is the non-dimensional temperature (via division by T_1) at the particular cycle state indicated by the subscript and $\tau = T_{3A}/T_2$. The fraction inside the curly brackets on the right-hand side of Equation 10.19 will appear frequently in the subsequent discussion and, henceforth, will be referred to as Γ (capital γ):

$$\Gamma \equiv \frac{\tau^{1/\gamma} - 1}{\tau - 1} \tag{10.20}$$

Even though they describe conceptually very different heat engines, both Atkinson and Humphrey cycle efficiencies are represented by the same formula, Equation 10.19, which can be recast into a more descriptive form:

$$\eta_A = 1 - \frac{1}{\tilde{q}} \cdot \left[\left(\frac{\gamma \cdot \tilde{q}}{\pi'^k} + 1 \right)^{1/\gamma} - 1 \right], \tag{10.21}$$

where $\tilde{q}$ is the total cycle heat addition (per unit mass flow) between states 2 and 3A normalized via division by $c_p \times T_1$. After some algebra, Equation 10.21 can be recast as

$$\eta_A - \eta_B = (1 - \eta_B) \cdot (1 - \gamma \Gamma), \tag{10.22}$$

where η_B is the efficiency of a Brayton cycle with overall cycle pressure ratio, π, the same as the precompression pressure ratio of the Atkinson cycle, π'. The right-hand side of Equation 10.22 is always positive so that η_A is always greater than η_B but only when Brayton π = Atkinson π'. When Brayton π is the same as Atkinson π (efficiency is denoted as η'_B), Equation 10.22 becomes

$$\eta_A - \eta'_B = (1 - \eta'_B) \cdot (1 - \gamma \tau^k \Gamma). \tag{10.23}$$

The right-hand side of Equation 10.23 is always negative so that η_A is always less than η'_B.

This is a very important point, which is usually ignored in discussions on cycle thermodynamics. Two ideal, air-standard heat engine cycles describing turbomachines can be compared on an apples-to-apples basis, *if and only if* their overall cycle pressure ratios are the same.

10.3.1 A New Ideal Cycle

A turbomachine (e.g., a gas turbine) is a *steady-flow* device, which can be naturally described by an air-standard cycle comprising four steady-flow (also known as *control volume*) processes:

1. Compression
2. Heat addition
3. Expansion
4. Heat rejection

The second process, heat addition, can be as follows:

1. Constant pressure
2. Constant volume

The natural and mechanically easy-to-achieve (or, more aptly, *easy-to-approximate*) heat addition is the constant-pressure one. As a *gedankenexperiment*, though, there is no reason why we cannot make it constant volume. The resulting cycle, while conceptually identical to the Atkinson cycle on a *T-s* diagram, is significantly different in performance. This will be elaborated upon later using fundamental thermodynamic considerations and numerical examples.

The cycle will be referred to as *Reynst-Gülen* (R-G) cycle in recognition of F. H. Reynst, who conceptually defined a gas turbine cycle with *pressure-gain* combustor (not necessarily at constant-volume) and numerically evaluated the performance [17]. Its formulation and derivation of its performance was first done by the author in 2009–2010 [16]. At the time, he was unable to locate a single published study to make the same derivation. This is still the case.

The *T-s* diagram representations of Brayton and R-G cycles are shown in Figure 10.7, based on the parameters and assumptions listed subsequently:

- State 1 is at ISO conditions −14.7 psia and 59°F (1.014 bara, 15°C)
- Brayton cycle pressure ratio, $PR = P_2/P_1 = 18{:}1$
- R-G cycle precompression pressure ratio, $PR' = P_2/P_1 = 6{:}1$
- $T_3 = 2{,}500°F$ (1,370°C)
- $\gamma = 1.40$, $R = 0.068577$ Btu/lb-R, $c_p = 0.24$ Btu/lb-R

Note that cycle pressure ratio and T_3 are approximately representative of a vintage F class gas turbine at ISO base load conditions. The cycles are chosen such that the following values for both are identical:

- Heat addition (q_{in})
- Overall cycle pressure ratio (PR)

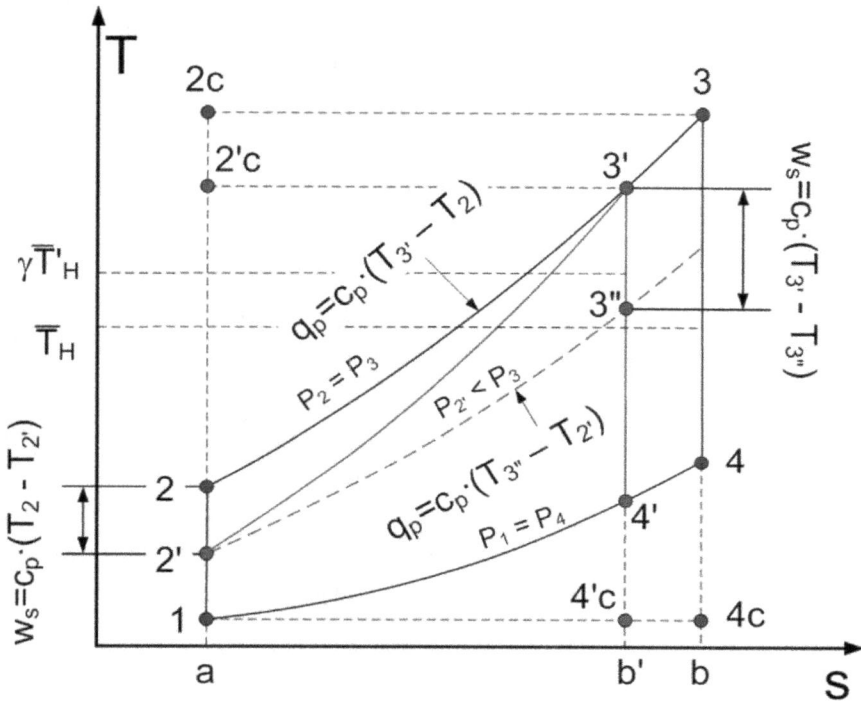

FIGURE 10.7   Brayton {1-2-3-4-1} and R-G {1-2'-3'-4'-1} cycles on T-s diagram.

Since the working fluid for both cycles is air as perfect gas, this is equivalent to the following relationship:

$$T_3 - T_2 = T_{3'} - T_{2'}.$$

(10.24)

Each cycle is characterized by two parameters:

1. Cycle total pressure ratio, PR, which is the same for isentropic compression and expansion of the Brayton cycle but not for the R-G cycle
2. Cycle maximum temperature, T_3 or $T_{3'}$, which is at the end of the heat addition and the start of isentropic expansion

The significant difference in the R-G cycle is the pressure rise during the constant-volume heat addition process.

The constant-volume heat addition process in the R-G cycle is described by the *second Tds equation*, Equation 2.2, introduced in Chapter 2. Both terms on the right-hand side of the equation are retained because $dP \neq 0$. The first term on the right-hand side of the equation describes the temperature rise of the working fluid via heat transfer at constant pressure, and the second term represents the temperature rise via the isentropic work done on the working fluid, which increases the temperature *and* pressure simultaneously. Dividing and multiplying the second term on the right-hand side with ds, the second Tds equation is rewritten subsequently as

$$dh = \left(T + v\frac{dP}{ds} \right) \cdot ds \qquad (10.25)$$

The exact differential dP is given by

$$dP = \left(\frac{\partial P}{\partial s} \right)_v \cdot ds + \left(\frac{\partial P}{\partial v} \right)_s \cdot dv \qquad (10.26)$$

For the constant-volume process $dv = 0$ so that one ends up with

$$\frac{dP}{ds} = \left(\frac{\partial P}{\partial s} \right)_v = \frac{P}{c_v} \qquad (10.27)$$

Combining Equations 10.25–10.27 and using the ideal gas equation of state, the result is

$$dh = T \cdot \left(1 + \frac{R}{c_v} \right) \cdot ds = T \cdot \gamma \cdot ds \qquad (10.28)$$

Integration between the states $2'$ and $3'$ gives

$$h_{3'} - h_{2'} = \gamma \cdot \int_{2'}^{3'} T \cdot ds = \gamma \cdot \overline{T}_H' \cdot \int_{2'}^{3'} ds = \gamma \cdot \overline{T}_H' \cdot (s_{3'} - s_{2'}) = q_{in} \qquad (10.29)$$

$$\gamma \cdot \overline{T}_H' = \frac{h_{3'} - h_{2'}}{s_{3'} - s_{2'}} = \frac{c_p \cdot (T_{3'} - T_{2'})}{c_v \cdot \ln(T_{3'}/T_{2'})} = \gamma \cdot \frac{(T_{3'} - T_{2'})}{\ln(T_{3'}/T_{2'})} \qquad (10.30)$$

$$\overline{T}_H' = \frac{(T_{3'} - T_{2'})}{\ln\left(\dfrac{T_{3'}}{T_{2'}} \right)} \qquad (10.31)$$

In particular, the details are as follows:

- Equation 10.30 gives the mean-effective temperature for the *entire* constant-volume heat addition process between states $2'$ and $3'$.
- Equation 10.31 gives the mean-effective temperature of the hypothetical heat transfer process at constant pressure between states $2'$ and $3'$.

For the two cycles shown in Figure 10.7, the heat addition, q_{in}, was identical for the constant-pressure and constant-volume processes so that Equation 10.30 becomes

$$\frac{\gamma \cdot \overline{T}_H'}{\overline{T}_H} = \frac{s_3 - s_2}{s_{3'} - s_{2'}} \tag{10.32}$$

As can be visually gathered from Figure 10.7, the entropy change in the CVHA process is smaller than the entropy change in the CPHA process. Thus, the right-hand side of Equation 10.32 is greater than unity, which provides the key to the principal advantage of the R-G cycle with CVHA over the Brayton cycle with CPHA with the same heat input and overall cycle pressure ratio:

The R-G cycle has a higher mean-effective heat addition temperature than the Brayton cycle:

$$\gamma \cdot \overline{T}_H' > \overline{T}_H \tag{10.33}$$

Using the cycle parameters listed in the caption of Figure 10.7, the actual numbers are as follows:

- $\gamma \cdot \overline{T}_H' = 1{,}760°F\ (960°C)$
- $\overline{T}_H' = 1{,}126°F\ (608°C)$
- $\overline{T}_H = 1{,}479°F\ (804°C)$

In other words, the heat addition during the R-G cycle takes place at a mean-effective temperature that is ~ 300°F (167°C) higher than the corresponding Brayton cycle with the same heat input and overall cycle pressure ratio. In this example, the R-G cycle also gets an *additional benefit* from the lower mean-effective heat rejection temperature; that is, using the formulations developed earlier in Chapter 3 for both cycles to evaluate the mean-effective temperature values, one gets the following:

- $\overline{T}_L' = 335°F\ (168°C)$
- $\overline{T}_L = 389°F\ (198°C)$

Using the relevant mean-effective cycle temperatures for R-G and Brayton cycles, it can be shown the former is better by about eight percentage points (i.e., 64.2% and 56.2%, respectively). For the R-G cycle, going through similar algebra as before, it can be shown that

$$\eta_{R-G} = 1 - \frac{\Gamma}{\theta_{2'}} \tag{10.34}$$

$$\eta_{R-G} = 1 - \frac{1}{\tilde{q}} \cdot \left[\left(\frac{\tilde{q}}{\pi'^k} + 1 \right)^{1/\gamma} - 1 \right]. \tag{10.35}$$

Comparing Equations 10.19 and 10.23, one can write

$$\eta_{R\text{-}G} - \eta_A = (1-\eta_B) \times (\gamma - 1) \times \Gamma \tag{10.36}$$

where η_B is the efficiency of a Brayton cycle with overall cycle pressure ratio the same as the precompression pressure ratio of R-G and Atkinson cycles. The right-hand side of Equation 10.36 is always positive so that $\eta_{R\text{-}G}$ is always higher than η_A. Note that, unlike the Atkinson cycle, the R-G cycle always has higher efficiency than the Brayton cycle, even at the same overall cycle *PR*.

Rewriting Equation 10.19 for the R-G cycle, and noting that $\tau = T_3/T_{2'}$, one gets

$$\eta_{R\text{-}G} - \eta'_B = (1-\eta'_B) \cdot (1 - \tau^k \cdot \Gamma). \tag{10.37}$$

The right-hand side of Equation 10.37 is always positive so that $\eta_{R\text{-}G}$ is always greater than η'_B.

In an R-G cycle, heat added to the cycle during the constant-volume flow process between states 2′ and 3′ manifests itself in two distinct forms:

1. *Heat* transfer at constant-pressure, *T·ds*, state 2′ to state 3″ *or* state 2 to state 3′ in Figure 10.7
2. Isentropic *Work* transfer, *v·dP*, state 3″ to 3′ *or* state 2′ to state 2 in Figure 10.7

In order to conceptualize the R-G cycle gas turbine, consider the drawing in Figure 10.8, in which the constant-volume heat addition process is represented by a motor-compressor assembly as described by Reynst [17]. The motor, which drives the compressor, is a heat engine and gives its heat into the compressed air instead of rejecting it to the surroundings [4]. The constant-volume, pressure-gain combustor of the hypothetical turbine converts about 18% of the fuel chemical energy to mechanical work. (In the alternate representation per Figure 10.7, i.e., the working fluid heated from state 2′ to state 3″ and compressed to state 3′, about 41% of the fuel chemical energy would be converted to mechanical work.)

In his work, Reynst envisioned a hybrid or turbo-compound engine where the combustor of the gas turbine is replaced by a two-stroke reciprocating engine, based on the three-process Lenoir cycle (missing the isentropic compression process), whose shaft drove a compressor. Reynst calculated the efficiency of that ideal constant-volume combustion engine as a function of its precompression pressure ratio, *PR′*, and T_1 as follows (Equation 12 on page 151 of Ref. [17]):

$$\eta = 1 - \frac{1}{\tilde{q}} \cdot \left[\left(\frac{\gamma \tilde{q}}{\pi'^k} + 1 \right)^{\frac{1}{\gamma}} - 1 \right],$$

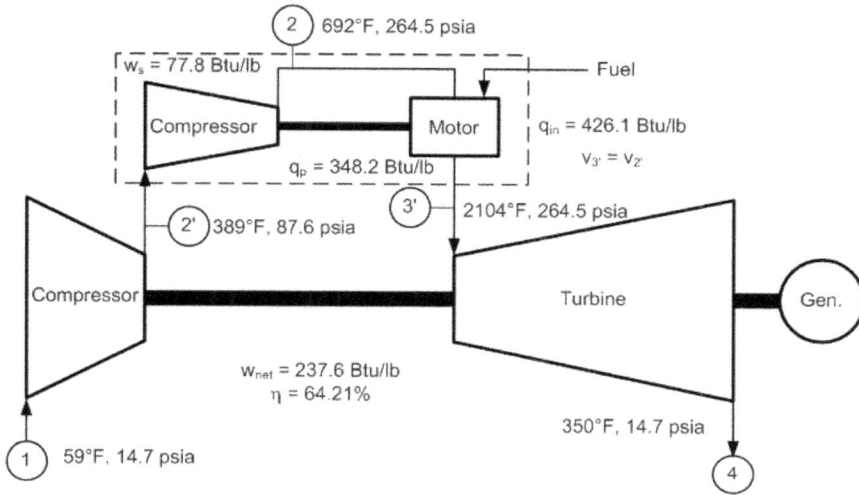

FIGURE 10.8 Hypothetical Reynst-Gülen cycle GT with constant-volume combustor replaced by a motor-compressor assembly per Ref. [17]. The state-points correspond to those shown in Figure 10.7.

which, of course, is Equation 10.21 derived earlier for the Atkinson cycle. In fact, Atkinson's and Reynst's ideal constant-volume heat addition engine are identical for the same non-dimensional heat input parameter, $\tilde{q}$ .However, whereas $\tilde{q}$ is about 10 for the compound engine of Reynst with constant-volume heat addition, Atkinson cycle $\tilde{q}$ for comparable pressure ratios and/or heat inputs as the base Brayton and R-G cycles is 2.4 to 4.2.

This is the reason why the cycle rigorously developed herein starting from the first principles is called as the *Reynst-Gülen* cycle. Reynst never made the step from his hypothetical device to a fundamental cycle as did the author in his paper [16]. Interestingly, by replacing the term $\gamma\tilde{q}$ in the parentheses on the right-hand side of Equation 10.21 with $\tilde{q}$, we end up with Equation 10.35, which is the efficiency of the R-G cycle.

What is the fuss about a theoretical cycle most likely destined to stay on paper one might be compelled to ask? This is not an unreasonable question at all. In order to answer it, the reader is directed to Chapter 9; that is, the concept of *technology factor* that is instrumental in establishing what is reasonably possible with existing state-of-the-art technology from a few key design parameters (in the case of gas turbines, cycle *PR* and *TIT*) using a fundamental model (in the case of gas turbines, the Brayton cycle).

It has been unequivocally established that for state-of-the-art F, G, H/HA, or J class heavy-duty industrial gas turbines, with reference to the *cold* air-standard Brayton cycle, the said technology factor is about 0.7 to 0.75.

Using fundamental thermodynamic arguments, it has been demonstrated earlier that the correct yardstick to be used for the same approach to be applied to a gas turbine with a pressure-gain combustor (which approximates the ideal constant-volume heat addition process) is the R-G cycle. It is not the Atkinson cycle (or the Humphrey cycle as it is referred to by many investigators due to some unknown reason).

In a recent paper by the author, this fact has been numerically illustrated using rigorous calculations [18]. In particular, for a typical E class gas turbine (PR = 13:1, TIT = 1,300°C), the technology factor is 0.68. In combined cycle configuration, the technology factor is 0.75 for a rating efficiency of 53.2% (from the *GTW Handbook*).

Let us say that we would like to modify this E class technology with pressure-gain combustion. What is the combined cycle efficiency that we can reasonably expect with existing technology (except the combustor)? Here is a two-step upgrade approach:

1. Keep the same E class compressor
2. Replace the combustor

The ramification of the first step is clear; precompression PR is set to 13:1. In the second step, we have a choice to make:

- Same heat input (i.e., same $\tilde{q}$ but higher TIT)
- Same TIT (i.e., smaller $\tilde{q}$ because $c_v < c_p$ and T_2 and T_3 are the same)

If one chooses the Atkinson and Humphrey and the second option, there is a small snag. Since the mean-effective heat addition temperature is the same as the base E class Brayton cycle (because T_2 and T_3 are the same), ideal combined cycle efficiency is exactly the same as the base E class GTCC; that is, 71.5%, for the 53.2% rating efficiency with a technology factor of 0.75 (because the mean-effective heat rejection temperature for the ideal combined cycle is T_0). Overall cycle PR is 34.1 with constant-volume heat addition pressure ratio of 2.6:1.

With the first option, Atkinson/Humphrey cycle TIT is 1,690°C with an overall cycle PR of 42.5:1 (constant-volume heat addition pressure ratio is 3.3:1). The ideal combined cycle efficiency is 74.9%. For gas turbines with similar TIT, combined cycle technology factor is about 0.8, which predicts a rating value of about 60%. This, of course, is not too appealing because it is actually *below* the existing state of the art with conventional combustors.

Let us pause here and consider the ramifications of using the *wrong* ideal cycle. There is no point in making an investment into pressure-gain combustion technology. The proverbial carrot is not even small; it is nonexistent.

For the same two-step upgrade approach, R-G cycle predicts the following:

- Overall cycle PR of 34.1: and ideal CC efficiency of 79.6% (same TIT)
- Overall cycle PR of 31.5:1 and ideal CC efficiency of 80.3% (same $\tilde{q}$, TIT = 1,482°C)

Projected CC rating efficiency with a technology factor of 0.8 is about 64%. The accuracy of this prediction will indeed be borne out by rigorous heat and balance simulations later.

10.3.2 A New Heat Engine

Present state of the art in land-based, heavy-duty industrial gas turbine (HDGT) technology is the J class, with 1,600°C *TIT* and *PR* of 23. Largest J class HDGTs are 50 Hz machines rated at nearly 500 MWe generator output and more than 42% thermal efficiency. Smaller 60 Hz machines are rated at about 350 MWe with similar efficiency. On a combined cycle basis, J class HDGTs have more than 61% thermal efficiency rating. As the basis of the real cycle analysis, a 60 Hz J class HDGT with 700 kg/s (nominal) airflow, *PR* of 23 and *TIT* of ~1,600°C is assumed. Thermoflow's Inc.'s THERMOFLEX software is used to model the gas turbine with appropriate chargeable and nonchargeable cooling flows (see Figure 10.9).

The bottoming cycle, comprising the heat recovery steam generator (HRSG), the steam turbine generator (STG) and the heat rejection system (once-through, open-loop water-cooled condenser with 40 mbar steam pressure), is modelled in Thermoflow's Inc.'s GT PRO software. A 2,400 psig (165 barg), 1,112°F (600°C) steam cycle is chosen with state-of-the-art STG technology and suitable exhaust annulus area. The combined cycle net performance is 505.3 MWe and 61.7% thermal efficiency with 174 MWe STG and 8.2 MWe plant auxiliary load (1.6% of the gross output). Gas turbine performance fuel heating to 410°F (210°C) with IP economizer feed water is used.

The base HDGT in Figure 10.9 is modified by replacing the combustor with a *pressure-gain combustor* (PGC) and adding a *booster compressor* (BC), as shown in Figure 10.10. The BC ensures that chargeable and nonchargeable cooling flows for the turbine stage 1 are delivered from the compressor discharge, which is now at a lower pressure. This is so because part of the compression duty is taken over by the *quasi* constant-volume combustion in the PGC. (Note that the specific volume ratio across a PGC is less than unity, e.g., about 0.6 for a pulse detonation combustor.) Thus, hot gas pressure at turbine stage 1 is in fact higher than that at the compressor exit.

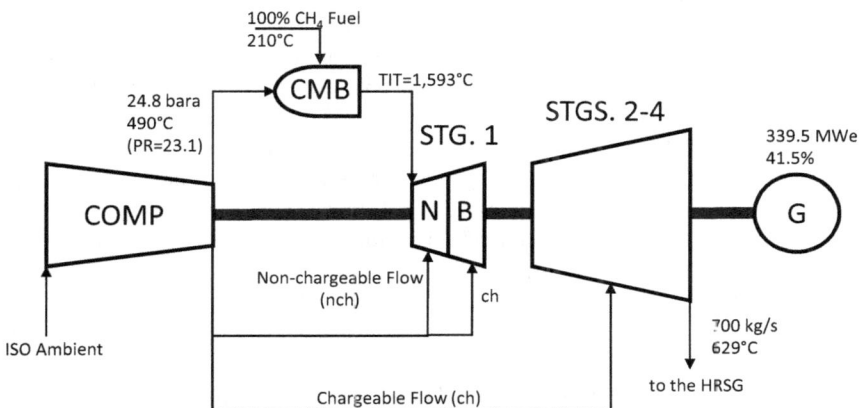

FIGURE 10.9 J class gas turbine model and performance.

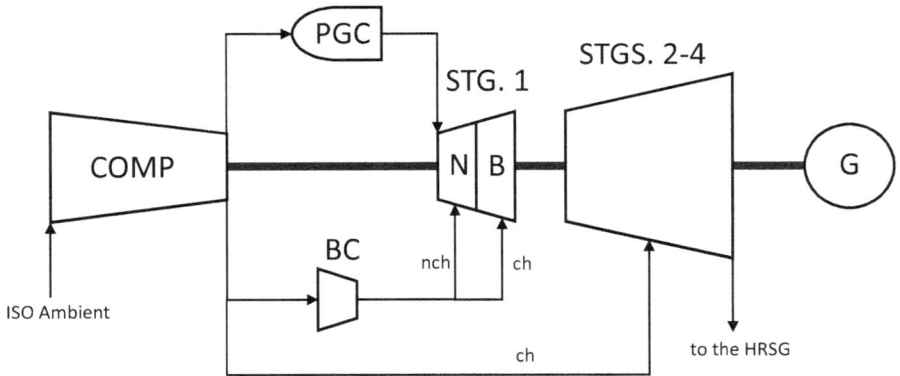

FIGURE 10.10   Gas turbine with pressure gain combustor.

Pulsed detonation combustion/combustor (PDC) is assumed to be the particular PGC process. The PDC is modelled in THERMOFLEX as a dummy compressor-combustor combination controlled by a script (to implement temperature and pressure rise via transfer functions published in Ref. [18] along with the fuel consumption via a heat balance). A pressure-temperature changer accounts for pressure and purge-air-dilution loss (5% and ~100°C, respectively) between the PDC and the turbine inlet (see Figure 10.11). The unfavorable impact of the pulsating gas flow, somewhat smoothed out in the transition piece between the PDC and stage 1 nozzle, is quantified by a two percentage point debit to the stage 1 isentropic efficiency.

The PDC cycle performance is calculated for different *TIT* and precompression *PR* combinations. Fuel compression is not included (i.e., pipeline pressure is assumed to be high enough). However, 0.5% transformer loss is accounted for. Booster compressor load is debited to the plant auxiliary load. Selected results are listed in Table 10.5.

As shown in Table 10.5, with *TIT* and cycle *PR* comparable to those representative of today's F class technology (~1,500°C), detonation combustion gas turbine is capable of almost 64% net CC efficiency. The 65% goal post is feasible with state-of-the-art J class *TIT*. Furthermore, as represented by the ratio of steam and gas turbine generator outputs, the performance increase is largely due to topping cycle (i.e., the gas turbine) improvement. Considering that each kilowatt from the bottoming cycle is nearly six to seven times as expensive as a kilowatt from the topping cycle, pressure gain combustion is a truly cost-effective method of achieving significant combined cycle efficiency improvement.

Pulsed detonation combustion has been investigated for more than a half century with an eye on aircraft propulsion applications. So far, significant engineering design challenges, such as noise, vibration, fatigue, and durability, prevented its successful commercialization [19–20]. The absence of weight and volume limitations (up to a point) in a land-based power generation application renders a satisfactory resolution of some of the most pressing problems possible (e.g., sturdy

Fuel

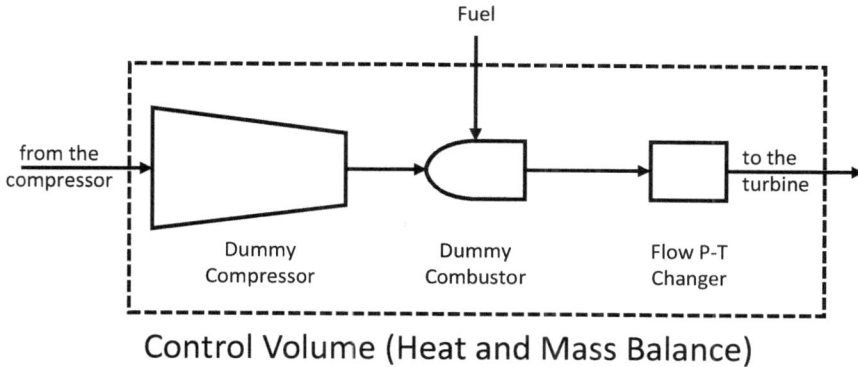

Control Volume (Heat and Mass Balance)

FIGURE 10.11   Implementation of the PDC in the THERMOFLEX model.

TABLE 10.5

Gas turbine with PDC – rigorous heat and mass balance model results
(TIP: Turbine Inlet Pressure).

	BASE	PRESSURE GAIN COMBUSTION			
GTG Output, kWe	339,531	438,104	418,721	398,958	405,541
Heat Input, kWth	819,518	892,497	857,284	821,360	848,679
GTG Efficiency	41.43%	49.09%	48.84%	48.57%	47.78%
T_{exh}, °C	629	612	593	573	595
m_{exh}, kg/s	700	702	701	700	701
Precompression PR	NA	13.6	13.6	13.6	11.9
P_2, bara	24.8	13.8	13.8	13.8	12.1
T_2, °C	490	371	371	371	346
PDC PR	NA	1.854	1.810	1.765	1.821
Cycle PR	23.1	25.2	24.6	24.0	21.7
TIP, bara	23.4	25.6	24.9	24.3	22.0
TIT, °C	1,593	1,593	1,538	1,482	1,482
STG Output, kWe	173,722	166,265	156,082	145,705	157,194
CC Gross Output, kWe	513,253	604,369	574,803	544,663	562,735
Plant Aux. Load, kWe	7,912	24,672	23,001	21,330	19,662
CC Net Output, kWe	505,341	579,697	551,802	523,332	543,072
CC Net Efficiency	61.7%	65.0%	64.4%	63.7%	64.0%

design of combustor components, space for deflagration-to-detonation (DDT)
transition). Whether the theoretical performance potential calculated herein via
Chapman-Jouget theory will survive a realistic design with myriad imperfections
(e.g., imperfect mixing of fuel and air, unsteady initial conditions, pressure, temperature, and composition variations downstream of the detonation wave, etc.)
remains an open question.

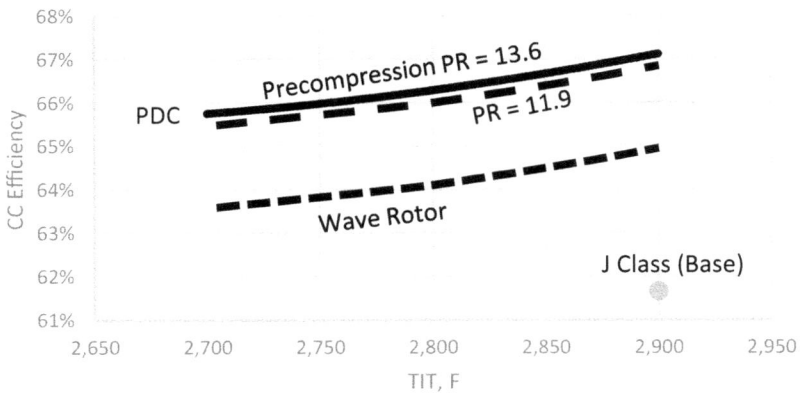

FIGURE 10.12   Combined cycle efficiency with PGC gas turbines.

Nevertheless, even with a more modest approach using a dynamic pressure exchange (wave rotor) PGC with the perfect CVC model for *TR-PR* prediction, significant CC performance potential is present (see Figure 10.12). In this case, 64+% CC efficiency is achievable with a *TIT* of 2,900°F (~1,600°C), same as today's state-of-the-art J class gas turbine.

REFERENCES

1. Frutschi, H., *Closed-Cycle Gas Turbines: Operating Experience and Future Potential*, New York: ASME Press, 2005.
2. Gas Turbine World, *2020 GTW Handbook, Volume 35*, Fairfield: Pequot Publishing, Inc., 2020.
3. Pritchard, J.H., H System™ Technology Update, GT2003–38711, *ASME Turbo Expo 2003*, June 16–19, 2003, Atlanta, GA.
4. Umemura, S., Akita, E., Tsukuda, Y., Akagi, K., and Iwasaki, Y., Development of 1500°C Class 501G Gas Turbine, *MHI Technical Review*, 35 (1), pp. 1–5, 1998.
5. Wilson, D.G., and Korakianitis, T., *The Design of High Efficiency Turbomachinery and Gas Turbines*, 2nd Edition, Upper Saddle River: Prentice-Hall, 1998.
6. Gülen, S.C., Étude on Gas Turbine Combined Cycle Power Plant—Next 20 Years, *Journal of Engineering Gas Turbines Power*, 138, p. #051701, 2015.
7. Bejan, A., Models of Power Plants that Generate Minimum Entropy While Generating Maximum Power, *America Journal of Physics*, 64 (8), pp. 1054–1059, 1996.
8. Facchini, B., Fiaschi, D., and Manfrida, G., Exergy Analysis of Combined Cycles Using Latest Generation Gas Turbines, *Journal of Engineering GTs & Power*, 122, pp. 233–238, 2000.
9. El-Masri, M., Exergy Analysis of Combined Cycles: Part 1 — Air-Cooled Brayton-Cycle Gas Turbines, *Journal of Engineering GTs & Power*, 109, pp. 228–236, 1987.
10. Caton, J.A., Thermodynamic Considerations for Advanced, High Efficiency IC Engines, *Journal of Engineering GTs & Power*, 136, pp. 101512–1, 2014.

11. Klimstra, J., and Hattar, C., Performance of Natural Gas-Fueled Engines Heading Towards Their Optimum, ICES 2006–1379, *ASME Internal Combustion Engine Division 2006 Spring Technical Conference*, May 8–10, 2006, Aachen.
12. Sutkowski, M., Wärtsila 18V50SG – The World's Biggest Four-Stroke Spark-Ignited Gas Engine, PTNSS–2011-SC-046, *PTNSS Congress—2011: International Congress on Combustion Engines Radom*, 2011, Poland.
13. Van Gerpen, J.H., and Shapiro, H.N., Second-Law Analysis of Diesel Engine Combustion, *Journal of Engineering Gas Turbines Power*, 112, pp. 129–137, 1990.
14. Mattson, J., Reznicek, E., and Depcik, C., Second-Law Heat Release Modeling of a Compression Ignition Engine Fueled with Blends of Palm Biodiesel, *Journal of Engineering Gas Turbines Power*, 138, pp. 091502–1, 2016.
15. Amann, C.A., Applying Thermodynamics in Search of Superior Engine Efficiency, *Journal of Engineering Gas Turbines Power*, 127, pp. 670–675, 2005.
16. Gülen, S.C., Gas Turbine with Constant Volume Heat Addition, *ASME Paper ESDA2010–24817, ASME 2010 10th Biennial Conference on Engineering Systems Design and Analysis*, July 12–14, 2010, Istanbul.
17. Reynst, F.H., *Pulsating Combustion—The Collected Works of F. H. Reynst* (Edited by M.W. Thring), Oxford: Pergamon Press, 1961.
18. Gülen, S.C., Pressure-Gain Combustion Advantage in Land-Based Electric Power Generation, *GPPF2017–0006, 1st Global Power and Propulsion Forum, GPPF 2017*, January 16–18, 2017, Zurich.
19. Lu, F.K., Braun, E.M., Massa, L., and Wilson, D.R., Rotating Detonation Wave Propulsion: Experimental Challenges, Modeling and Engine Concepts, *AIAA 2011–8043, 47th Joint Propulsion Conference & Exhibit*, 2011, San Diego, CA.
20. Roy, G.D., Frolov, S.M., Borisov, A.A., and Netzer, D.W., Pulse Detonation Propulsion: Challenges, Current Status, and Future Perspective, *Progress in Energy and Combustion Science*, 30, pp. 545–672, 2004.

11 External Combustion Engines

I sell here, Sir, what all the world desires to have—power.
    —Matthew Boulton (about James Watt's steam engine)

The author had to agonize over the problem of selecting the title of this chapter quite a bit. The problem is similar to that discussed in the introduction to Chapter 10 (i.e., the dichotomy between *heat engines* and *energy conversion devices*). In Chapter 10, when making the distinction between the internal combustion engine (an energy conversion device) and the ideal air-standard cycle describing its operation, the problem was swept aside rather easily. In this chapter, the subject matter (implicitly) includes heat engines without combustion. Another apt title for the chapter would be "Closed-Cycle Devices". (Because both the gas turbine and RICE discussed in Chapter 10 are open-cycle devices in actual operation – never mind the fact that the term *open cycle* is an *oxymoron*, whereas the term *closed cycle* is a *tautology*.) Herein, we turn to an old paper (from the 1960s) by Lauck et al. [1], who divided the power generation devices into two classifications: *Q-engines* and *E-engines*. Q-engines are engines (devices) whose method of operation requires that the internal energy of the fuel first be converted to heat and then transferred to a working fluid which in turn goes through a cycle. Thus, the working fluid for Q-engines both in theory and practice executes a cycle. In addition to the steam turbines (see Section 11.1 later on), according to the authors, Q-engines include thermoelectric and thermionic generators (the electrons are the fluid here); hot-air engines, such as the Stirling engine; closed-cycle gas turbines; and solar batteries and engines.

E-engines, as described by Lauck et al. (ibid.), are devices whose method of operation does not require that the internal energy of the fuel be first converted to heat. Consequently, the authors conclude that, since the internal energy of the fuel is not converted to heat and then transferred to the working fluid, *E-engines are concerned both in practice and in theory with a process and not a cycle*. Examples of this type are internal combustion engines, including spark-ignition reciprocating engines, compression-ignition reciprocating engines, open-cycle gas turbines, fuel cells, batteries, and living organisms.

Fair enough. There is no argument here to the earlier statement (in italics) as far as it goes, but as discussed in the introduction to Chapter 10, the theory part is not that clear-cut. Granted, the underlying theory describing the actual processes does not define a cycle per se, but there is an *ideal proxy* that precisely describes the operation of the device in question; that is, the air-standard Brayton and Otto cycle for gas turbines and RICE, respectively. In any event, the whole point of this introductory discussion is that this chapter deals with two types of Q-engines (i.e., the steam turbine

DOI: 10.1201/9781003247418-13

and the closed-cycle gas turbine). Since the term *Q-engines* would be rather cryptical to most readers, the more recognizable term *external combustion engines* is chosen. The reader should be cognizant of the fact that either device can receive heat from a process *that does not include combustion.*

11.1 STEAM TURBINE – RANKINE CYCLE

The ideal thermodynamic cycle describing the operation of the steam turbine is the Rankine cycle. As is the case with the ideal Brayton cycle, the working fluid, H_2O, completes a closed loop during the process. In contrast to the gas turbine Brayton cycle, however, this is indeed the case for the real system as well. The working fluid (i) stays the same (but it undergoes phase change from liquid to vapor and then back to liquid) and (ii) starts and ends at the same thermodynamic state. The big difference between gas and steam turbine cycles is that cycle heat addition process is an integral part of the heat engine in the former. (This is why the gas turbine is classified as an internal combustion engine.) In the Rankine cycle, however, cycle heat addition process forms the basis of a major piece of equipment (i.e., the boiler), which has to be studied on its own and this can be done outside the cycle itself. Hence, the conventional steam plant operating in a Rankine cycle is an external combustion engine, wherein the steam turbine is only one of the three major pieces of equipment (the other two are the boiler and the condenser/cooling tower, i.e., the heat sink).

Consequently, whereas ideal Brayton cycle analysis provides direct information on the heat engine itself (i.e., the gas turbine), ideal Rankine cycle analysis does not provide direct information on the steam turbine. Like all practical heat engine cycles, the Rankine cycle (see the *T-s* diagram in Figure 11.1) is an attempt to approximate the ideal Carnot cycle with (i) isothermal heat addition and rejection and (ii)

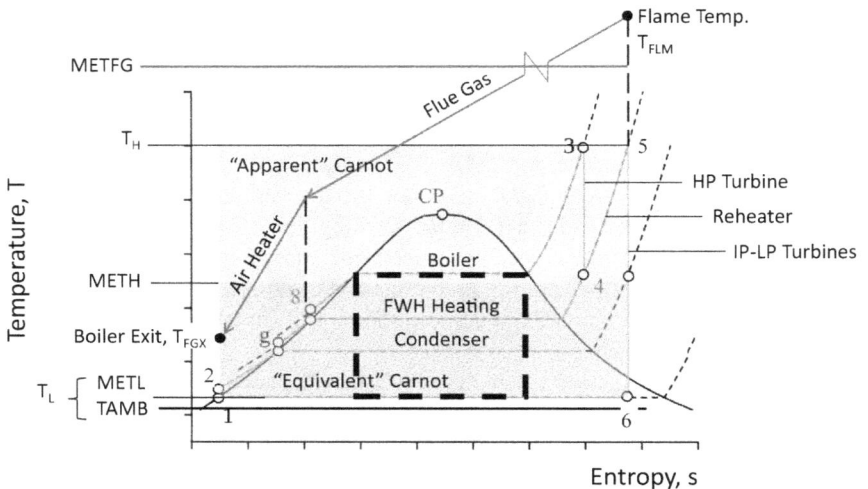

FIGURE 11.1 Temperature-entropy diagram of Rankine steam cycle with reheat and feed water heating.

isentropic compression and expansion. In fact, the cycle that is being aspired to is the apparent Carnot cycle operating between T_H and T_L, which is the Carnot target of the Rankine cycle shown in the figure and whose efficiency is given by

$$\eta_C = 1 - \frac{T_L}{T_H} \tag{11.1}$$

Note that the cycle diagram in Figure 11.1 is a qualitative sketch; that is, it is not based on precise thermodynamic property data. The intention is to illustrate the relative positions of key cycle temperatures and their (logarithmic) averages. What is depicted is a very basic cycle with only one feed water heater (FWH) and subcritical main steam pressure. In contrast, modern steam cycles with supercritical main steam pressure have eight or nine FWHs and showing them all on the T-s diagram would make it unintelligible without adding anything to an understanding of the fundamental concepts. Also shown in the diagram is a dashed rectangle inside the H_2O phase dome. This is a commonly used Carnot cycle proxy in explaining the Rankine cycle to laypersons. It is hoped that the following discussion will make clear that this is an erroneous construct. (For example, interested reader can verify readily that, for a subcritical cycle, METH is *less than* the saturation temperature at main steam pressure.)

Even if one can design a perfect cycle with zero losses and isentropic pumps and turbines, the resulting efficiency is much less than the Carnot target given by Equation 11.1. (This can be easily proven by using a commercially available heat balance tool to perform the relatively straightforward calculation.) In fact, the efficiency of such a *perfect* cycle is equal to the efficiency of the equivalent Carnot cycle defined as follows:

$$\eta_{CE} = 1 - \frac{\overline{T}_L}{\overline{T}_H} \tag{11.2}$$

where $\overline{T}_H$ and $\overline{T}_L$ are cycle mean-effective heat addition and heat rejection temperatures, METH and METL in Figure 11.1, respectively (see Section 3.2.1). By virtue of the constant-pressure–temperature heat rejection process via condensation, Rankine cycle $\overline{T}_L$ is indeed equal to the real temperature T_L. However, $\overline{T}_H$ is a hypothetical temperature, which, for a hypothetical isothermal heat addition process between states 8 and 5 in Figure 11.1, results in the same amount of heat addition, which is the sum total of non-isothermal main and reheat heat addition processes. Thus, starting from the general expression for an arbitrary number of constant-pressure heat addition processes ($i = 1, 2, \ldots, N$)

$$\overline{T}_H = \frac{\displaystyle\sum_{i=1}^{N} \Delta h_i}{\Delta s_{overall}} \tag{11.3}$$

where h is enthalpy and s is entropy (from a suitable equation of state using pressure, P, and temperature, T, e.g., *The ASME Steam Tables*), for the specific reheat boiler in Figure 11.1, one arrives at

$$\overline{T}_H = \frac{(h_3 - h_8) + (h_5 - h_4)}{(s_5 - s_8)}. \tag{11.4}$$

For typical subcritical, *supercritical* (SC) and *ultra-supercritical* (USC) cycles, the relationship between METH and cycle's main and reheat steam temperatures is shown in Figure 11.2. In high-efficiency designs, reheat steam temperature is higher than main steam temperature. The reason is to enable low condenser pressures without excessive moisture in the LP turbine last stages (i.e., higher efficiency with reduced erosion). (This can be visually deduced easily from the simple *T-s* diagram in Figure 11.1.) It also increases the load range with minimal moisture erosion of the LP last stage blades. Thus, one can use the average of main and reheat steam temperatures if they are not the same.)

Let us consider an advanced steam turbine cycle with 595°C main and reheat steam temperatures and condenser pressure of 50 mbar (saturation temperature is 33°C). The apparent Carnot cycle efficiency is

$$\eta_C = 1 - (33 + 273) / (595 + 273) = 0.647 \text{ or } 64.7\%.$$

The equivalent Carnot cycle efficiency is

$$\eta_{CE} = 1 - (33 + 273) / 672 = 0.545 \text{ or } 54.5\%$$

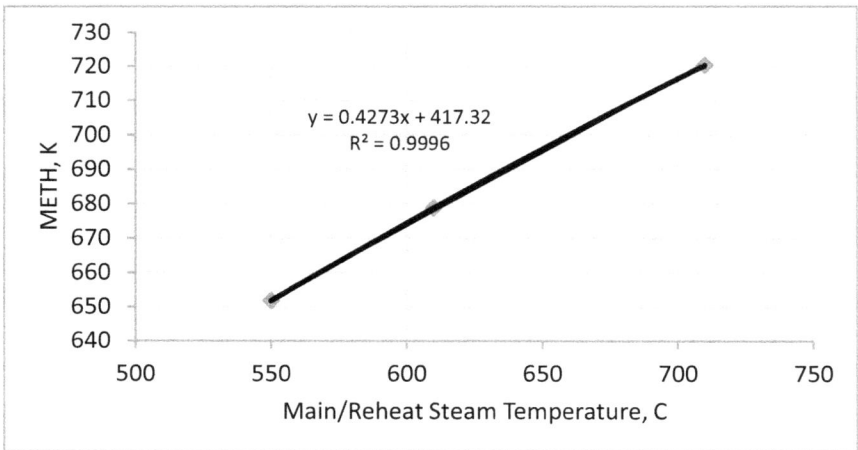

FIGURE 11.2 Rankine steam cycle mean-effective heat addition temperature (for 250 bara steam). For each 50 bara increase in steam pressure, add 6 K to METH.

(Note that both efficiencies are ideal efficiencies.) The ratio of the equivalent and apparent Carnot efficiencies, which is 0.84 from the numbers cited earlier, defines the cycle factor (*CF*), which is a measure of the deviation of cycle heat transfer processes, one or both, from the isothermal ideal. For the purpose of generality, it is preferable to use a standard temperature for $\overline{T}_L$ and the logical choice is the ISO ambient temperature of 15°C. This convention eliminates one design specification from the discussion/analysis; that is, condenser pressure (steam turbine backpressure), which is highly dependent on site conditions, project economics, and prevailing regulations. In that case, $\eta_C = 66.8\%$ and $\eta_{CE} = 57.1\%$ so that $CF = 0.855$. In conclusion, assuming a *CF* of 0.85 is sufficiently accurate for most practical purposes.

The foregoing analysis was exclusively based on the steam cycle and steam temperatures. Also shown in Figure 11.1 are three boiler temperatures:

- Burner flame temperature, T_{FLM}
- Boiler flue gas exit temperature, T_{FGX}
- Mean-effective average of the two, *METFG*

Typical flame temperature is around 2,000°C (3,600°F). Flue gas temperature at the boiler exit is around 130°C (266°F). Logarithmic mean of these two temperatures is 808°C. Rewriting Equations 11.1 and 11.2 with T_{FLM}, *METFG*, and T_{AMB}, Carnot efficiencies and *CF* are recalculated as

$$\eta_C = 1 - \frac{T_{AMB}}{T_{FLM}} = 1 - \frac{288}{2000 + 273} = 87.3\%,$$

$$\eta_{CE} = 1 - \frac{T_{AMB}}{METFG} = 1 - \frac{288}{808 + 273} = 73.4\%,$$

$$CF = 73.4 / 87.3 = 0.84.$$

One would argue that this is the true measure of the steam cycle's thermodynamic potential, and the true T_H is T_{FLM}. This argument is fundamentally correct. However, for the high-level thermodynamic analysis described herein, burner flame and gas exit temperatures are not readily available numbers. Obtaining these numbers requires in-depth boiler thermodynamic and heat transfer analysis. This involves not only the heat transfer between hot combustion gas from the furnace and the heater coils (i.e., evaporator, superheaters, and reheaters) but also heat recovery via combustion air heating (to increase the boiler efficiency), flue gas recirculation, and so on. (See chapters 4 and 22 in Ref. [2] for detailed boiler combustion calculations.) For the second law analysis, it is easier to leave this tedious exercise out and take care of it later using a boiler efficiency (typically, 0.90–0.95) in the first law roll-up of efficiencies.

Thermal efficiency of the actual steam cycle with myriad losses (e.g., heat and pressure losses in pipes and valves, turbine expansion efficiency, etc.) as a fraction of its Carnot equivalent sets the *TF* for the Rankine steam cycle. Today's state of the art in SC and USC design can achieve a *TF* of 0.80 to 0.85 at METH range of 670 to

675 K (steam cycle only, i.e., steam turbine generator (STG) output minus boiler con-
densate/feed pump motor input, excluding boiler efficiency and other plant auxiliary
loads). For quick estimates, one can assume that each 30 K in METH is worth 0.01
(equivalent to one percentage point) in *TF*. Modern utility boiler LHV efficiencies are
in low to mid-'90s (percent). Plant auxiliary load is largely a function of heat rejection system and the air quality control system (e.g., flue gas desulfurizer, etc.). Both
systems are highly dependent on existing environmental regulations and permits.
Optimistic values are 5 to 6% of STG output but can be higher. Electric motor-drive
boiler condensate and feed pumps consume about 3–4% of STG output. In some
designs, the large boiler feed pump (BFP) is driven by a mechanical drive steam
turbine (in a steam power plant rated at 600 MWe, the BFP consumes more than 20
MWe). This reduces electric motor power, but steam extracted to run the BFP steam
turbine reduces STG output; that is, the bottom-line plant thermal performance is
not affected much. Going with the earlier example, assuming that *TF* = 0.82, boiler
efficiency is 92%, and remaining house load and transformer losses constitute 2% of
gross output, conventional steam plant efficiencies (net LHV) become

$$\eta_{cycle} = 0.82 \times 0.571 = 0.468 \text{ or } 46.8\%,$$

$$\eta_{gross} = 0.92 \times 0.468 = 0.431 \text{ or } 43.1\%, \text{ and}$$

$$\eta_{net} = 0.431 \times (1 - 0.02) = 0.422 \text{ or } 42.2\%.$$

For the entire plant, TF_{net} is 0.422/0.571 = 0.74, and the product of *TF* and *CF* is
0.74 × 0.855 = 0.63, which can be considered an overall Carnot factor of conventional
steam cycle technology on a net plant efficiency basis. On a cycle-only basis, state-of-
the-art Carnot factor range is (0.80 to 0.85) × 0.855 = 0.68 to 0.73.

Let us consider an advanced USC cycle with 300 bar main steam pressure and 700°C
main and reheat steam temperatures (i.e., η_C = 70.4%). From Figure 11.2, METH is about
725 K (with pressure effect) so that η_{CE} = 60.3% and *CF* = 0.603/0.704 = 0.857. Let us use
TF = 0.87 (0.85 plus 0.02 for higher METH), and plant efficiencies are found as

$$\eta_{cycle} = 0.87 \times 0.603 = 0.525 \text{ or } 52.5\%,$$

$$\eta_{gross} = 0.92 \times 0.525 = 0.483 \text{ or } 48.3\%, \text{ and}$$

$$\eta_{net} = 0.483 \times (1 - 0.02) = 0.473 \text{ or } 47.3\%.$$

Thus, for the entire plant, TF_{net} = 0.473/0.603 = 0.785. For *TF* = 0.90 (note that
TF = 1.0 means perfect (isentropic) steam turbine with no leakages and exhaust
losses), η_{net} = 49.0% (TF_{net} = 0.49/0.603 = 0.812). If the boiler efficiency is 95%,
η_{net} = 50.6% (TF_{net} = 0.506/0.603 = 0.84).

11.1.1 Exergy Analysis

Let us start with a state-of-the-art gas turbine combined cycle (GTCC) with General Electric's 7HA.02 gas turbine (60 Hz). In order to have a realistic assessment, site ambient conditions are set to a hot climate such as one would expect to find, say, in Texas, USA. Cycle heat rejection system is based on a dry air-cooled condenser (ACC) operating at 244 mbar (7.2 in. Hg). The system is modeled in Thermoflow Inc.'s GT PRO software. The cycle schematic is shown in Figure 11.3.

Steam turbine heat and mass balance diagram is shown in Figure 11.4. Stream property data for calculating stream exergies is summarized in Table 11.1. Note that the exergy reference is 1 atm and 25°C (corresponding to $h_0 = 45.08$ kJ/kg and $s_0 = 0.0877$ kJ/kg-K). Gas turbine (GT) exhaust exergy calculated by the program is 230,713 kW (1 atm 25°C, H_2O as vapor).

The following findings are based on the data in Table 11.1:

- Steam turbine expansion power output is 153,784 kW
- Corresponding irreversibility is 12,018 kW (7.81%)
- Steam turbine generator (STG) output after mechanical and electrical losses is 151,367 MW
- Corresponding irreversibility is 14,436 (9.54%)
- Referenced to GT exhaust exergy, STG irreversibility is 6.26%
- STG irreversibility breakdown is as follows:

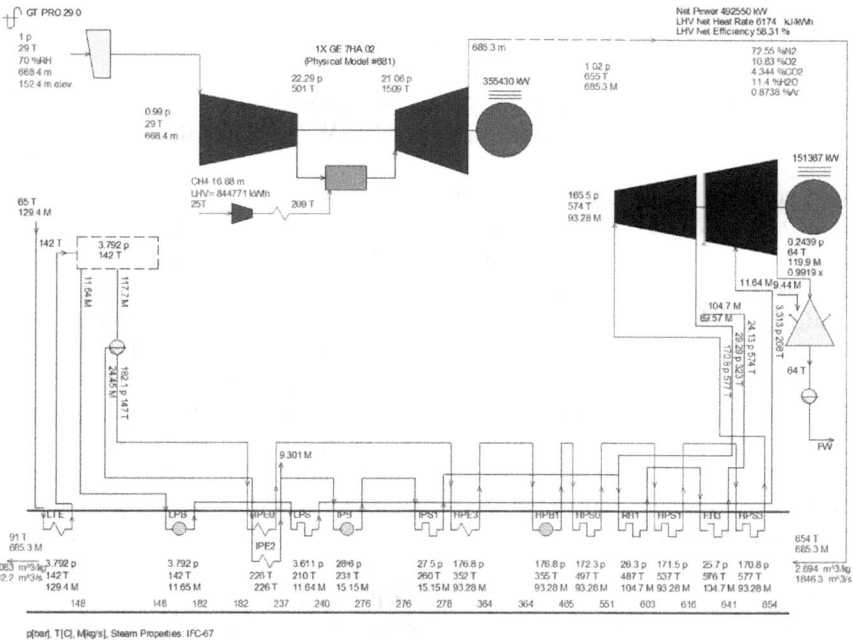

FIGURE 11.3 GTCC example with GE's 7HA.02 gas turbine.

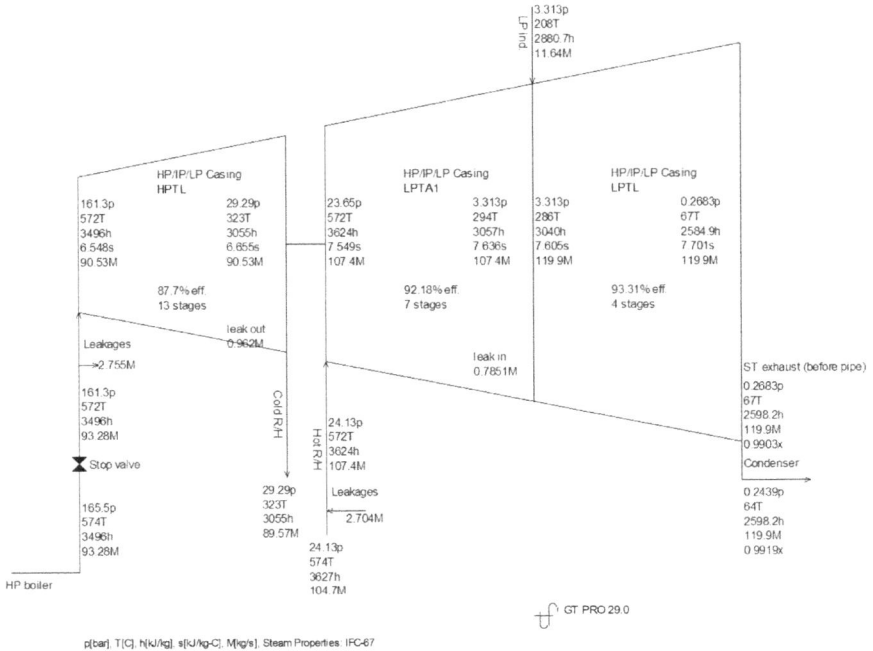

FIGURE 11.4   Steam turbine heat and mass balance (HMB) schematic.

- HP turbine, 2,863 kW (19.8% of the total)
- IP turbine, 2,769 kW (19.2% of the total)
- LP turbine, including exhaust loss, 6,386 kW (44.2% of the total)
- Mechanical and electrical losses, 2,418 kW (16.8% of the total)

Based on the STG output, bottoming cycle gross exergetic efficiency is 151,367/230,713 = 65.61%. Accounting for the cycle pumps and STG auxiliary losses, 3,705 kW, the net bottoming cycle output is 151,367 − 3,705 = 147,662 kW for the net bottoming cycle exergetic efficiency of 147,662/230,713 = 64%.

Admittedly, these are not stellar numbers. However, this is the stark reality. ISO conditions (1 atm, 15°C, 60% relative humidity) do not exist in many places. Furthermore, due to ever-stringent environmental regulations prompted by water scarcity, in many places in the world, but especially in the US, water-cooled cycle heat rejection systems, once-through (i.e., open loop with a readily available cooling water source) or with a mechanical cooling tower, are very difficult to be permitted. The developer/owner must either invest into a *zero liquid discharge* (ZLD) system, which is extremely finicky to operate (at extra CAPEX and parasitic power consumption to boot), or resort to an ACC, which is expensive, consumes a lot of fan power, and hampers the steam turbine via high back pressure.

In order to illustrate the impact of the aforementioned facts, let us look at the GTCC performance with the same gas turbine but at ISO ambient conditions and with an open-loop water cooled condenser operating at 59 mbar back pressure (1.2 in. Hg):

TABLE 11.1

Steam Turbine HMB – Stream Data

	P, bar	T, °C	h, kJ/kg	M, kg/s	S, kJ/kg-K	a, kJ/kg	A, kW	H, kW	Power, kW	ΔA, kW	Irrev., kW
Before stop valve	165.47	573.9	3,496.49	93.283	6.5378	1,551.80	144,757				
After stop valve	161.33	572.4	3,496.49		6.5485	1,548.61	0				
-Valve stem leak 1			3,496.49	0.348		1,548.61	539				
-Valve stem leak 2			3,496.49	0.051		1,548.61	79				
-HP inlet leak 1			3,496.49	2.355		1,548.61	3,647				
HPT inlet	161.33	572.4	3,496.49	90.528	6.5485	1,548.61	140,193				
GROUP IN (HPTL)	161.33	572.4	3,496.49	90.528	6.5485	1,548.61	140,193				
GROUP OUT	29.29	323.1	3,054.79	90.528	6.6546	1,075.28	97,343		39,987	42,850	2,863
-HP exit leak 1			3,054.79	0.785		1,075.28	844				
-HP exit leak 2			3,054.79	0.177		1,075.28	190				
HPT exit	29.29	323.1	3,054.79	89.566	6.6546	1,075.28	96,308				
Hot reheat	24.13	573.9	3,626.9	104.719	7.5439	1,382.24	144,747				
+Valve stem leak 1			3,496.49	0.348		1,382.24	481				
+HP inlet leak 1			3,496.49	2.355		1,382.24	3,255				
Before stop valve	24.13	572.4	3,623.61	107.422	7.54	1,380.12	148,255				
After stop valve	23.65	572.2	3,623.61	107.422	7.5492	1,377.37	147,960				
GROUP IN (LPTA1)	23.65	572.2	3,623.61	107.422	7.5492	1,377.37	147,960				
GROUP OUT	3.31	294.3	3,057.05	107.422	7.6357	785.02	84,329		60,862	63,631	2,769
before LP induction	3.31	294.3	3,057.05	107.422	7.6357	785.02	84,329				
LP induction	3.31	208.2	2,880.75	11.645	7.2987	709.20	8,259				
+HP exit leak 1			3,054.79	0.785			0				
GROUP IN (LPTL)	3.31	285.9	3,039.91	119.853	7.6053	776.95	93,119	364,342			
GROUP OUT	0.268	66.58	2,584.88	119.853	7.7014	293.26	35,149	309,806			
After Leaving Loss	0.268	66.58	2,598.24	119.853	7.7407	294.91	35,346	311,407			
To condenser	0.244	64.44	2,598.24	119.853	7.784	282.00	33,798	311,407	52,935	59,321	6,386

- STG output is 182,392 kW (cf. 151,367 kW above, 17% lower)
- Gross bottoming cycle exergetic efficiency is 182,392/224,265 kW (GT exhaust exergy) = 81.3%
- Net bottoming cycle exergetic efficiency is (182,392 − 3,796)/224,265 = 79.6%

Now, these are very respectable numbers. More details on GTCC exergy analysis will be provided in Chapter 12.

Clearly, the LP turbine irreversibility makes the largest contribution to the STG irreversibility, accounting nearly for 45% of it. Referenced to the GT exhaust exergy, its contribution to the bottoming cycle exergy loss is 6,386/230,713 = 0.0277 or 2.8%. HP and LP turbine irreversibilities are comparable in magnitude at 2,863 and 2,769 kW, respectively. LP turbine irreversibility has two contributors:

- Friction and leakage losses during expansion
- Exhaust loss (see the exhaust loss curve in Figure 11.5)

In general, the exhaust loss is a combination of two losses; that is, leaving loss (LL) and hood loss. The leaving loss is the kinetic energy of steam leaving the last stage, which cannot be utilized because there is no following turbine stage. The hood loss is the pressure loss between the exhaust of the last stage and the inlet of the condenser.

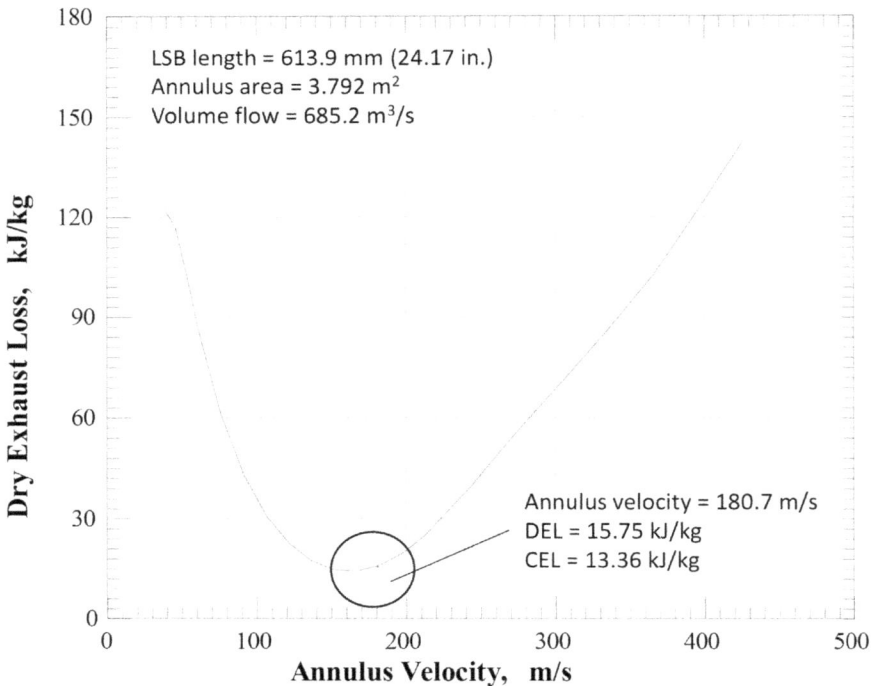

FIGURE 11.5 Steam turbine exhaust loss curve (from GT PRO output); LSB: last stage bucket, DEL: dry exhaust loss, CEL: corrected exhaust loss.

At low load points with low steam velocity, one can add the turn-up loss to those two. For a thorough coverage of exhaust loss calculation, the reader is referred to the 1963 paper by Spencer, Cotton, and Cannon [3]. (It should be noted that the exhaust loss curve in Figure 11.5 includes only the leaving loss.)

The expansion path of LP steam in the LP turbine is shown on an enthalpy-entropy (Mollier) diagram in Figure 11.6. Corresponding state-point data is summarized in Table 11.2. From Figure 11.5, the leaving loss is calculated as 119.853 kg/s × 13.36 kJ/kg = 1,601.2 kW. The isentropic work is 119.853 kg/s × (3,309.91 − 2,552.33) = 58,447 kW so that the isentropic efficiency is 54,537/58,447 = 93.3%. The lost work is thus calculated as 58,447 − 54,537 = 3,910 kW. Combined with the leaving loss, this adds up to 3,910 + 1,601 = 5,511 kW.

As shown in Table 11.1, application of the SSSF conservation of exergy to the LP turbine control volume between state-points 1 and c results in 6,386 kW for the total LP turbine irreversibility $\dot{I}$. Between state-points 1 and 2, $\dot{I}$ is calculated as 3,434 kW. Similarly, application of the SSSF conservation of exergy to the leaving loss and hood loss processes, the irreversibility terms are found as 1,404 kW and 1,547 kW, respectively. Thus, while the lost work is calculated as 5,511 kW above, total irreversibility (before the hood loss) is found as 3,434 + 1,404 = 4,838 kW. The difference, 5,511 − 4,838 = 673 kW, is the lost work opportunity because the irreversibilities push the expansion end point up (higher enthalpy and entropy at the control volume exit).

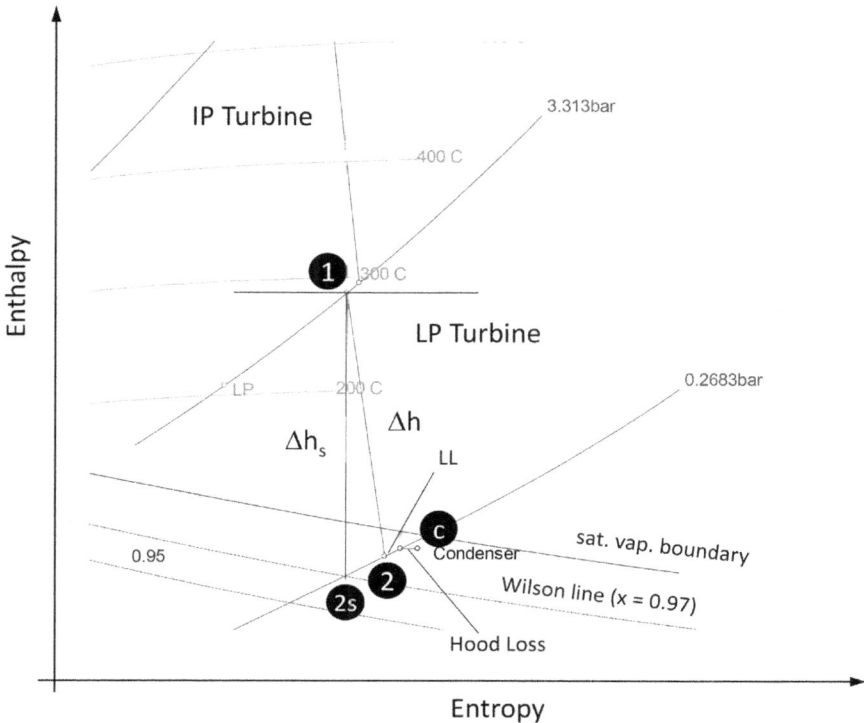

FIGURE 11.6 Steam turbine expansion path (from GT PRO).

TABLE 11.2
State-Point Data for Steam Expansion Path in Figure 11.6

	P, bar	T, °C	h, kJ/kg	s, kJ/kg-K	a, kJ/kg
1	3.31	285.9	3,039.91	7.6053	776.95
2	0.268	66.58	2,584.88	7.7014	293.26
2s	0.268	66.58	2,552.33	7.6053	289.23
c	0.244	64.44	2,598.24	7.7840	282.00

11.2 CLOSED-CYCLE GAS TURBINE

The ideal air-standard gas turbine (Brayton) cycle heat rejection process between state-points 4 and 1 is imaginary. This is why the standard gas turbine cycle is an open cycle (to be precise, as already noted earlier, an *oxymoron*); the working fluid, which itself varies in composition via combustion, does not complete a full loop. In that sense, using the term *cycle* for an actual gas turbine is erroneous in the first place. A true gas turbine cycle with a single working fluid, which stays constant in composition, would be a closed cycle. Obviously, in this case, the term *cycle* is scientifically correct, but the moniker *closed* is superfluous. Nevertheless, this is the widely accepted terminology one has to live with.

Readers interested in the theory and history of closed-cycle gas turbines is referred to the one-and-only monograph on the subject matter, namely, the excellent book by Hans Ulrich Frutschi (Ref. [1] in Chapter 10). Hans Frutschi was assisted in writing the book and its translation into English by several venerable industry veterans; that is, Peter Rufli, Hans Wettstein (who wrote the foreword), late Septimus van der Linden, and Axel von Rappard. He had actually led and/or assisted in the development of almost all closed-cycle gas turbine power plants in Europe by the Escher Wyss AG.

The basic patent for the cycle was registered in Switzerland in 1935 by Ackeret (of ETH Zurich) and Keller (of *Escher Wyss* (EW) in Zurich). This is why the cycle is frequently referred to as the Ackeret-Keller or the AK cycle. The first test installation, AK36, with air as the working fluid was in EW factory in Zurich in 1939. In 1950s and 1960s, under licenses provided by EW, several AK cycle power plants were built in the UK, France, Germany, Austria, and Japan. The largest, commissioned in 1972 in Vienna, was rated at 30 MWe. The range of the remaining AK cycle power plants with air as the working fluid were 2 to 17 MWe (ibid.).

Several closed-cycle concepts with helium as the working fluid were developed primarily for nuclear power plant applications in mind. Helium with the lowest neutron capture cross-section (it is chemically and neutronically inert or *non-fertile*[1]) was considered as an ideal coolant for high-temperature nuclear reactors and fast breeders. Supercritical carbon dioxide and its mixture with helium were also considered as working fluids for similar applications and many studies were done on direct cycles with either working fluid [4].

In the 1950s and 1960s, with their ability to utilize difficult fuels, such as coal, steel mill off-gases, and peat, closed-cycle gas turbines achieved limited commercial success. Their inherent size disadvantage (just imagine the size of a modern H class gas turbine power plant with an air mass flow rate approaching 1,000 kg/s operating in a closed cycle with near-atmospheric pressure at the compressor inlet) was minimized by charging the working fluid to high pressures. Charging not only helped to keep the size of piping and heat exchangers down from a volumetric flow perspective, but it also enhanced the heat transfer coefficients as well. Furthermore, since the velocity triangles in the rotating components changed very little with charging, part load control of closed-cycle turbines with nearly constant efficiency via pressure control (i.e., letting working fluid into or out of the cycle) was a clear advantage.

Nevertheless, the inability of closed-cycle machines to match ever-increasing turbine inlet temperatures, combined with the simplicity and agility of industrial gas turbines, which are similar to jet engines in construction and thermal cycle, ended their run in the early seventies. In a nutshell, a cost-effective closed-cycle gas turbine design is out of question at high temperatures requisite for acceptable efficiencies.

Sometime around 2010 and thereafter, however, trade publications and archival journals were inundated with articles and papers filled with hyperbole and lofty claims about the closed-cycle *supercritical CO_2* turbines and their merits with no regard to reality dictated by fundamental thermodynamic principles. In principle, a closed-cycle power generation system with supercritical CO_2 as its working fluid (henceforth, SCO2) is an excellent *bottoming cycle* to complement a *topping cycle* with low-grade exhaust energy (e.g., an aeroderivative gas turbine or reciprocating gas-fired engine). It is also a good fit to advanced nuclear reactors; for example, Generation IV reactors, such as LMR (liquid metal reactor). In particular, a sodium-cooled fast-breeder LMR is ideally suitable to SCO2 Brayton cycle with ~500°C at the exit of the intermediate heat exchanger.[2] Similarly, an SCO2 turbine, small enough to be put at the top of a solar tower along with a solar receiver, is an intriguing option for *concentrated solar power* (CSP). Even in those good fit scenarios, things are not as simple as the proponents of the technology claim. But there is at least a conceivable and realistic path to a feasible, cost-effective performance, which would be difficult and/or costly to achieve with, say, a steam cycle. For a critical look at the SCO2 cycles, the reader is referred to an article by the author [5] and chapter 22 of his monograph ([4] in Chapter 1). Herein, we will investigate the true potential of the SCO2 closed-cycle gas turbine via three ideal cycle calculations (i.e., in descending order of efficiency predictions):

- Carnot cycle (via cycle maximum and minimum temperatures, TMAX and TMIN, respectively)
- Equivalent Carnot cycle (via mean-effective cycle heat addition and rejection temperatures, METH and METL, respectively)
- Ideal cycle with a real gas equation of state (EOS), zero pressure, heat, and friction losses, and 100% component isentropic efficiencies

The first two can easily be identified on the temperature-entropy (T-s) diagram of any heat engine cycle. The T-s diagram of the intercooled-recuperated SCO2 cycle is shown (in a generic manner) in Figure 11.7. The Carnot cycle efficiency is easy to evaluate; that is, for T_4 = TMAX = 1,150°C, and T_1 = TMIN = 15°C (ISO ambient temperature), it is given by

Temperature, °C

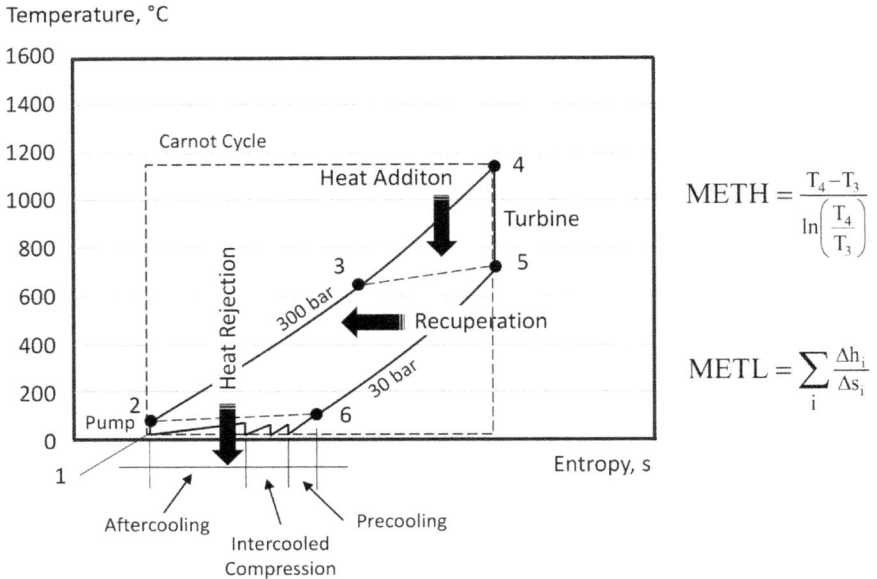

$$\text{METH} = \frac{T_4 - T_3}{\ln\left(\frac{T_4}{T_3}\right)}$$

$$\text{METL} = \sum_i \frac{\Delta h_i}{\Delta s_i}$$

FIGURE 11.7 Temperature-entropy (*T-s*) diagram of the SCO2 cycle with recuperation and intercooled compression.

$$\eta_c = 1 + \frac{15 + 273.15}{1{,}150 + 273.15}. \tag{11.5}$$

(For TMAX = 1,500 K, the result is η_c = 80.8%.) For the calculation of METH and METL, we will use the *air-standard* cycle approach with the following assumptions: (i) ideal gas EOS; (ii) constant specific heat, c_p (i.e., *perfect* gas); and (iii) isothermal compression with recuperation. With these assumptions, it follows that $T_3 = T_5$ and $T_2 = T_1 = T_6$ = TMIN. Consequently, without resorting to tedious enthalpy/entropy change evaluations for pre-, inter-, and after-cooling, METL = TMIN = 288 K, and METH is the logarithmic average of T_3 and T_4 = TMAX = 1,423 K. Turbine exhaust temperature, T_5, is evaluated using the isentropic formula

$$T_5 = T_4 \times \text{PR}^{-k}, \quad k = 1 - \frac{1}{3}. \tag{11.6}$$

In standard textbook calculations, the value of the specific heat ratio, γ, is taken as 1.4 for air as perfect gas. For supercritical CO_2, in the region of interest, c_p varies between 1.17 and 1.31 kJ/kg-K. Using an average value of 1.24 kJ/kg-K and molecular weight of 44 kg/kmol, γ is calculated as 1.18. Using this value in Equation 11.6, we find that $T_5 = T_3 = 1{,}002$ K so that METH is evaluated to be 1,200 K (927°C). Using this value in the denominator of the fraction on the right-hand side of Equation 11.5, the *equivalent* Carnot efficiency is found to be η_{ec} = 76.0%. (For TMAX = 1,500 K and PR = 10.9:1, METH = 1,257 K, and the *equivalent* Carnot efficiency, η_{ec} = 77.1%.)

Successful *Carnotization* of the SCO2 cycle with a relatively modest cycle *PR* is evident from these results. (In comparison, the efficiency of the standard Brayton cycle is a function of cycle *PR* and γ only, which would amount to a paltry 30.5%.) For a conventional, simple recuperated SCO2 cycle with indirect heating, *PR* = 3:1, and TMAX = 760°C, METH is calculated as 951 K (γ = 1.18) so that η_{ec} = 69.7% (vis-à-vis the Carnot efficiency, η_c = 72.1%). However, for the recompression cycle with a flow split fraction of ψ = 0.6, the *equivalent* Carnot efficiency becomes

$$\eta_{ec} = 1 - 0.6\frac{288}{951} = 81.8\%. \tag{11.7}$$

which is higher than the *conventional* Carnot efficiency. The reason for this seemingly impossible finding is the reduction in cycle heat rejection because only a *fraction* of the cycle working fluid is involved in that process. Without flow split, the equivalent Carnot efficiency would simply be η_{ec} = 1 − 288/951 = 69.7% because the working fluid is *fully* involved in both heat addition *and* rejection processes. A full derivation of Equation 11.7 can be found in the monograph by Gülen [7]. This is a clear and simple-to-understand demonstration of the effectiveness of reduction in cycle heat rejection (via flow split and re-compression) on cycle efficiency.

Finally, let us look at an ideal cycle calculation in Thermoflow Inc.'s THERMOFLEX flowsheet simulation software with real gas properties (using the REFPROP (2018) package for SCO2 property calculation) with zero losses, 100% recuperative heat exchange effectiveness, and isentropic components. Cycle heat and mass balance data (700 kg/s SCO2 flow, T_4 = 1,500 K and PR = 10.9:1) data is shown in Figure 11.8. Cycle heat input is 564,614 kWth, and heat rejection is −22,870 − 174,681 = −197,551 kWth for cycle work output of 564,614 − 197,551 = 367,063 kW and cycle efficiency of 367,063/564,614 = 65.0%; that is, nearly five percentage points *lower* than its air-standard counterpart. Note that in the ideal cycle calculations, the working fluid is 100% CO_2, which is *not* the case in the actual cycle with combustion. The goal therein is to keep the CO_2 purity to at least 97%(v) to derive the benefit of quasi-isochoric pumping of dense phase CO_2.

A two-stage compression with intercooling adds only 0.3 percentage points to the efficiency. A more worthwhile modification would be split-flow recompression, whose THERMOFLEX heat and mass balance with ψ = 0.5 is shown in Figure 11.9. From the data shown in the figure, the net output is 338,546 kW and the cycle

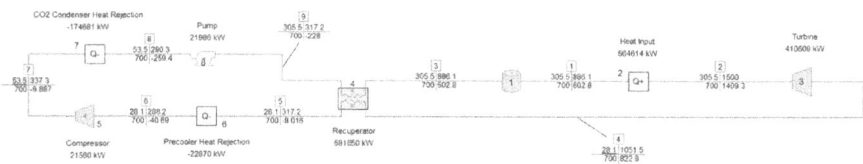

FIGURE 11.8 Ideal SCO2 cycle with recuperation (THERMOFLEX model with REFPROP).

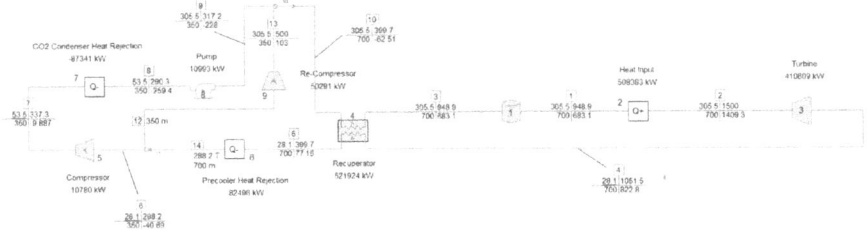

FIGURE 11.9   Ideal SCO2 cycle with split-flow ($\psi = 0.5$) and recompression.

efficiency is 66.6%; that is, 1.6 percentage points better than that for the simple recuperation cycle in Figure 11.8.

For an SCO2 cycle with recuperation and isothermal compression, *TIT* (TMAX) = 1,500 K and *PR* = 10.9:1, the following is a summary:

- Carnot efficiency is 80.8% (*ultimate* theoretical limit)
- Equivalent Carnot efficiency is 77.1% (theoretical limit with perfect gas assumption)
- Ideal cycle limit with real gas model is 65% (65.3% with intercooled compression)
- Ideal cycle limit with split flow ($\psi = 0.5$) and recompression is 66.6%

So far, SCO2 power cycles have not progressed beyond being an academic exercise. There is no commercial design, engineering, procurement, construction, and field operation experience to provide a framework for proper evaluation of the findings earlier. The only means available to put the ideal cycle efficiency numbers into a meaningful perspective is the theory and experience gained in Brayton-Rankine (i.e., gas and steam turbine) combined cycle. Air-standard Brayton-Rankine combined cycle (CC) efficiency is a function of Brayton (gas turbine) cycle *PR* and TMAX (i.e., gas turbine *TIT*). Horlock (1995) has shown that the two parameters are related and there is an optimum cycle *PR* for given *TIT* that maximizes CC efficiency. The author confirmed that assertion and established the relationship between *TIT* and cycle *PR* at the point where gas turbine specific power output is maximized by using published OEM rating data for heavy-duty industrial gas turbines (Ref. [4] in Chapter 1). For a given *TIT* (T_3 in Brayton cycle), the optimum cycle *PR* is given by (for $T_1 = 288.15$ K)

$$PR_{opt} = \left(\frac{T_3}{288.15}\right)^{1/2\kappa}, \quad \kappa = 0.3093\left(\frac{T_3}{1873.15}\right)^{0.2939}. \tag{11.8}$$

Using P_{opt} from Equation 11.8, combined cycle efficiency can be found as (for $\gamma = 1.4$)

$$\eta_{cc} = 1 - \frac{288.15}{METH}, \quad METH = \frac{T_3 - T_2}{\ln\left(\frac{T_3}{T_2}\right)}, \quad T_2 = 288.15\,PR_{opt}^{0.2857}. \tag{11.9}$$

Air-standard cycle efficiencies of the SCO2 cycle (with recuperation, isothermal heat rejection, and compression, and $\gamma = 1.18$) and Brayton-Rankine CC are plotted as a function of TIT (T_4 in Figure 11.7 for the SCO2 cycle, T_3 for the CC) in Figure 11.10. On a theoretical (ideal) cycle basis, the SCO2 cycle at 1,150–1,223°C TIT (PR 10.9:1) matches the performance of the Brayton-Rankine combined cycle at 1,700°C gas turbine TIT (PR = 23.5:1). This is indeed a testament to the thermodynamic advantage of recuperation and isothermal compression (both cycles have isothermal heat rejection) as well as the superiority of SCO2 vis-à-vis air as cycle working fluid (represented by γ).

On an ideal cycle basis, as illustrated by the graphic in Figure 11.10, there is a clear efficiency advantage for the SCO2 power cycle vis-à-vis gas-steam turbine combined cycle. Undoubtedly, there will be significant changes when a concept is translated from the drawing board to the field. In other words, the efficiencies in Figure 11.10 do not reflect the *reality*. Thus, the next task for the scientist/engineer is to quantify the gap between theory and practice (i.e., the *reality*). To do that, we examine the ISO base load rating performances for heavy-duty industrial gas turbine combined cycles published in the annual handbook of the *Gas Turbine World* (*GTW*) magazine, which is a leading trade publication. Simple and combined cycle gas turbine performance data, in terms of net output and LHV efficiency (with 100% CH_4 fuel), are provided directly by the OEMs for ISO ambient conditions and with minimal auxiliary loads. A study of the published net rating performances established the auxiliary load implicit in the listed values as 1.6% of gross (generator) output [6]. For net cycle performance, one must account for cycle power consumers; that is, condensate and feed pumps, typically 1.9–2.0% of steam turbine generator output or about 0.5–0.6% of gross output (ibid.). Utilizing the *2016 GTW Handbook* data and making the adjustment to convert net performance to gross or cycle performance (generator output minus cycle pump power consumption), a comparison of real and theoretical performances can be made as shown in Figure 11.11.

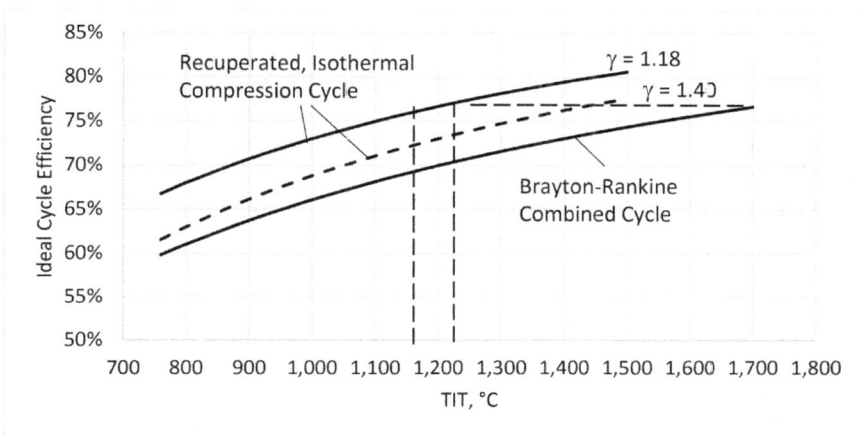

FIGURE 11.10   Ideal cycle comparison.

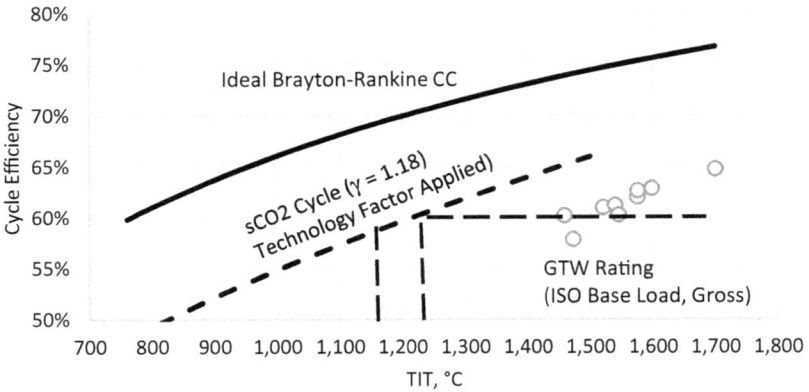

FIGURE 11.11   Comparison of theory and state-of-the-art technology.

Using the data in Figure 11.11, one can establish a *technology factor* (*TF*) that quantifies the state-of-the-art technology (represented by published rating efficiency) as a fraction the theory (represented by the air-standard cycle efficiency). For the gas-steam turbine (Brayton-Rankine) combined cycle, *TF* is found as 0.820+0.018 in the limited range covering advanced F, G, H, and J class gas turbine *TIT*s, 1,450 to 1,700°C. As a function of *TIT* (in °C), *TF* on a gross/cycle basis is represented by the linear curve-fit equation (R^2 = 0.8835, removing two outliers)

$$\text{TF} = 0.1337\left(\frac{\text{TIT}}{1000}\right) + 0.6175. \tag{11.10}$$

Modern gas and steam turbines represent the pinnacle in prime mover (heat engine) technology at the time of writing, drawing upon the latest developments in metallurgy, 3D aerodynamics, combustion, electro-mechanics, rotor-dynamics, and digital controls. Connected with a three-pressure, reheat HRSG, they form the backbone of the most efficient fossil fuel–fired electric power generation technology: the *gas turbine combined cycle* (GTCC) with rated efficiencies peaking at about 64% net LHV and CO_2 emissions less than 350 kg/MWh. This achievement is the result of a long period of intense research and commercial development going back to late 1960s and early 1970s and involving billions of dollars in investment, millions of man-hours spent in design, development, and operation, and millions of hours of fired operation in the field by thousands of units. It is patently unrealistic to expect better *TF* for a first-of-a-kind (FOAK) technology than that achieved by the state-of-the-art GTCC in the third decade of the twenty-first century.

The technology factor in Equation 11.10 is applied to the ideal SCO2 cycle in Figure 11.10, and the result is also plotted in Figure 11.11. While the SCO2 cycle still shows an advantage vis-à-vis GTCC (also note that it implicitly includes CO_2 capture with oxycombustion), it seems that the latter might have a fighting chance with H or J class *TIT*s even with the performance penalty incurred by the addition of *post-combustion carbon capture* (PCC). This will require a closer look at the oxycombustion SCO2 cycle with all the proverbial baggage; that is, plant auxiliary loads, turbine cooling, and direct heat addition (combustion). From Figure 11.11, predicted *gross/cycle* efficiency of an oxycombustion cycle with 1,500 K *TIT* and 10.9:1 cycle pressure ratio is about 60%. As noted earlier, for this cycle, Carnot and equivalent Carnot efficiencies are 80.8% and 77.1%, respectively, implying a *cycle factor* of CF = 77.1/80.8 = 0.954. The *technology factor* for 1,500 K *TIT* from Equation 11.0 is found as *TF* = 0.782. Thus, achievable gross/cycle efficiency is 0.782 × 77.1% = 60.3%. It is imperative to understand that this is *not* the *plant net thermal efficiency*, which is found *after* subtracting plant's own (auxiliary) power consumption and step-up transformer losses. Assuming a (quite optimistic) 2% for the latter, predicted plant net thermal efficiency would be 60.3% × (1 − 2%) = 59.1% (*excluding* the fuel and O_2 compressors and the ASU – more on those later in the paper). More optimistically, using the average *TF* value of 0.82, we find that 0.82 × 77.1% × (1 − 2%) = 62%. This is a number based *solely* on *fundamental thermodynamic* considerations without a proper cycle heat and mass balance simulation based on realistic hardware design assumptions. The next step is to do just that. The reader is referred to the paper by Gülen et al. for the details and findings [8].

NOTES

1 A fertile material can be converted into a fissile material by neutron absorption and subsequent nuclei conversions. An example is U-234, which can be converted into U-235.
2 See the *Gas Turbine World* March–April 2007 article "Closed cycle nuclear plant rated at 165 MW and 41% efficiency" by S. van der Linden for a recuperated helium Brayton cycle turbine envisioned for a pebble bed modular reactor.

REFERENCES

1. Lauck, F., Uyehara, O.A., and Myers, P.S., *An Engineering Evaluation of Energy Conversion Devices*, SAE Trans., 71, Paper 630, 1963.
2. Tomei, G.L. (Editor), *Steam – Its Generation and Use*, 42nd Edition, Charlotte: The Babcock & Wilcox Company, 2015.
3. Spencer, R.C., Cotton, K.C., and Cannon, C.N., *A Method for Predicting the Performance of Steam Turbine Generators . . . , 16500 kW and Larger*, 1974 revised version of ASME Paper No. 62-WA-209, 1974.
4. Lee, J.C., Campbell, Jr., J., and Wright, D.E., Closed-Cycle Gas Turbine Working Fuels, *ASME Paper 80-GT-135, Gas Turbine Conference & Products Show*, March 10–13, 1980, New Orleans, LA.
5. Gülen, S.C., Supercritical CO2 — What Is It Good For? *Gas Turbine World*, September/October Issue, pp. 26–34, 2016.
6. Gülen, S.C., Importance of Auxiliary Power Consumption for Combined Cycle Performance, *Journal of Engineering GTs & Power*, 133, p. #041801, 2011.
7. Gülen, S.C., *Gas & Steam Turbine Power Plants—Applications in Sustainable Power*, Cambridge: Cambridge University Press, 2022.
8. Gülen, S.C., Taher, M., and Lyddon, L.G., *Oxy-Fuel Combustion Supercritical CO_2 Power Cycle – A Critical Look*, GPPS-TC-2022–0134, GPPS Chania22, September 12–14, 2022.

12 Gas Turbine Combined Cycle

Two are better than one.

—Ecclesiastes 4.9–10

This is how we ended Chapter 10: Consider the triangular area {1-4-4C-1} in Figure 3.4, which quantifies the lost work associated with cycle heat rejection. What if we could place another power cycle in there, which has a mean-effective heat addition temperature of $\overline{T}_L$ and mean-effective heat rejection temperature of T_1? This is the thought underlying the combined cycle concept, which is best illustrated graphically on a *T-s* diagram as shown in Figure 12.1 using the second law–base considerations.

The following observations are made in Figure 12.1:

- The starting point is the gas turbine Brayton cycle {1-2-3-4-1}, which *approximates* the ideal heat engine (Carnot) cycle operating between T_3 and T_1. It can be represented by an *equivalent* Carnot cycle operating between $\overline{T}_H$ and $\overline{T}_L$
- Heat energy rejected by the Brayton cycle can be utilized by another, for the time being *unspecified*, cycle {1-4-4C-1}, which can be represented by another equivalent Carnot cycle operating between $\overline{T}_L$ and T_1.
- Due to their respective positions in the *T-s* diagram, the first cycle (i.e., the Brayton cycle) is referred to as the *topping* cycle. The second cycle is referred to as the *bottoming* cycle.
- The two cycles form a *combined* cycle {1-2-3-4C-1}, which is represented by the equivalent Carnot cycle operating between $\overline{T}_H$ and T_1.

The efficiency of that ideal bottoming cycle would be given by the following equation:

$$\eta_{BC} = 1 - \frac{T_1}{\overline{T}_L} \tag{12.1}$$

From the cycle *T-s* diagram in Figure 12.1, the ideal *combined* efficiency with ideal Brayton (topping) and ideal bottoming cycles can be deduced to depend on the ratio of $\overline{T}_H$ and the ambient temperature T_1, which, after rearranging the terms a bit leads us to the combined cycle efficiency:

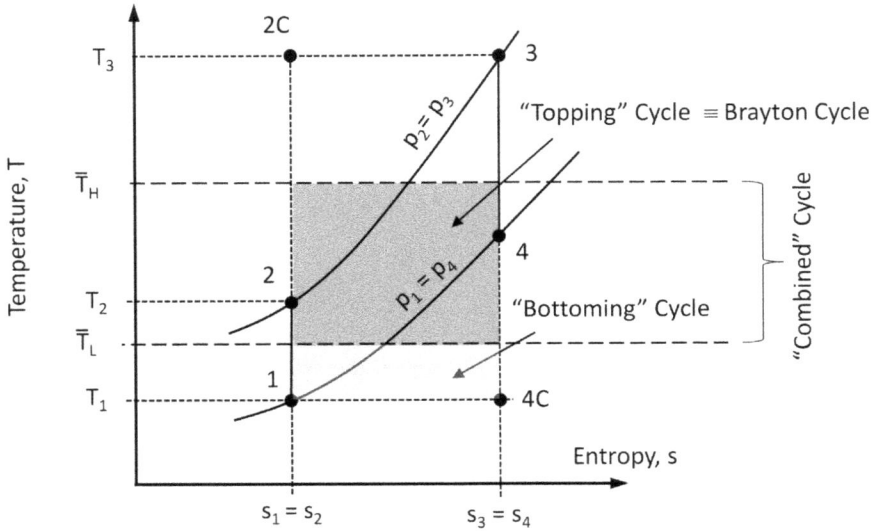

FIGURE 12.1 Combined cycle concept on *T-s* diagram.

$$\eta_{CC} = 1 - \frac{T_1}{\overline{T}_H} \qquad (12.2)$$

$$\eta_{CC} = 1 - \frac{\ln\left(\dfrac{\tau_3}{PR^k}\right)}{\tau_3 - PR^k}, \qquad (12.3)$$

where $PR = P_2/P_1$ and $\tau_3 = T_3/T_1$.

Note that taking the derivative of η_{cc} in Equation 12.3 with respect to *PR* and setting it to zero will not give us the optimum *PR* for maximum combined cycle efficiency. This is so because Equation 12.3 does not contain *information* about two cycles. Efficiency predicted by Equation 12.3 will increase indefinitely until $\overline{T}_H \sim T_3$ and $\overline{T}_L \sim T_1$; that is, for all practical purposes, one would end up with a Carnot cycle generating zero power output (i.e., a purely mathematical limiting case of no practical value).

However, a visual examination of Figure 12.1 clearly points to two extremes:

- Very high *PR*, as discussed before, leading to a very thin topping cycle (i.e., practically, no bottoming cycle)
- Very low *PR*, leading to a disappearing topping cycle so that we are left with the bottoming cycle

At one extreme, cycle efficiency is at a maximum, but cycle power output is essentially zero. At the other extreme, cycle efficiency is low, but cycle power output is positive. Clearly, somewhere in between these two trends are balanced so that we end up with an *optimal* combined cycle efficiency. This optimum can be determined using a simple approach, as described below (originally proposed in a paper by Horlock [1]).

Ignoring the miscellaneous topping and bottoming cycle losses and minor inputs, a simplified version for the combined cycle efficiency can also be written as

$$\eta_{CC} = \eta_{TC} + (1 - \eta_{TC}) \cdot \eta_{BC}, \tag{12.4}$$

where subscripts *TC* and *BC* denote topping and bottoming cycles, respectively. Taking the derivative of both sides with respect to *PR* and setting the combined cycle efficiency derivative to zero to find its maximum, noting that the bottoming and topping cycle efficiencies are approximately the same in magnitude, we find that

$$\frac{\partial \eta_{TC}}{\partial PR} \approx - \frac{\partial \eta_{BC}}{\partial PR}. \tag{12.5}$$

This correlation states that the maximum combined cycle efficiency occurs at the point where the rate of increase of the topping cycle efficiency with *PR* is the same as the rate of decrease of the bottoming cycle efficiency. The rate of increase of the topping cycle efficiency with increasing *PR* can be found from the Brayton cycle efficiency

$$\frac{\partial \eta_{TC}}{\partial PR} = \alpha \frac{k}{PR^{k+1}}, \tag{12.6}$$

where α is the technology factor applied to the topping cycle (see Chapter 10). Similarly, the rate of decrease of the bottoming cycle efficiency with increasing *PR* is found as

$$\eta_{BC} = \beta \left(1 - \frac{\ln X}{X - 1} \right), \quad X = \frac{\tau_3}{PR^k}, \tag{12.7}$$

where β is the technology factor applied to the bottoming cycle. At this point, we can go ahead and state that a steam *Rankine* cycle is an ideal candidate for the bottoming cycle because of the following:

- Current gas turbine technology with exhaust temperatures around 1,150°F or higher constitutes an ideal heat source for generating high-pressure steam at 1,000–1,100°F in a *heat recovery steam generator* (HRSG).
- Heat rejection from the steam turbine condenser at constant pressure and temperature is the perfect fit to the fourth ideal Carnot cycle process; that is, *isothermal* cycle heat rejection.

Taking the derivative of Equation 12.7, we obtain

$$\frac{\partial \eta_{BC}}{\partial PR} = \beta \frac{k}{PR} \cdot \frac{\left(1 - \dfrac{X \ln X}{X-1}\right)}{X-1}. \qquad (12.8)$$

Using the technology factors $\alpha = 0.7$ and $\beta = 0.8$, solving for *PR* in Equation 12.5 with the help of Equations 12.6 and 12.8, it is found that the optimal *PR* for maximum combined cycle efficiency is quite close to the optimum topping (Brayton) cycle *PR* maximizing the topping cycle output (see Section 10.1).

12.1   BOTTOMING CYCLE – SINGLE PRESSURE, REHEAT

Let us now look closer at the bottoming cycle, which we already know to be a Rankine steam cycle, particularly, a reheat cycle. Modern combined cycles with advanced-class gas turbines are equipped with three-pressure, reheat (3PRH) bottoming steam cycles. In multiple-pressure steam generation, the goal is to minimize irreversibility (exergy destruction) in the HRSG, which can be visually assessed by the gap between gas and steam in a heat release diagram (e.g., see Figure 6.8 in Section 6.4). For more details, refer to the monograph by the author [2]. For the purpose of illustrating the basic principles, however, the single-pressure cycle is most suitable.

Let us specify a state-of-the-art, single-pressure, reheat steam cycle as follows:

- 2,500 psia high-pressure (HP) steam at 1,112°F
- 400 psia intermediate pressure (IP) reheat steam at 1,112°F
- 0.8 psia condenser pressure
- No losses, isentropic pump and turbine

The system is depicted on a temperature-entropy diagram in Figure 12.2. The gas turbine is specified as a J/HA class gas turbine with 1,153°F exhaust temperature (state-point a in Figure 6.6) and 1,500 lb/s exhaust flow. The gas is cooled to 180°F in an HRSG to make steam (state-point b in Figure 12.2).

We identify the *three* different ways to quantify the thermodynamic potential of the system in Figure 12.2 utilizing the mean-effective heat addition and heat rejection concepts. For the Rankine steam cycle in Figure 12.2, there are *three* possible $\overline{T}_H$ definitions:

1. Between gas turbine exhaust and stack temperatures (i.e., T_a and T_b, respectively)
2. Between gas turbine exhaust and ambient temperatures (i.e., T_a and T_0, respectively)
3. Between steam and water temperatures at state-points 3, 3R and 2

Furthermore, there are *two* possible $\overline{T}_L$ definitions:

1. Condenser steam temperature, $T_1 = T_4$
2. Ambient temperature, T_0

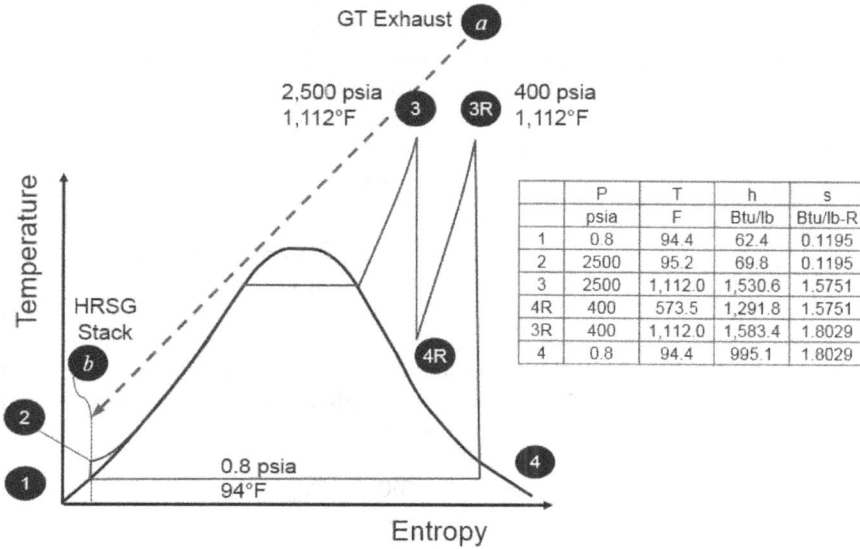

	P	T	h	s
	psia	F	Btu/lb	Btu/lb-R
1	0.8	94.4	62.4	0.1195
2	2500	95.2	69.8	0.1195
3	2500	1,112.0	1,530.6	1.5751
4R	400	573.5	1,291.8	1.5751
3R	400	1,112.0	1,583.4	1.8029
4	0.8	94.4	995.1	1.8029

FIGURE 12.2   Steam Rankine bottoming cycle.

We calculate $\bar{T}_H$ according the second definition earlier as follows (using the gas properties in THERMOFLEX):

$$\bar{T}_H = \frac{(h_a - h_0)}{(s_a - s_0)} = \frac{(286.7 - -4.4)}{(0.2894 - -0.0083)} - 460 = 518°\,F$$

The Carnot equivalent efficiency implied by this mean-effective heat addition temperature and heat rejection at ambient temperature (i.e., the first possible $\bar{T}_L$ definition identified earlier) is given by

$$\eta_1 = 1 - \frac{59 + 460}{518 + 460} = 46.9\%. \tag{12.9}$$

Note that Equation 12.9 is the translation of the gas turbine exhaust gas exergy into an efficiency term. This can be easily seen by rewriting Equation 12.9 as

$$\eta_1 = 1 - T_0 \frac{(s_a - s_0)}{(h_a - h_0)} \tag{12.10}$$

$$\eta_1 = \frac{(h_a - h_0) - T_0(s_a - s_0)}{(h_a - h_0)}. \tag{12.11}$$

We immediately recognize the numerator of Equation 12.11 as the exergy of the gas turbine exhaust gas (remember: the ambient is the dead state). The denominator of Equation 12.11 is the maximum possible heat transfer from the exhaust gas. Thus, Equations 12.10 and 12.11 give us the *theoretically possible maximum bottoming cycle efficiency.*

Rankine cycle $\overline{T}_H$, which is the third possible definition identified earlier, is calculated as follows using the ASME steam tables:

$$\overline{T}_H = \frac{(h_3 - h_2) + (h_{3R} - h_{4R})}{(s_4 - s_2)} = \frac{(1531 - 70) + (1583 - 1292)}{(1.803 - 0.12)} - 459.67 = 581°F$$

Since $\overline{T}_L$ for the Rankine cycle is the steam condensing temperature of 94°F, which is the second possible $\overline{T}_L$ definition identified earlier, another definition of ideal (i.e., Carnot equivalent) Rankine cycle efficiency is

$$\eta_2 = 1 - \frac{94 + 460}{581 + 460} = 46.8\%. \tag{12.12}$$

According to Figure 12.2, heat taken from the gas turbine exhaust gas is

$$\dot{Q}_{exh} = \dot{m}_{exh}(h_a - h_b). \tag{12.13}$$

Similarly, heat given to the Rankine cycle working fluid (i.e., water/steam) is

$$\dot{Q}_{stm} = \dot{m}_{stm}((h_3 - h_2) + (h_{3R} - h_{4R})). \tag{12.14}$$

The ratio of the two is defined as the heat recovery effectiveness:

$$\eta_{HR} = \frac{\dot{Q}_{stm}}{\dot{Q}_{exh}} \tag{12.15}$$

The actual bottoming cycle work is given by

$$\dot{W}_{BC} = TF \cdot \dot{m}_{exh} a_{exh} \tag{12.16}$$

From the numerator of the term on the right-hand side of Equation 12.11, we find the exhaust gas exergy as

$$a_{exh} = (h_a - h_0) - T_0(s_a - s_0), \text{ or} \tag{12.17}$$

$$a_{exh} = \eta_1(h_a - h_0). \tag{12.18}$$

Substituting Equation 12.18 into Equation 12.16, we obtain

$$\dot{W}_{BC} = TF \cdot \eta_1 \cdot \dot{m}_{exh}\left(h_a - h_0\right). \tag{12.19}$$

Note that the term in parentheses on the right-hand side of Equation 12.19 is the *maximum possible heat transfer from the gas turbine exhaust gas per unit mass*:

$$\dot{Q}_{exh,max} = \dot{m}_{exh}\left(h_a - h_0\right) \tag{12.20}$$

Based on Equations 12.12 and 12.14, *maximum possible* bottoming cycle work is

$$\dot{W}_{stm,max} = \eta_2 \cdot \dot{m}_{stm}\left(\left(h_3 - h_2\right) + \left(h_{3R} - h_{4R}\right)\right). \tag{12.12}$$

Consequently, *actual* bottoming cycle work can be found via another technology factor, say, *TF'*:

$$\dot{W}_{BC} = TF' \cdot \eta_2 \cdot \dot{m}_{stm}\left(\left(h_3 - h_2\right) + \left(h_{3R} - h_{4R}\right)\right) \tag{12.22}$$

$$\dot{W}_{BC} = TF' \cdot \eta_2 \cdot \dot{Q}_{stm} \tag{12.23}$$

$$\dot{W}_{BC} = TF' \cdot \eta_2 \cdot \eta_{HR} \cdot \dot{Q}_{exh} \tag{12.24}$$

Combining Equations 12.19 and 12.22, the second technology factor is found as

$$TF' = \frac{TF \cdot \eta_1 \cdot \dot{m}_{exh}\left(h_a - h_0\right)}{\eta_2 \cdot \dot{m}_{stm}\left(\left(h_3 - h_2\right) + \left(h_{3R} - h_{4R}\right)\right)} \tag{12.25}$$

$$TF' = TF \frac{\eta_1}{\eta_2} \cdot \frac{1}{\eta_{HR}} \cdot \frac{\left(h_a - h_0\right)}{\left(h_a - h_b\right)}. \tag{12.26}$$

To a very good approximation, $\eta_1 \sim \eta_2$, typical heat recovery effectiveness is about 0.9, and the ratio of the gas enthalpy differences is ~1.11 so that *TF'* ~ 1.24·*TF*. As will be shown in the next section, state-of-the-art technology at gas turbine exhaust temperature of ~1,150°F corresponds to *TF* ~ 0.75 so that *TF'* is ~0.93.

Continuing with the example, from Equation 12.20, we calculate the heat transfer from the exhaust gas as

$$\dot{Q}_{exh} = 1,500 \times \left(287.6 - 26\right) = 391,050\,\text{Btu}/\text{s}.$$

Finally, using Equation 12.24, the bottoming cycle net (actual) output is found as

$$\dot{W}_{BC} = TF' \cdot \eta_2 \cdot \eta_{HR} \cdot \dot{Q}_{exh}, \text{or}$$

$$\dot{W}_{BC} = TF \cdot \eta_1 \cdot \frac{(h_a - h_0)}{(h_a - h_b)} \cdot \dot{Q}_{exh}$$

$$\dot{W}_{BC} = 0.75 \cdot 46.9\% \cdot 1.11 \cdot (391,050 \cdot 1.05506) = 162,000 \, kWe$$

where 1.05506 is the unit conversion factor.

12.2 BOTTOMING CYCLE – STATE OF THE ART

We established that the exergy of the gas turbine exhaust gas is a thermodynamic property and gives the maximum possible (theoretically, that is) bottoming cycle output. (Note that this is essentially a thermodynamic law; specifically, the second law in disguise. In other words, no specification of the bottoming cycle, working fluid, etc. is required a priori.) What can actually be achieved in the field with realistic hardware design is only a fraction of that theoretical maximum. That fraction is the technology factor. Using the rating data published in the *Gas Turbine World 2016 Handbook*, the author calculated the average technology factor of steam bottoming cycles as 0.75 ± 0.03. Starting from Equation 12.16,

$$TF = \frac{\dot{W}_{BC}}{\dot{m}_{exh} a_{exh}} \text{ with}$$

$$a_{exh} [\text{Btu / lb}] = 0.001628 \cdot T_{exh}^{1.60877} [\text{in }^{\circ}F] \tag{12.27}$$

In using the rating data for *TF* calculation, several assumptions and/or approximations are made implicitly. First, $\dot{W}_{BC}$ is found from the listed steam turbine generator output by a reduction of 3.5% to account for cycle feed pumps. Second, gas turbine exhaust temperatures used in Equation 12.27 to calculate the exhaust exergy are those for simple cycle ISO base load ratings. Those ratings are typically with zero inlet and exhaust losses (or values commensurate with simple cycle installations). For a combined cycle application, a three-pressure HRSG imposes at least 10 to 12 inches of water column additional back pressure on the gas turbine. This would result in 4–6 degrees rise in the gas turbine exhaust temperature. Rating data provided by the OEMs is not always clear on the exact assumptions used in the performance calculations (which also includes commercial margins adding to the uncertainty). Thus, splitting hairs at this point is futile.

The scatter in calculated *TF* values (there were eight data points, covering the advanced-class gas turbines by three major OEMs) was significant, covering a range of {0.71, 0.78} with a standard deviation of 0.03. This is also a reflection of different assumptions used by the OEMs in their gas turbine and bottoming cycle performance assumptions. Since the details provided by the OEMs are rather sparse, trying to resolve discrepancies is simply not possible. As one would expect, there is a strong dependence of the bottoming cycle technology factor on exhaust gas temperature; that is, higher quality energy leads to better bottoming cycle efficiencies. The technology factor is represented by the following curve-fit formula:

$$TF = 0.647 \frac{TEXH[^\circ F]}{1,000}, \quad R^2 = 0.999.$$

(12.28)

12.3   EXERGY ANALYSIS

In the preceding chapter, we have looked at the exergy analysis of a steam turbine generator (STG) in a modern gas turbine combined cycle (GTCC) power plant. Now, we will carry out the exergy analysis of the bottoming cycle of that plant at ISO base load rating conditions. Once again, the GTCC is modeled in Thermoflow Inc.'s GT PRO simulation software. The cycle diagram is shown in Figure 12.3. As shown in the diagram, net GTCC efficiency is 61.5% (LHV).

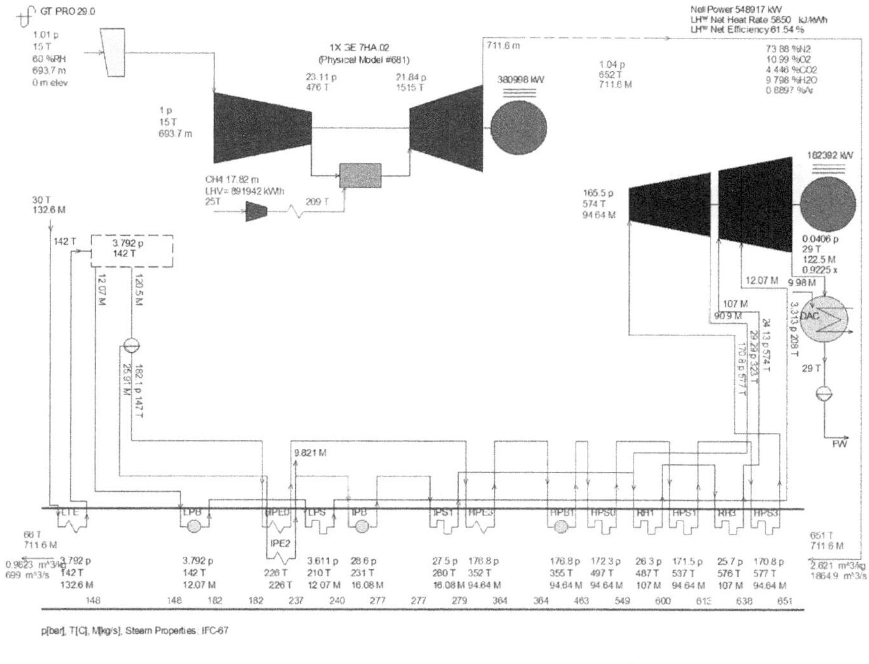

FIGURE 12.3   GTCC example with GE's 7HA.02 gas turbine (ISO base load).

TABLE 12.1
Bottoming Cycle Exergy Balance

GT exhaust exergy (in)	244,179	100%
HRSG irreversibility	26,901	11.02%
STG irreversibility	19,079	7.81%
Condenser lost work	13,099	5.36%
HRSG stack gas exergy (out)	2,708	1.11%
Miscellaneous losses	1,893	0.78%
Exergy transfer to fuel gas heater	1,718	0.70%
Balance	178,780	73.22%

Bottoming cycle exergy breakdown is summarized in Table 12.1. Note the following:

- Gas turbine exhaust exergy is calculated using Equation 14.8 (15°C dead state or reference temperature)
- HRSG stack gas exergy is calculated using Equation 14.14
- Condenser irreversibility and exergy transfer are calculated using Equations 7.7–7.10 (why GT PRO calculation is not used was explained in Chapter 7)
    - Heat rejection from steam to the cooling water is 278,038 kWth
    - $\dot{I}_C$ is calculated as 8,360 kW
    - $\dot{E}_C$ is calculated as 4,739 kW
- HRSG and STG irreversibilities are taken from the GT PRO output
- Miscellaneous losses (i.e., losses from desuperheating, mixing, throttling, and small water streams) are also taken from the GT PRO output
- STG output is 182,392 kWe
- Mechanical and electrical losses plus STG auxiliary load add up to 3,796 kWe
- Thus, net bottoming cycle output is 178,596 kWe

The breakdown in Table 12.1 identifies three big hitters in terms of exergy losses: HRSG, STG, and the condenser. At 41 mbar (1.2 in. Hg) condensing pressure with open-loop cooling, there is no realistic opportunity for condenser exergy loss alleviation via engineering design. One is simply hostage to the site conditions and/or the prevailing environmental rules and regulations. As far as the HRSG is concerned, there are also limited options available to the designer, all a function of increased size and cost (i.e., reduced pinch and approach temperature deltas). Adding a fourth pressure level is possible on paper, but it is extremely unlikely to be cost-effective.

The performance in Figure 12.3 is based on the following HRSG specs:

- HP/IP/LP pinch temperature deltas of 15°F/15°F/12°F
- HP/IP economizer approach temperature deltas of 7°F/10°F
- Casing radiant heat loss of 0.75%
- Minimum allowable superheater approach temperature difference (i.e., gas inlet temperature minus steam exit temperature) of 26°F for HP and reheat superheaters, 30°F for IP/LP superheaters

The HRSG design is revised as follows:

- HP/IP/LP pinch temperature deltas of 10°F/10°F/10°F
- Casing radiant heat loss of 0.25%
- Minimum allowable superheater approach temperature difference 20°F across the board

When the performance is rerun with this larger HRSG, the following is found:

- HRSG irreversibility is 25,196 kW (i.e., 6.34% lower than in Table 12.1)
- GTCC net LHV efficiency is 0.2 percentage points higher (i.e., 61.7%)

Let us now look at the STG design. The performance in Figure 12.3 is based on HP, IP, and LP section efficiencies of 87.74%, 92.34%, and 90.15%, respectively. The STG design is revised with HP, IP, and LP section efficiencies of 91.0%, 93.0%, and 93.0%, respectively. When the performance is rerun with this more efficient STG, the following is found:

- STG irreversibility is 15,501 kW (i.e., 18.75% lower than in Table 12.1)
- HRSG irreversibility is 27,035 kW (i.e., 0.5% higher than in Table 12.1)
- GTCC net LHV efficiency is 0.4 percentage points higher (i.e., 61.9%)

The slight increase in HRSG irreversibility is due to the changes in the distribution of heat transfer among the superheaters. For example, higher HP section efficiency leads to colder cold reheat steam temperature entering the HRSG and increasing the duty of the reheat superheater.

In a 2008 paper, Gülen and Smith examined the bottoming cycle exergy breakdown of the GTCC cycles (Ref. [1], Chapter 6). The HRSG design in the cited paper was based on 15°F evaporator pinch deltas (0.5% casing heat loss), and steam turbine HP/LP/IP efficiencies were set to 89%, 90%, and 91%, respectively. Condenser pressure was 1.5 in. Hg (51 mbar). A comparison of key design assumptions is in Table 12.2. The bottoming cycle exergy breakdowns from that work and the current one are compared in Table 12.3.

The data listed in Table 12.3 identifies the two key exergy loss buckets (i.e., higher condenser pressure and higher HRSG stack temperature). Both are design choices so that the bottoming cycle technology represented in the 2008 paper and in Figure 12.3 are essentially the same.

TABLE 12.2
Comparison of Design Assumptions

	Figure 12.3	Ref. [1] Chapter 6
HRSG HP/IP/LP pinch deltas, °F	15/15/12	15/15/15
HRSG casing loss	0.75%	0.50%
HRSG stack temperature, °F	150	175
Steam turbine HP/IP/LP efficiencies, %	87.74/92.34/90.15	89/90/91
Condenser pressure, in. Hg	1.20	1.50

TABLE 12.3
Comparison of Bottoming Cycle Exergy Breakdown

	Figure 12.3	Ref. [1] Chapter 6
GT exhaust exergy (in)	100.00%	100.00%
HRSG irreversibility	11.02%	10.10%
STG irreversibility	7.81%	7.70%
Condenser lost work	5.36%	7.00%
HRSG stack gas exergy (out)	1.11%	2.60%
Miscellaneous losses	0.78%	1.50%
Exergy transfer to fuel gas heater	0.70%	0.80%
Balance	73.22%	70.30%

Calculation of the HRSG irreversibility for a modern 3PRH HRSG with many inlet/outlet water/steam and gas streams (as can be appreciated by looking at the schematic in Figure 12.3) is a tedious undertaking (albeit straightforward). This is why that one would be well advised to use a software package such as GT PRO. Gülen and Smith (ibid.) show that there is an exact formula to calculate the HRSG irreversibility (e.g., see Section 6.4), but estimating the mean-effective (average) steam/water and gas temperatures in it is the *chore* (for the lack of a better term). The reader can use the transfer function provided in that paper for front-end quick and dirty calculations. Thus, some of the difference between the HRSG irreversibility values in Table 12.3 can be traced back to the inexactness of the transfer function used in the cited paper. Consider that HRSG plus miscellaneous losses add up to the following:

- 11.02% + 0.78% = 11.80% for the example herein
- 10.10% + 1.50% = 11.60% for the cited work

(In Gülen and Smith (ibid.), miscellaneous losses category was just a catch-all for "anything else very tedious to calculate individually.")

One last thing to try is to improve the steam cycle, which is typically referred to by a combination of HP (main steam) pressure and HP and reheat steam temperatures. For example, the steam cycle in Figure 12.3 is a 2,400/1,065/1,065 (psia and °F) cycle, which corresponds to 165.5/574/574 in SI units (bara and °C). Presently, the state of the art in steam temperatures is 600°C (1,112°F). It can be shown that the HP steam pressure has a small impact on bottoming cycle and GTCC efficiency (e.g., see Gülen and Jacobs [2]). Rerunning the GT PRO model with the new advanced steam cycle of 180/600/600 in SI units, we find that the GTCC efficiency is 61.77% LHV (0.27 percentage points improvement). Since the gas turbine is the same, the improvement must be fully attributed to the bottoming cycle. Furthermore, since the gas turbine exhaust exergy does not change, the improvement must be traced back to the bottoming cycle irreversibility reduction. How do higher steam pressure and temperatures help the bottoming cycle in terms of irreversibility reduction? In order to answer this question, we have to look at a comparison of the exergy breakdown for the two cases (e.g., see Table 12.4). Clearly, the main contributor is the reduction in HRSG irreversibility.

TABLE 12.4

GT Exhaust Exergy Breakdown for Base and Advanced Steam Cycles

	Base Steam Cycle	Advanced Steam Cycle
Steam cycle, bara/°C/°C	165.5/574/574	180/600/600
GT exhaust exergy (in)	100.00%	100.00%
HRSG irreversibility	11.02%	10.31%
STG irreversibility	7.81%	7.69%
Condenser lost work	5.36%	5.31%
HRSG stack gas exergy (out)	1.11%	1.19%
Miscellaneous losses	0.78%	0.78%
Exergy transfer to fuel gas heater	0.70%	0.70%
Balance	73.22%	74.03%

In order to explain this finding, we have to resort to the formula for HRSG irreversibility (i.e., Equation 6.19), which is reproduced subsequently for convenience.

$$\dot{I}_{HRSG} = \frac{T_0}{\overline{T}_{stm}} \left(1 - \frac{\overline{T}_{stm}}{\overline{T}_{gas}} \right) \dot{Q}_{rec}. \tag{6.9}$$

What we achieved by increasing the steam temperatures (mainly) and HP steam pressure is to increase the mean-effective temperature of steam/water inside the HRSG, $\overline{T}_{stm}$. Gas turbine exhaust gas temperature does not change. HRSG stack gas temperature is primarily driven by the LP evaporator pressure, which is kept constant. (The underlying theory for this is explained in minute detail in the author's monograph on combined cycles, Ref. [6] in Chapter 6.) Nevertheless, a difference of 1.7°F is observed in the HRSG stack gas due to the small change in steam production. Thus, $\overline{T}_{gas}$ is unchanged for all practical purposes. Consequently, as can be easily seen from Equation 6.19, increasing $\overline{T}_{stm}$ leads to decreasing $\dot{I}_{HRSG}$.

Finally, let us check what the OEM is saying on their bottoming cycle performance. Using the rating data from the *Gas Turbine World*'s *2022 GTW Handbook*, a side-by-side comparison of simple and combined cycle performances is presented in Table 12.5 and Table 12.6. Note that *GTW* simple cycle data is based on certain inlet/exit loss, fuel gas performance heating (or none), and other assumptions. When the machine runs in combined cycle, changes are to be expected, especially at the back end; that is, higher exhaust loss, 440°F (227°C) fuel gas heating, and so on. Furthermore, *GTW* combined cycle data for the 1 × 1 × 1 option does not provide a breakdown of gas and steam turbine generator outputs (a single shaft unit). Thus, they are estimated from the multi-shaft 2 × 2 × 1 unit rating and listed gross output.

Clearly, the ISO base load rating data provided by the OEM is not only overly optimistic (i.e., very low auxiliary power consumption, which should at least have been 1.6%, excluding the transformer losses) but also does not pass the proverbial smell test using the second law analysis. Specifically, on a gross (i.e., STG output only) basis, bottoming cycle exergetic efficiency is 197.5/249.1 = 79.3% (cf. 74.7% in our base sample). Is

TABLE 12.5
GE 7HA.02 Simple Cycle Data

	GTW 2022	Figure 12.3
Output, MW	384	381
Efficiency	43.30%	42.72
Exhaust flow, lb/s (kg/s)	1,609.4 (730.0)	1,568.7 (711.6)
Exhaust temperature, F (C)	1,202 (650)	1,206.1 (652.3)
Exhaust gas exergy, MW	249.1	244.2

TABLE 12.6
GE 7HA.02 Combined Cycle Data (*: Guess)

	GTW 2022	Figure 12.3
Gross output, MW	580.1	563.4
Net output, MW	573.0	548.9
Net efficiency	63.4%	61.54%
GT output, MW	382.6*	381.0
ST output, MW	197.5*	182.4
Aux. load, % of gross	1.24%	2.64%

that even possible? Note that our earlier analysis came up with the following performance deltas (GTCC net LHV efficiency basis and using two significant digits):

- 0.17 percentage points for larger HRSG
- 0.39 percentage points for more efficient steam turbine
- 0.23 percentage points for advanced steam conditions

Assuming that these improvements are additive (which, to a good approximation, they are), we can project an improvement of 0.17 + 0.39 + 0.23 = 0.79 percentage points in net LHV efficiency points for 61.54% + 0.79% = 62.33% net LHV. We are still short 63.40 − 62.33 = 1.07 percentage points. Of that delta, 0.89 percentage points can be attributed to the extra (and realistic, one must add) auxiliary load book-kept by the GT PRO model (including transformer losses). That leaves 0.22 percentage points of net GTCC (LHV) efficiency to be explained by better bottoming cycle beyond those we *hypothesized* earlier.

In general, each percentage point of improvement in bottoming cycle exergetic efficiency (on a gross basis) is worth anywhere between 0.25 to 0.28 percentage points in net GTCC efficiency. In the case of the present example, 0.79 percentage points in net LHV efficiency points corresponds to about 3.0 percentage points in bottoming cycle (gross) exergetic efficiency. The performance stairsteps are shown in Table 12.7. Note that the final push to match *GTW Handbook* GTCC efficiency required a bump of 0.67 percentage points in the bottoming cycle gross exergetic efficiency.

The exergy balance for the bottoming cycle shown earlier can be distilled into a simple formula for rapid estimation of the steam turbine contribution for a given gas

TABLE 12.7

GTCC Performance Stairsteps; A: Plant Aux Load Adjustment, B: +0.79 Net Efficiency Points (Explained in the Text), C: Final Adjustment

	GTW 2022	Figure 12.3	A	B	C
GT exhaust exergy, MW	249.1	244.2	244.2	244.2	244.2
GT output, MWe	382.6	381.0	381.0	381.0	381.0
ST output, MWe	197.5	182.4	182.4	189.7	191.3
BC gross exergetic efficiency	79.27%	74.70%	74.70%	77.69%	78.36%
Gross CC output, MWe	580.10	563.40	563.40	570.71	572.34
Auxiliary load, % of STG	3.59%	7.95%	3.59%	3.59%	3.59%
Heat consumption, MWth	903.8	891.9	891.9	891.9	891.9
Net CC output, MWe	573.0	548.9	556.8	563.9	565.5
Net CC (LHV) efficiency	63.40%	61.54%	62.43%	63.22%	63.40%

turbine (GT) and technology choice. Evidently, the key parameter is the GT exhaust temperature (*TEXH*), which sets the theoretically possible maximum (i.e., the entitlement), and the percentage of that entitlement value that can be converted to net electric power via technically and economically feasible design choices. Thus, based on the premise that there is a well-designed or optimal steam bottoming (Rankine) cycle design for a given GT exhaust gas stream, different technology curves in the form of *f(TEXH)* can be generated. In principle, many technology levels can be selected based on specific assumptions. Such a formula has been developed by Gülen and Smith (Ref. [1] in Chapter 6, see Figure 12.4).

The chart in Figure 12.4 gives a reasonably accurate predictive tool for estimating the bottoming cycle performance for a given GT exhaust temperature. Note that there is a spectrum of bottoming cycle configurations that covers the range of possible GT exhaust temperatures. Temperatures below 538°C (1,000°F) are not enough to support a feasible reheat design because there is not enough energy to heat the steam twice to the same temperature. At the high end, for sufficiently high temperatures, there is a transition from three-pressure to two-pressure because the exhaust gas energy dedicated to HP steam production, superheating, and economizing, squeezes the IP section out. At still higher temperatures, typically 871°C (1,600°F) or higher, the temperature pinch transitions from the LP evaporator to the HRSG economizer inlet. This can be visualized by imagining the exhaust gas cooling line in Figure 6.7 in Chapter 6 as a cantilever beam rotating around the evaporator pinch point as the exhaust gas temperature increases. At a certain value of *TEXH*, the hot and cold heat release lines for the economizer section will become parallel so that further increase will pinch them together at the HRSG inlet. At that point, the LP evaporator is also squeezed out and the system reverts to a single-pressure system.

A curve that represents the family of well-designed bottoming cycles and implicitly contains the preceding design spectrum considerations is also given in Figure 12.4. This optimal curve goes through the data points representing the bottoming cycle exergetic efficiencies extracted from the *GTW 2006 Handbook* data and is adequately described by the following quadratic formula with *TEXH* in degrees °F:

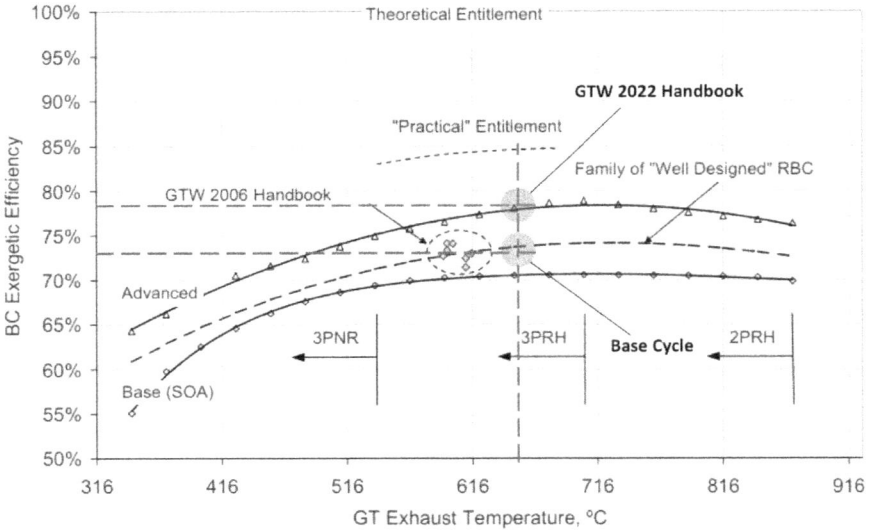

FIGURE 12.4 Gas turbine combined cycle, steam bottoming (Rankine) cycle technology curves. Note that 593°C = 1,100°F; each 100°C increment is 212°F.

$$\varepsilon_{BC} = 0.2441 + 0.0746\,t - 0.00279t^2, \quad t = \frac{\text{TEXH}}{100}. \tag{12.29}$$

Equation 12.29 is representative of a well-designed bottoming cycle based on the technology of early 2000s. For modern power plants (in the third decade of the twenty-first century) with advanced-class gas turbines, a good first estimate is adding four percentage points (or 0.40 as a fraction) to the value calculated using Equation 12.29.

12.4   WHY THE SECOND LAW (EXERGY ANALYSIS)?

As far as the proverbial bottom line are concerned, there is nothing that cannot be accomplished by using only the first law of thermodynamics (see Section 14.4 for an especially enlightening example of this assertion). The technology curves in Figure 12.4 can be reconstructed from a first law perspective to result in bottoming cycle *thermal* efficiency. The power of second law or exergy approach comes from the following two perspectives:

1. It reveals the true potential of individual components. The best example is the condenser, which accounts for a huge energy loss from the bottoming cycle (> 30% of the GT heat consumption in modern 3PRH plants), whereas from a work potential point, it is rather puny (i.e., less than 1% of the GT heat consumption). However, it also reveals that, although a very small player from a total combined cycle system point of view, from a bottoming cycle system perspective, the condenser is essentially the most important lost work component.

2. It provides a readily available yardstick toward the entitlement, (i.e., the theoretically possible maximum performance). From a first law perspective, one can certainly find that for a given GT by setting all losses to zero and all efficiencies to 100%. However, this would require a detailed heat balance calculation, and the final answer is totally uninteresting (e.g., 40%, 45%, etc.). However, the second law provides the answer unambiguously and unassailably by just a single thermodynamic property calculation and sets the 100% effectiveness, an intuitive upper limit or entitlement value. And just by evaluating the steam turbine power output, the designer can immediately assess the current status from an entitlement perspective (both theoretical and practical) and can also judge how difficult or easy the improvement steps toward that goal are.

REFERENCES

1. Horlock, J.H., Combined Cycle Power Plants—Past, Present, and Future, *ASME Journal of Engineering Gas Turbines Power*, 117, pp. 608–616, 1995.
2. Gülen, S.C., and Jacobs III, J.A., *Optimization of Gas Turbine Combined Cycle*, Las Vegas, NV: Power-Gen International, 2003.

13 Peripheral Systems

The environment is everything that isn't me.

—Albert Einstein

Decarbonization is the heart of *energy transition*, which refers to the global energy sector's shift from fossil fuel–based energy production and consumption to renewable energy sources, like wind and solar, supplemented by energy storage. Decarbonization is the reduction of carbon dioxide emissions through the use of different means. Carbon dioxide (CO_2) is a potent *greenhouse gas* (GHG), contributing significantly to global warming and, thus, plays a major role in climate change. In the United States, transportation, industry, and electricity are three economic sectors with roughly equal shares in GHG emissions. By far the dominant mechanism of CO_2 generation is combustion of hydrocarbon (fossil) fuels (i.e., coal; petroleum-derived liquid fuels, such as diesel and gasoline; and natural gas). Thus, as dictated by simple logic, the first and foremost action to be taken is elimination of fossil fuels from the picture. There are several solutions in this respect:

1. Increase the share of carbon-free resources, such as wind, solar, nuclear, and hydro, in the electricity generation portfolio
2. Switch from coal to natural gas to, eventually, hydrogen-manufactured using carbon-free technologies (e.g., *electrolysis* of water powered by wind or solar PV)

The second logical step is to prevent the release of CO_2 into the atmosphere while the fossil fuels are still in the generation portfolio. This is the premise of the technologies under the umbrella term of *carbon capture, utilization, and sequestration* (CCUS). The term *capture* refers to the removal of CO_2 either from the fuel itself (pre-combustion) or from the flue gas (post-combustion). Utilization and sequestration refer to using the captured CO_2 in other economic activities, such as *enhanced oil recovery* (EOR), and its eventual storage in underground reservoirs.

Carbon capture technologies expand the thermal power plant based on heat engines via addition of complex systems, which are bona fide plants on their own. Two such systems are the chemical process plant to strip CO_2 from the stack gas using a solvent and the air separation plant (better known as the *air separation unit*, ASU) to produce oxygen for oxy-combustion of natural gas. These two technologies and electrolysis are covered in this chapter using the second law approach.

DOI: 10.1201/9781003247418-15

13.1 CARBON CAPTURE

To date, post-combustion CO_2 removal from the stack gases via deployment of aqueous amine-based chemical absorption technology is the only commercially available option, which is applicable to new units as well as to the retrofit of the existing plants. The only commercially available absorbents active enough for recovery of dilute CO_2 at low partial pressures are aqueous solutions of *alkanolamines*, such as monoethanolamine (MEA), diethanolamine (DEA), methyl-diethanolamine (MDEA), and the newly developed sterically hindered amines (e.g., piperazine).

As shown in Figure 13.1, the post-combustion capture (PCC) system consists of two main components:

- An *absorber* in which the CO_2 is removed
- A *regenerator (stripper)* in which the CO_2 is released in a concentrated form and the solvent is recovered

Prior to the CO_2 removal, the flue gas (at around 90°C at the heat recovery steam generator (HRSG) stack for the most efficient GTCC power plants) is typically cooled to about 40°C and then treated to reduce particulates that cause operational problems and other impurities, which would otherwise cause costly loss of the solvent (e.g., in a direct contact cooler or quench tower). The amine solvent absorbs the CO_2 (together with traces of NOx) by chemical reaction to form a loosely bound

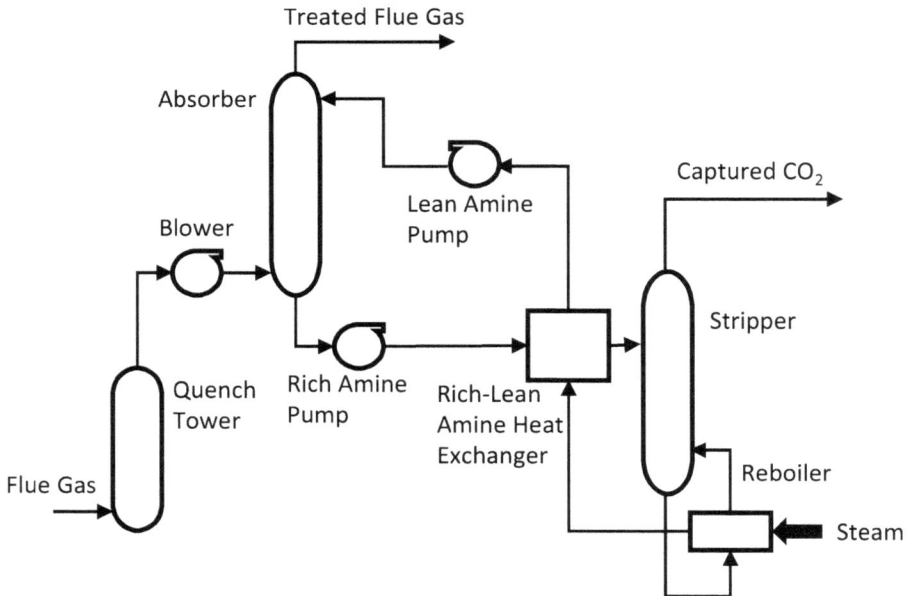

FIGURE 13.1 Highly simplified schematic diagram of CO_2 capture from the power plant flue gas via aqueous amine-based absorption.

compound. A booster fan (blower) is requisite to overcome the pressure loss in the capture plant and is a significant (parasitic) power consumer.

The largest penalty imposed on the power plant output by the PCC system is due to the large amount of heat required to regenerate the solvent. The temperature level for regeneration is normally around 120°C. This heat is typically supplied by steam extracted from the bottoming cycle and reduces steam turbine power output and, consequently, net efficiency of the GTCC significantly. As is the case for all other carbon capture technologies, electrical power is consumed to compress the captured CO_2 for transportation to the storage site and injection into the storage reservoir.

Energy required for the absorbent regeneration in the stripper is supplied by a kettle-type reboiler utilizing the steam extracted from the bottoming cycle of the power block. The duty of the reboiler has three constituents: (1) the heat of reaction for the desorption of the CO_2, (2) the energy required to generate the stripping vapor (i.e., steam), and (3) the sensible heat required to raise the temperature of the incoming rich amine to the stripper operating temperature. Thus,

$$\dot{Q}_{reb} = \dot{Q}_{rx} + \dot{Q}_{str} + \dot{Q}_{sens}. \qquad (13.1)$$

This energy is supplied by the condensing steam (from the power block) in the reboiler; that is,

$$\dot{Q}_{reb} = \dot{m}_{stm} h_{fg} \left(P_{stm} \right), \qquad (13.2)$$

where h_{fg} is the latent heat of condensation of steam at pressure, P_{stm}. For proper evaluation of the terms on the right-hand side of Equation 13.1, one needs to model it in a chemical process simulation software, such as *Aspen Plus* or *ProMax*. Such software packages are extremely expensive, with hefty maintenance fees and steep learning curves. In short, they are only available to those who are employed by private or public R&D organizations, engineering companies, and OEMs. Nevertheless, it is still possible to make reasonably good estimates starting from the first principles using an Excel spreadsheet. By far the best method is based on the *minimum separation work (MSW)* principle, which is described in minute detail with step-by-step examples in the superb monograph by Wilcox [1]. Specifically, the method is based on the evaluation of *partial molal Gibbs free energy* of gas streams in the process. The underlying theory will be introduced in the next section.

13.1.1 APPLICATION EXAMPLE

The minimum thermodynamic work required to separate CO_2 from a gas mixture in an isothermal and isobaric process is equal to the negative of the difference in the Gibbs free energy between the final initial states shown in the simplified capture process diagram in Figure 13.2 (after Figure 1.10 on p. 23 in Wilcox [1]). A rigorous derivation of this statement (mathematically expressed in Equation 2.53) from the first principles is provided in Section 2.3.2.1.

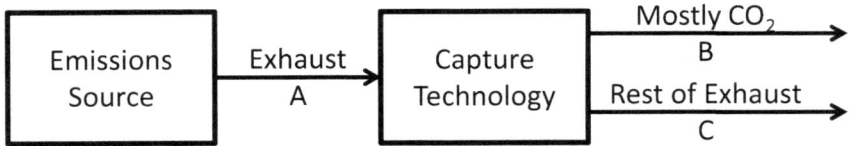

FIGURE 13.2   Carbon capture process inlet/outlet material streams.

Specifically, minimum separation work (*MSW*) can be found from

$$MSW = (G_B + G_C) - G_A \tag{13.3}$$

$$MSW = (\dot{n}_B \mu_B + \dot{n}_C \mu_C) - \dot{n}_A \mu_A \tag{13.4}$$

For stream A, as an example, using Equation 2.65 for each component, the chemical potential can be expressed as

$$\mu_A = \bar{g}_{CO2}^{\circ} + \bar{R}T \ln \frac{y_{CO2,A} P}{P_{ref}} + \sum_j \left(\bar{g}_j^{\circ} + \bar{R}T \ln \frac{y_{j,A} P}{P_{ref}} \right)_{j \neq CO2}. \tag{13.5}$$

For simplicity, assume that $P = P_{ref}$ so that

$$\mu_A = \bar{g}_{CO2}^{\circ} + \bar{R}T \ln y_{CO2,A} + \sum_j \left(\bar{g}_j^{\circ} + \bar{R}T \ln y_{j,A} \right)_{j \neq CO2}. \tag{13.6}$$

Equation 13.6 can be written in a similar way for streams B and C as well. The resulting three equations for three streams A, B, and C can then be substituted into Equation 13.4. It is easy to show that the terms associated with Gibbs free energy of formation will cancel each other out. For example,

$$\dot{n}_{CO2,B} \, \bar{g}_{CO2}^{\circ} + \dot{n}_{CO2,C} \, \bar{g}_{CO2}^{\circ} - \dot{n}_{CO2,A} \, \bar{g}_{CO2}^{\circ} = X,$$

$$\left(\dot{n}_{CO2,B} + \dot{n}_{CO2,C} - \dot{n}_{CO2,A} \right) \bar{g}_{CO2}^{\circ} = X, \text{ but}$$

$$\dot{n}_{CO2,B} + \dot{n}_{CO2,C} = \dot{n}_{CO2,A}, \text{ so that}$$

$$\left(\dot{n}_{CO2,B} + \dot{n}_{CO2,C} - \dot{n}_{CO2,A} \right) \bar{g}_{CO2}^{\circ} = 0. \tag{13.7}$$

Another simplification is bookkeeping of stream components other than CO_2 as a single group instead of individually. For example, Equation 13.6 can be rewritten as

$$\mu_A = \left(\bar{g}_{CO2}^{\circ} + \sum_j \bar{g}_{j \neq CO2}^{\circ} \right) + \bar{R}T \ln y_{CO2,A} + \bar{R}T \ln \left(1 - y_{CO2,A} \right). \quad (13.8)$$

As noted earlier, the terms in parentheses on the RHS of Equation 13.8 will cancel each other out when substituted into Equation 13.4. Consequently, we end up with terms containing only the molar flow rates and mole fractions; that is,

$$MSW = \bar{R}T_{proc} \left[\begin{array}{c} \left(\dot{n}_{CO2,B} \ln y_B + \dot{n}_{CO2,C} \ln y_C - \dot{n}_{CO2,A} \ln y_A \right) + \\ \left(\left(\dot{n}_B - \dot{n}_{CO2,B} \right) \ln \left(1 - y_B \right) + \left(\dot{n}_C - \dot{n}_{CO2,C} \right) \ln \left(1 - y_C \right) - \left(\dot{n}_A - \dot{n}_{CO2,A} \right) \ln \left(1 - y_A \right) \right) \end{array} \right], \quad (13.9)$$

where $\bar{R} = R_{unv} = 8.314$ J/mol-K and T_{proc} is the separation process operating temperature (*not* the flue gas temperature at the source of emission, e.g., HRSG stack). Furthermore, we note that

- $\dot{n}_A$, $\dot{n}_B$ and $\dot{n}_C$ are the total number of moles in streams A, B, and C, respectively
- $\dot{n}_{CO2,A}$ $\dot{n}_{CO2,B}$ and $\dot{n}_{CO2,C}$ are the total number of moles of CO_2 in streams A, B, and C, respectively
- y_A, y_B, and y_C are the mole fractions of CO_2 in streams A, B, and C, respectively

Equation 13.9 is equivalent to Equation 1.11 on p. 24 of Wilcox [1]. While the equation is relatively straightforward, its evaluation with actual data is quite tedious. For convenience, the following Min_Sep_Work is an Excel VBA function calculating the *MSW* as given by Equation 13.9.

```
Function Min _ Sep _ Work(sTemp As Single, sMdot As Single,
sCapRate As Single, sPurity As Single, vComp As Variant)
As Single
        ' 1  'N2
        ' 2  'O2
        ' 3  'H2O
        ' 4  'CO2
        ' 5  'Ar
        ' 6  'SO2
        ' 10 'Air
        ' 7  'CH4
    Dim sGasMolWeight As Single, sMoleFlowRate As Single
    ' A: Exhaust
    ' B: Mostly CO2
    ' C: Rest of Exhaust
    Dim n_A_CO2 As Single, n_B_CO2 As Single, n_C_CO2
        As Single
```

```
Dim n_A_Others As Single, n_B_Others As Single,
  n_C_Others As Single
Dim y_A_CO2 As Single, y_B_CO2 As Single, y_C_CO2
  As Single
Dim y_A_Others As Single, y_B_Others As Single,
  y_C_Others As Single
Dim i As Integer
Dim sTemp_K As Single
Dim G_A As Single, G_B As Single, G_C As Single
Const Rgas As Single = 8.314 'kJ/kmol-K
sTemp_K = (sTemp - 32) / 1.8 + 273.15 'temp. in
  Kelvins
sGasMolWeight = mWgas(vComp, 10) 'lbm/lbmole or kg/
  kmole
sMoleFlowRate = (sMdot * 0.4536) / sGasMolWeight
  'kmole/s
y_A_CO2 = vComp(4)
y_A_Others = 1 - y_A_CO2
n_A_CO2 = y_A_CO2 * sMoleFlowRate
n_A_Others = sMoleFlowRate - n_A_CO2
n_B_CO2 = sCapRate * n_A_CO2
n_B_Others = n_B_CO2 / sPurity - n_B_CO2
y_B_CO2 = n_B_CO2 / (n_B_CO2 + n_B_Others)
y_B_Others = 1 - y_B_CO2
n_C_CO2 = (1 - sCapRate) * n_A_CO2
n_C_Others = sMoleFlowRate - (n_B_CO2 + n_B_Oth-
  ers) - n_C_CO2
y_C_CO2 = n_C_CO2 / (n_C_CO2 + n_C_Others)
y_C_Others = 1 - y_C_CO2
G_A = Rgas * sTemp_K * (n_A_CO2 * Log(y_A_
  CO2) + n_A_Others * Log(y_A_Others))
G_B = Rgas * sTemp_K * (n_B_CO2 * Log(y_B_
  CO2) + n_B_Others * Log(y_B_Others))
G_C = Rgas * sTemp_K * (n_C_CO2 * Log(y_C_
  CO2) + n_C_Others * Log(y_C_Others))
Min_Sep_Work = G_B + G_C - G_A 'in kW
End Function
```

Evaluation of *MSW* as a function of $y_{CO2,A}$ and T_{proc} shows that *MSW* increases as follows:

- As the concentration of CO_2 in stream A decreases
- As the process temperature, T_{proc}, increases

This is a distinct disadvantage for carbon capture from the HRSG stack gas in a GTCC power plant with $y_{CO2,A}$ around 0.04 (4% CO_2 by volume). In coal-fired power plants, boiler stack gas CO_2 concentration is around 0.14 (14% CO_2 by volume).

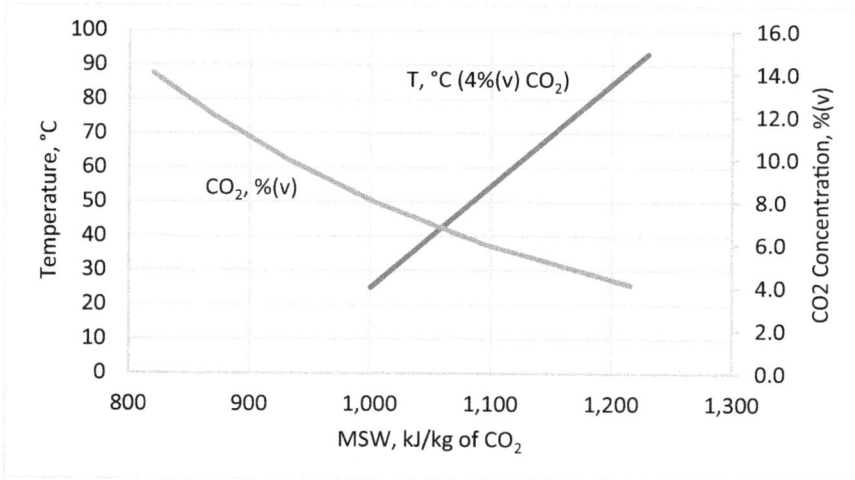

FIGURE 13.3 Minimum separation work (MWE), correlation with CO_2 content of the flue gas and process temperature.

Note that *MSW* is the theoretical minimum value for the capture process. In other words, it is analogous to the Carnot equivalent efficiency of an actual heat engine (e.g., a gas turbine) or the exhaust gas exergy of a gas turbine. As such, it is unachievable in practice. To convert it into a realistic estimate, we need a technology factor This technology factor is the second law (exergetic) efficiency defined by Wilcox on pp. 26–27 in his book and plotted in Figure 1.12 on p. 27 of Ref. [1]. There are only two data points for post-combustion, amine-based CO_2 scrubbing using chemical absorption; one for a coal-fired power plant (low-sulfur, Appalachian bituminous coal); and the other for a natural gas–fired GTCC:

	CO_2, mole fraction	2nd Law Efficiency, %
PCC	0.120	27.4
GTCC	0.038	21.5

The paucity of data is not conducive to establish a firm correlation between flue gas CO_2 concentration and the said technology factor. For a combined cycle power plant, as a first guess, it can be taken as 0.22 so that the *actual separation work (ASW)* is found as

$$ASW = \frac{MSW}{0.22}. \qquad (13.10)$$

To get a feel for the magnitudes involved, consider that, for an F class gas turbine combined cycle, rated at 400 MWe net, the following data is available:

- Flue gas flow rate, 573.4 kg/s (HRSG Stack)
- Flue gas temperature, 89.1°C
- Flue gas MW, 28.33 kg/kmol

- Flue gas molar composition:

  - Nitrogen, 74.13%
  - Oxygen, 11.711%
  - Water vapor, 9.148%
  - Carbon dioxide, 4.118%
  - Argon, 0.893%

For 90% capture and 99.9% CO_2 purity, using Min_Sep_Work (i.e., Equation 13.9) with $T = 89.1°C$ (i.e., flue gas is not cooled), we obtain 8.8 MW for MSW. Using Equation 13.10, actual separation work is found as $8.8/0.22 = 40.5$ MW. Auxiliary power consumption for CO_2 compression and conditioning is 17 to 19 MJ per kmol of captured CO_2. This can be estimated as follows. We start by estimating the isentropic compression work, w_{is}, from

$$w_{is} = N \frac{R_{unv} T_0}{k} \left(PR^{\frac{k}{N}} - 1 \right)$$

(13.11)

where N is the number of compressor casings with intercooling in between, PR is the overall compressor ratio (typically, 150, i.e., compressing CO_2 from roughly 1 bar to 150 bar), and $k = 1 - 1/\gamma$, with γ being the specific heat ratio of CO_2 (1.289). With $N = 3$, Equation 13.11 returns a value of 15 MJ/kmol, which translates into 17–18 MJ per kmol depending on the isentropic efficiency (i.e., compressor technology).

Continuing with the calculation, captured CO_2 molar flow rate is

$$(1264.1 / 28.33) \times 0.04118 \times 0.90 / 2.2046 = 0.75 \text{ kmol} / \text{s}$$

so that total compression power requirement is

$$0.75 \text{kmol} / \text{s} \times 17.5 \text{MJ} / \text{kmol} \approx 13 \text{MW}.$$

Another 2 MJ/kmol should be allocated to amine recirculation and other capture unit pumps; that is, the balance of plant (BOP) of the capture block.

Typically, the PCC plant is connected to the power plant via a flue gas duct, whose total length, cross-sectional area, and layout are site-specific. A booster fan is requisite to push the flue gas and compensate for myriad pressure losses in the system. Without a detailed design at hand, allocation of 10 MJ/kmol for booster fan duty is reasonable. Thus, total BOP and booster fan power requirement of the capture plant is found as

$$0.75 \text{kmol} / \text{s} \times (10 + 2) \text{MJ} / \text{kmol} = 9 \text{MW}.$$

Finally, total capture penalty is estimated as

$$40.5 + 13 + 9 \approx 62.5 \text{MW},$$

which amounts to $62.5/400 = 15.5\%$ of the plant rated output with no capture. Thus, as a front-end estimate, equipping a GTCC power plant with amine-based carbon capture system results in 15–16% reduction in plant net output. Since the fuel consumption of the gas turbines are not affected, this translates into 15–16% *increase* in plant net heat rate or, equivalently, 15–16% (in relative terms) *reduction* in net thermal efficiency. In other words, for a net 60% H class GTCC, carbon capture retrofit brings down the net thermal efficiency to about 51%.

A key number to be used in amine-based carbon capture calculations is the stripper reboiler energy requirement, which determines the extraction steam flow from the bottoming cycle and, as a result, lost steam turbine power output. Based on the numbers used for this example, captured CO_2 flow rate is

$$0.75 \text{kmol} / \text{s} \times 44 \text{kg} / \text{kmol} = 33 \text{kg} / \text{s}.$$

Actual separation work is calculated as 40.5 MW, which is the lost steam turbine work. Typical conversion efficiency is 35–36%. This means that the net energy input to the reboiler (the left-hand side of Equation 13.1) is

$$\dot{Q}_{reb} = 40.5 / .35 = 115.7 \text{MW th}.$$

Using the saturated vapor/steam and liquid enthalpies at 3.5 bara to calculate h_{fg}, requisite steam flow is found from Equation 13.2 as

$$115,700 / (2731.5 - 582) = 53.8 \text{kg} / \text{s}.$$

Furthermore, heat of desorption can be found as

$$115,700 \text{ kWth} / 33 \text{ kg} / \text{s} = 3,500 \text{ kJ per kg of captured } CO_2 \text{ (about } 1,500 \text{ Btu} / \text{lb)}.$$

This number is indeed about right for 30% (by weight) MEA solvent.

13.2   AIR SEPARATION UNIT

Cryogenic air separation is an established technology to manufacture liquid or gaseous oxygen and nitrogen (and argon) for industrial use. While there are other technologies to separate O_2 and N_2 from air, the multi-column cryogenic distillation process is presently the most mature and cost-effective technology. In power generation, oxygen is used in gasifiers producing the syngas (e.g., the IGCC power plant) or as the oxidizer in oxy-combustion (e.g., the supercritical CO_2 oxy-combustion Allam cycle). In the IGCC power plant, N_2 produced by the *air separation unit* (ASU) is used as a diluent in the gas turbine combustor.

A cryogenic ASU comprises five key processes:

- Air compression (intercooled, centrifugal process compressor with aftercooler)
- Air pretreatment or purification (to remove process contaminants, e.g., water, carbon dioxide, and hydrocarbons)

- Air cooling to cryogenic temperatures (about 200°C below the ambient):
  - Expansion through turbines and/or valves
  - Heat exchange between product streams and feed air (to cool air to cryogenic temperatures)
- Distillation (separation into N_2, O_2, and Ar)

These five processes can be combined in several different configurations to generate the three product streams in different quantities and purities (e.g., 95% pure O_2). The final selection depends on a trade-off between product requirements, capital cost, and parasitic power consumption of the ASU compressors. Another factor is the degree of integration between the ASU and the process and/or power blocks (e.g., utilization of N_2 as a diluent in the gas turbine combustor). There are three types of ASU processes (referred to as *cycles*) distinguished by the method of pressurizing the product streams or the air feed pressure [2]:

- Compression cycle
- Pumped liquid cycle
- Low and elevated pressure cycle

In the compression cycle, low pressure (subatmospheric) gaseous O_2 from the heat exchanger is compressed in a centrifugal compressor to the desired product pressure. Product O_2 can be in the liquid phase but at the expense of higher compression power required by the distillation system. This requires further optimization of system to for heat recovery between refrigerated and vaporizing streams in the heat exchangers. Liquid O_2 can either be pumped to an intermediate pressure lower than the final product pressure or to the full product pressure at the ASU battery limits.

In low-pressure (LP) ASU cycle, feed air pressure is between 5 and 6 bar (absolute). Nitrogen is rejected as a byproduct at atmospheric pressure from the system. In elevated pressure (EP) ASU cycle, N_2 is produced at higher pressures, which is the case when it is used in the accompanying power or process block. Thus, in EP cycles, feed air is compressed to pressures in excess of 6 bar.

In a supercritical CO_2 (SCO2) cycle with oxy-combustion (for high-efficiency power generation with inherent CO_2 capture ability), high-purity (at least 99.5%) O_2 is needed at a significantly elevated pressure; for example, around 300 bar or slightly higher (see Section 11.2). Producing O_2 at such a high purity and high pressure and at large quantities (e.g., 140 t/h for a nominal 300 MW plant, which comes to about 3,350 t/d) is a significant challenge. An ASU comprising a two-column system adopted from one described by Cockett and Smith [3] with an argon recovery column as a side column of the upper column is shown in Figure 13.4 (see Figure 6 on p. 182 of the cited work).

In this double-column Linde process, cooled and partially liquefied air is introduced at the bottom of the lower column. The condenser of the higher-pressure lower column drives the reboiler of the lower pressure upper column, which allows the condensation of N_2 in the lower column. (In the model depicted in Figure 13.4, the condenser and reboiler are linked via an energy stream. In reality, it is a combination of the two as a single piece of equipment known as the *vaporizer*.) Nitrogen could

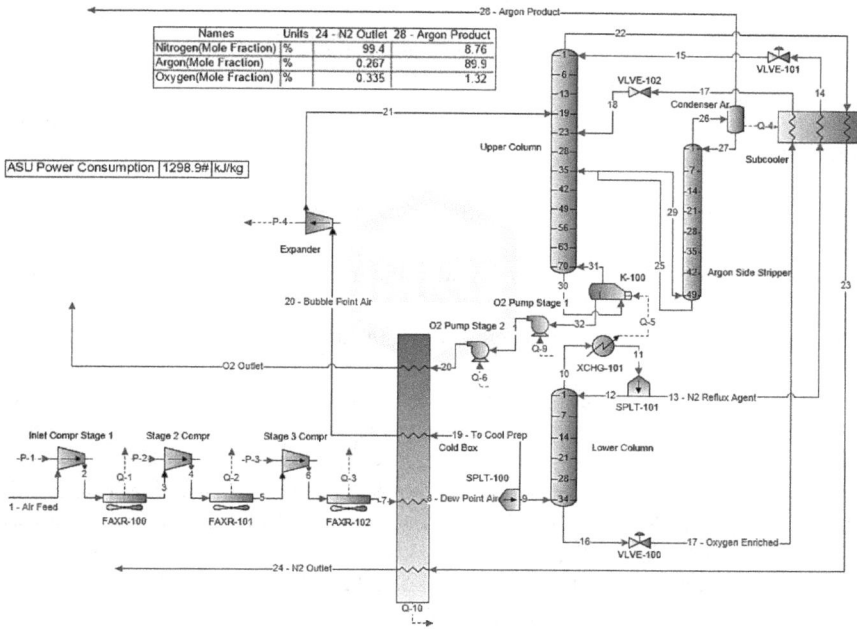

Names	Units	24 - N2 Outlet	28 - Argon Product
Nitrogen(Mole Fraction)	%	99.4	8.76
Argon(Mole Fraction)	%	0.267	89.9
Oxygen(Mole Fraction)	%	0.335	1.32

FIGURE 13.4 Double-column cryogenic ASU based on the Linde process [3].

not be condensed by cooling against liquid O_2 in the upper column reboiler at equal pressures since the normal boiling point of N_2 is lower than that of O_2. The upper column operates at 1.7 bar, while the lower column operates at 6.21 bar.

Part of the liquid N_2 from the condenser-reboiler falls down to the lower column as reflux. The remainder is sent to the top of the upper column through an expansion or let-down valve. The stream from the bottom of the lower, containing about 37% O_2, is also fed to the upper column through a let-down valve. The upper column is the heart of the process with a composition variation from its top to its bottom so that relatively pure N_2 and O_2 product streams are drawn from its top and bottom ends, respectively. The *argon side column* (ASC) configuration allows for higher purity and higher recovery of an O_2 product, as well as a N_2 product and an Ar product. For a more detailed description of the cryogenic separation process in Figure 13.4, the reader is referred to article by Chen [4].

For systems such as SCO2 oxy-combustion or IGCC power plants, the most important ASU parameter is the parasitic power consumption of the unit. This can be precisely determined from a rigorous system model using a process flowsheet simulation tool such as ProMax [5]. However, a reasonably accurate estimate can also be made starting from the first principles similar to what is done in Section 13.1. Once again, we can conceive the ASU as a van't Hoff box, as shown in Figure 13.5. Consequently, the minimum separation work can be found as the change in the Gibbs free energy of product and inlet air streams. The resulting equation is analogous to Equation 13.9 in Section 13.1, containing only the molar flow rates and mole fractions; that is,

$$\text{MSW} = \bar{R}T_{proc} \begin{bmatrix} \left(\dot{n}_{O2,B} \ln y_B + \dot{n}_{O2,C} \ln y_C - \dot{n}_{O2,A} \ln y_A\right) + \\ \left(\left(\dot{n}_B - \dot{n}_{O2,B}\right)\ln\left(1-y_B\right) + \left(\dot{n}_C - \dot{n}_{O2,C}\right)\ln\left(1-y_C\right) - \left(\dot{n}_A - \dot{n}_{O2,A}\right)\ln\left(1-y_A\right)\right) \end{bmatrix} (10.13)$$

where $\bar{R} = R_{unv} = 8.314$ J/mol-K and T_{proc} is the separation process operating temperature. Furthermore,

- $\dot{n}_A$, $\dot{n}_B$ and $\dot{n}_C$ are the total number of moles in streams A, B, and C, respectively
- $\dot{n}_{O2,A}$ $\dot{n}_{O2,B}$ and $\dot{n}_{O2,C}$ are the total number of moles of O_2 in streams A, B, and C, respectively
- y_A, y_B, and y_C are the mole fractions of O_2 in streams A, B, and C, respectively

Let us assume that the inlet air composition is given as 78.11%(v), 20.96%(v), and 0.93%(v), for nitrogen, oxygen, and argon, respectively. Molecular weight is calculated as 28.95 kg/kmol (thus, gas constant, $R = 259.81$ J/kg-K). For unit flow rate of air (stream A) in kg/s, one can calculate that the normal volume flow rate is 2,787 Nm³/h ($T_{ref} = 273.15$ K, $P_{ref} = 101,325$ Pa), and its molar flow rate is 0.0345 kmol with 0.027 kmol/s, 0.0072 kmol/s, and 0.0003 kmol/s for N_2, O_2, and Ar, respectively. Corresponding mass flow rates are 0.7555 kg/s, 0.2317 kg/s, and 0.0128 kg/s, respectively. The following assumptions are made for the separation process:

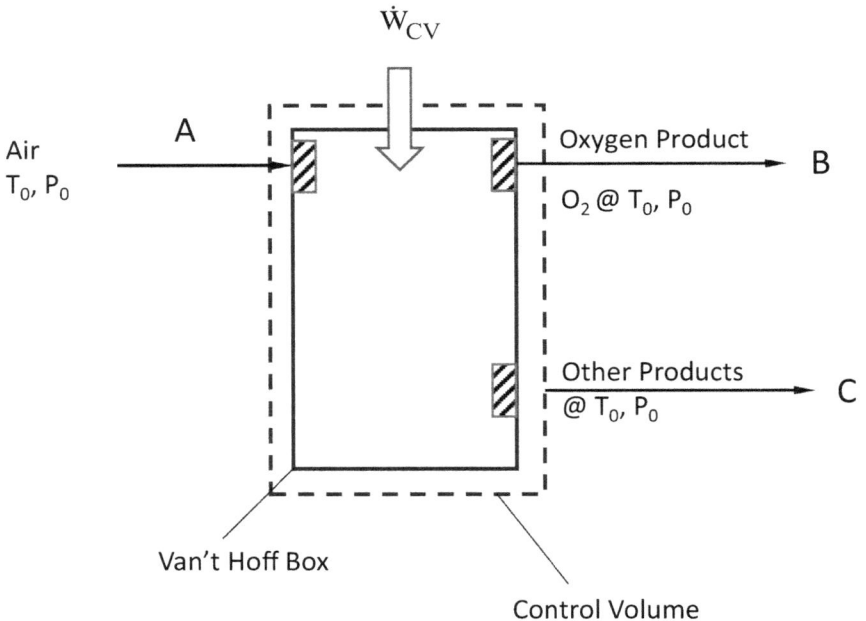

FIGURE 13.5   ASU representation as van't Hoff box.

- $T_0 = 288.15$ K, $P_0 = 101{,}325$ Pa
- Recovery factor is 90%
- O_2 purity 99.5%(v) (with 0.5%(v) Ar)

Based on these assumptions and the input data, the following calculations can be made:

- $0.9 \times 0.2317 = 0.2085$ kg/s O_2 in stream B
- $0.995 \times 32 + 0.005 \times 40 = 32.04$ kg/kmol MW of stream B
- $0.2085/32 = 0.0065$ kmol/s molar flow rate of O_2 in stream B
- $(0.0065 \times 32.04 - 0.2085)/(40 - 32.04) = 0.00003$ kmol/s molar flow rate of Ar in stream B
- $0.00003 \times 40 = 0.0013$ kg/s of Ar in stream B
- $0.2085 + 0.0013 = 0.2098$ kg/s total stream B
- $0.2098/32.04 = 0.00655$ kmol/s total stream B
- $1.0000 - 0.2098 = 0.7902$ kg/s total stream C
- $0.2317 - 0.2085 = 0.0232$ kg/s O_2 in stream C
- $0.0128 - 0.0013 = 0.0115$ kg/s Ar in stream C
- $0.7555/28 = 0.02698$ kmol/s N_2 in stream C
- $0.0232/32 = 0.00072$ kmol/s O_2 in stream C
- $0.0115/40 = 0.00029$ kmol/s Ar in stream C
- $0.02698 + 0.00072 + 0.00029 = 0.02799$ kmol/s total stream C
- $0.02698/0.02799 = 0.96383$ mole fraction of N_2 in stream C
- $0.00072/0.02799 = 0.02586$ mole fraction of O_2 in stream C
- $0.00029/0.02799 = 0.01031$ mole fraction of Ar in stream C

Substituting the values calculated earlier into Equation 13.12, *MSW* is calculated as 33.9 kW or 33.9/0.2085 = 162.8 kJ/kg of separated O_2. A typical technology factor for the separation process is 0.2 [4] so that the *actual* separation work, *ASW* = 162.8/0.2 = 813.9 kJ/kg of separated O_2. For an ideal recovery factor (*RF*) of unity, *MSW* and *ASW* are 180.5 (41.8 kW) and 902.5 kJ/kg, respectively. Normal volume flow rate of O_2 is calculated as 584 Nm³/h so that *ASW* = 41.8/584/0.2 = 0.3576 kWh/Nm³.

Typical separation process requires compression to about 6 bar, which should also be accounted for in the final number for ASU parasitic power consumption. The ideal compression work is found as (for $\gamma = 1.4$ and $k = 0.2857$)

$$1.005 \text{kJ}/\text{kg} - \text{K} \times 288.15\text{K} \times \left(6^\wedge 0.2857 - 1\right) = 193.6 \text{kJ}/\text{kg}.$$

Actual compression work is 50% higher [4] so that we have $1.5 \times 193.6 = 290.4$ kJ/kg. Thus, for the ASU with *RF* = 1.0 and O_2 purity of 99.5%, the total estimated specific power requirement is 902.5 + 290.4 = 1,192.9 kJ/kg. This is quite close to the number calculated using a rigorous process simulation model, about 1,300 kJ/kg.

13.3 ELECTROLYSIS

Hydrogen (H_2) is expected to play a major role in decarbonization of the electricity sector. While combustion of H_2 does not generate CO_2 emissions, its production must also be carbon-free to fulfill its promise as a green fuel. Hydrogen production via electrolysis of water is a technology that fits the bill if the electrolyzer is powered by carbon-free resources, such as wind, solar PV, or nuclear.

Electrolysis is a technique that uses direct electric current (DC) to drive an otherwise non-spontaneous chemical reaction. The electrolysis of one mole of water produces one mole of hydrogen gas and one-half mole of oxygen gas in their normal diatomic forms. The reaction formula is as follows:

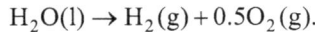

$$H_2O(l) \rightarrow H_2(g) + 0.5O_2(g).$$

If the process takes place at *standard temperature and pressure* (STP) – that is, 298.15 K (25°C) and 1 atmosphere pressure (1.01325 bar) – the process and the pertinent thermodynamic data for the chemical reaction is as shown in Figure 13.6.

The reactions taking place at the *anode* (oxidation) and *cathode* (reduction) at STP are as follows:

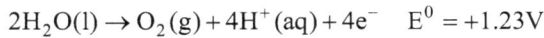

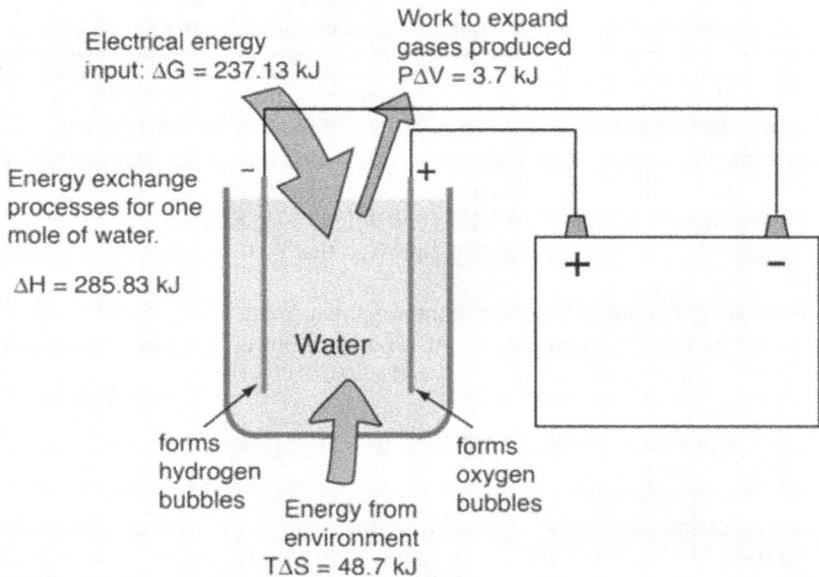

$$2H_2O(l) \rightarrow O_2(g) + 4H^+(aq) + 4e^- \quad E^0 = +1.23V$$

FIGURE 13.6 Schematic description of the simplified water electrolyzer.

$$2H^+(aq) + 2e- \rightarrow H_2(g) \quad E^0 = 0.00V$$

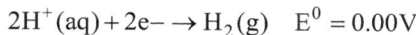

Consequently, for the electrolysis cell, one can write the following equality:

$$\text{cell } E^0 = \text{athode } E^0 - \text{anode } E^0 = -1.23V.$$

The *Gibbs free energy* for the electrolysis reaction is given by

$$\Delta G^0 = -n \cdot F \cdot E^0$$

where the *Faraday constant*, $F = 96{,}485.3321233°C/mol$, and $n = 4$ (i.e., two H_2 molecules formed from two H_2O molecules). Thus, $\Delta G^0 = 4 \times 96.485 \times 1.23 = 474.48$ kJ per 2 moles of water or 237.24 kJ/mol of water. The enthalpy of water at STP is −285.83 kJ/mol so that the enthalpy change of the reaction is $\Delta H^0 = 285.83$ kJ/mol. (Note that, by definition, the enthalpies of H_2 and O_2 at STP are 0 kJ/mol.) From the definition of the Gibbs free energy

$$\Delta G^0 = \Delta H^0 - T\Delta S^0$$

$$237.24 = 285.83 - T\Delta S^0$$

$$T\Delta S^0 = 285.83 - 237.24 = 48.59 kJ / mol.$$

Note that the entropy of water, H_2, and O_2 at STP is 69.91 J/K-mol, 130.68 J/K-mol, and 205.14 J/K-mol, respectively. Thus, the entropy change for the electrolysis reaction is

$$\Delta S^0 = 130.68 + 205.14/2 - 69.91 = 163.64 J / K - mol$$

so that

$$T\Delta S^0 = 298.15K \times 163.64 J / K - mol / 1{,}000 = 48.7 kJ / mol.$$

In the electrolysis process shown in Figure 13.6, two things happen: (1) dissociation of H_2O and (2) expansion of gaseous H_2 and O_2. The electrolysis does not happen on its own; that is, it is not a spontaneous reaction. In order for it to take place, energy input is required. This energy input translates into a change in the enthalpy of the system. Using the fundamental thermodynamic relationship $H = U + PV$, the change in the enthalpy can be written as (ignoring the superscript 0 for simplicity)

$$\Delta H = \Delta U + \Delta PV + P\Delta V = \Delta U + P\Delta V.$$

(Since the process takes place at constant pressure, $\Delta P = 0$.) The second term on the right-hand side of the equation earlier is the work done by the system; that is,

$$W = P\Delta V = 101,325\text{Pa} \times 1.5\text{mole} \times 22.4 \cdot 10^{-3}\,\text{m}^3 \, / \, \text{mole} \times (298.15\text{K} \, / \, 273.15\text{K})$$
$$= 3,715\text{J} = 3.715\text{kJ}.$$

(Note that the electrolysis process generates 1 mole H_2 and 0.5 mole O_2 gas, i.e., a total of 1.5 mole of gas from 1 mole of H_2O. At 273.15 K and 1 atm, 1 mole of gas occupies a volume of 22.4 liter.) Consequently, the change in the internal energy of the system is found from

$$\Delta U = \Delta H - P\Delta V = 285.83\text{kJ} - 3.72\text{kJ} = 282.1\text{kJ}\left(\text{per mol of } H_2O\right).$$

In conclusion, the change in enthalpy represents the necessary energy to accomplish the electrolysis It is equal to the change in the internal energy of the system *after* subtracting the work done during the expansion of the gases produced. Where does this energy (i.e., $\Delta H = 285.83$ kJ/mol) come from? As shown in Figure 13.6, it comes in two forms: electrical energy and heat. The latter, $\Delta Q = T\Delta S = 48.7$ kJ/mol, is provided from the environment at temperature $T = 298.15$ K. The remainder (i.e., the electrical energy) is supplied by the battery, and it is equal to the change in the Gibbs free energy:

$$\Delta G = \Delta H - T\Delta S = 285.83 - 48.7 = 237.1\text{kJ} \, / \, \text{mol}.$$

This is the minimum theoretical energy requisite for the electrolysis of water to take place. Thus, 237.1 MJ is required to produce 1 kmol (i.e., 2 kg of H_2) from 1 kmol (i.e., 18 kg of H_2O). Alternatively, 118.55 MW power is required to produced H_2 at a rate of 1 kg/s, which translates into 118.55/3.6 = 32.93 ~ 33 kWh/kg of H_2. This is the theoretical yardstick that one must use to gauge the claims made by electrolyzer manufacturers. For instance, claiming to have an electrolyzer that can electrolyze water to make hydrogen with 33 kWh/kg energy consumption is akin to having built a heat engine with the efficiency of a Carnot engine (i.e., an impossibility). In terms of heat engines (e.g., the advanced-class gas turbines), the state of the art in terms of a technology factor is about 0.70 to 0.75. In other words, at the present stage of human technological development, we can make real heat engines, whose thermal efficiency is 70 to 75% of that achievable in a theoretical Carnot engine operating between the same temperature reservoirs. Thus, what one should expect from a mature electrolyzer technology is at best 33/0.75 = 44 kWh/kg. A more likely number is to be around 0.33/0.7 = 47 kWh/kg.

REFERENCES

1. Wilcox, J., *Carbon Capture*, New York: Springer, 2012.
2. Smith, A.R., and Klosek, J., A Review of Air Separation Technologies and their Integration with Energy Conversion Processes, *Fuel Processing Technology*, 70, pp. 115–134, 2001.

3. Cockett, A.H., and Smith, K.C., *The Chemistry of The Monatomic Gases*, New York: Pergamon Press, 1975.
4. Chen, G., Calculate the Power of Cryogenic Air Separation Units, *Chemical Engineering Progress*, pp. 26–33, 2020.
5. *ProMax Version 5.0*, Bryan, TX: Bryan Research & Engineering, LLC.

14 Supplementary Material

To a great mind, nothing is little.
—Sherlock Holmes; Sir Arthur Conan Doyle (1887), *A Study in Scarlet*

This is the section typically referred to as the appendix. By definition, an appendix contains *supplementary material* that is not an essential part of the main body of the text itself but which may be helpful in providing a more comprehensive understanding of the subject matter, or it is information that is too cumbersome to be included in the body of the paper. Strictly speaking, the text (i.e., the previous chapters of the book, in this case) must be able to stand alone without the supplementary material; that is, it must contain all information including tables, figures, and results necessary to understand the subject matter. Style manuals typically recommend that the information in the appendix is *non-essential*. In other words, if it were removed, the reader would still be able to comprehend the significance, validity, and implications of what the author is trying to convey in the main body of the work.

In this case, the author decided to call this chapter "Supplementary Material" instead of "Appendix" with the assumption that it is intended to complete and enhance the material covered in the preceding chapters. Put another way, in this author's opinion, the material presented in this chapter is *essential*. They are combined into a standalone section at the end in order to make the narrative in the main body of the book flow more smoothly.

14.1 PROPERTY CALCULATIONS

14.1.1 IDEAL GAS PROPERTIES

Ideal gas properties of several key compounds and stable elements are provided herein. The data is taken from the appendix of the textbook by Moran and Shapiro (Ref. [1] in Chapter 2). The data is provided in tabular form (see Tables 14.1, 14.2, 14.3, 14.4, 14.5, 14.6, and 14.7). The original data covers a range of 260 K to 1,700 K (1,427°C) except H, which covers a range of 0 K to 3,250 K. Zero-enthalpy and entropy reference is 0 K. Herein, interpolated values for the common reference temperatures, 15°C and 25°C, have been added to the tables for convenience. Furthermore, an extrapolated value for 1,600°C is also added to the tables (except H).

The data in the tables can be readily represented by curve-fits, as shown in Figure 14.1, for oxygen (O_2), using the data in Table 14.1. For Excel applications, it is probably better to implement the table as is and use simple interpolation or extrapolation to find the enthalpy/entropy values for given temperatures.

For simple combustion calculations, air can be represented as a mixture of O_2 (21% by volume) and N_2 (79% by volume). Equilibrium products of stoichiometric

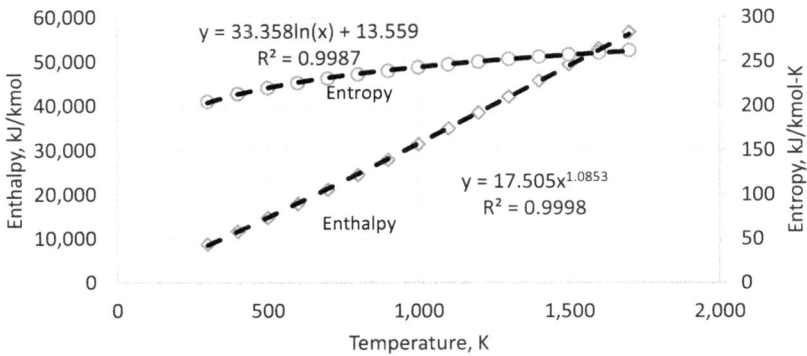

FIGURE 14.1 Ideal gas enthalpy and entropy of O_2.

TABLE 14.1

Ideal Gas Enthalpy and Entropy (at 1 atm) of O_2

T, K	$\bar{h}$, kJ/kmol	$\bar{s}^\circ$ kJ/kmol/K
288.15	8,383	204.200
298.15	8,681	205.055
300	8,736	205.213
400	11,711	213.765
500	14,770	220.589
600	17,929	226.346
700	21,184	231.358
800	24,523	235.81
900	27,928	239.823
1,000	31,389	243.471
1,100	34,899	246.818
1,200	38,447	249.906
1,300	42,033	252.776
1,400	45,648	255.454
1,500	49,292	257.965
1,600	52,961	260.333
1,700	56,652	262.571
1,873.15	63,020	266.580

combustion reaction are H_2O, CO_2, and N_2 (from air, does not react with O_2). In case of combustion with excess air, the products will also include unreacted O_2.

Ideal gas entropy of CH_4 can be calculated using its specific heat, which is represented by the following polynomial (in kJ/kmol-K) from Moran and Shapiro (Ref. [1] in Chapter 2, Table A-12, valid from 300 to 1,000 K):

$$\frac{\overline{c}_p}{R_{unv}} = \alpha + \beta T + \gamma T^2 + \delta T^3 + \varepsilon T^4,$$

(14.1)

TABLE 14.2
Ideal Gas Enthalpy and Entropy (at 1 atm) of N_2

T, K	$\bar{h}$ kJ/kmol	$\bar{s}^\circ$ kJ/kmol/K
288.15	8,377	190.688
298.15	8,669	191.527
300	8,723	191.682
400	11,640	200.071
500	14,581	206.63
600	17,563	212.066
700	20,604	216.756
800	23,714	220.907
900	26,890	224.647
1,000	30,129	228.057
1,100	33,426	231.199
1,200	36,777	234.115
1,300	40,170	236.831
1,400	43,605	239.375
1,500	47,073	241.768
1,600	50,571	244.028
1,700	54,099	246.166
1,873.15	60,177	249.994

TABLE 14.3
Ideal Gas Enthalpy and Entropy (at 1 atm) of H_2O

T, K	$\bar{h}$ kJ/kmol	$\bar{s}^\circ$ kJ/kmol/K
288.15	9,564	187.773
298.15	9,903	188.748
300	9,966	188.928
400	13,356	198.673
500	16,828	206.413
600	20,402	212.92
700	24,088	218.61
800	27,896	223.693
900	31,828	228.321
1,000	35,882	232.597
1,100	40,071	236.584
1,200	44,380	240.433
1,300	48,807	243.877
1,400	53,351	247.241
1,500	57,999	250.45
1,600	62,748	253.513
1,700	67,589	256.45
1,873.15	75,876	261.666

TABLE 14.4
Ideal Gas Enthalpy and Entropy (at 1 atm) of CO_2

T, K	$\bar{h}$ kJ/kmol	$\bar{s}^{\circ}$ kJ/kmol/K
288.15	8,964	212.575
298.15	9,358	213.706
300	9,431	213.915
400	13,372	225.225
500	17,678	234.814
600	22,280	243.199
700	27,125	250.663
800	32,179	257.408
900	37,405	263.559
1,000	42,769	269.215
1,100	48,258	274.445
1,200	53,848	279.307
1,300	59,522	283.847
1,400	65,271	288.106
1,500	71,078	292.114
1,600	76,944	295.901
1,700	82,856	299.482
1,873.15	93,045	305.895

TABLE 14.5
Ideal Gas Enthalpy and Entropy (at 1 atm) of CO

T, K	$\bar{h}$ kJ/kmol	$\bar{s}^{\circ}$ kJ/kmol/K
288.15	8,377	196.727
298.15	8,669	197.568
300	8,723	197.723
400	11,644	206.125
500	14,600	212.719
600	17,611	218.204
700	20,690	222.953
800	23,844	227.162
900	27,066	230.957
1,000	30,355	234.421
1,100	33,702	237.609
1,200	37,095	240.663
1,300	40,534	243.316
1,400	44,007	245.889
1,500	47,517	248.312
1,600	51,053	250.592
1,700	54,609	252.751
1,873.15	60,746	256.614

TABLE 14.6
Ideal Gas Enthalpy and Entropy (at 1 atm) of H

T, K	$\bar{h}$ kJ/kmol	$\bar{s}°$ kJ/kmol/K
288.15	8,178	129.764
298.15	8,468	130.599
300	8,522	130.754
400	11,426	139.106
500	14,350	145.628
600	17,280	150.968
700	20,219	155.204
800	23,171	159.44
900	26,149	162.777
1,000	29,154	166.114
1,100	32,191	168.898
1,200	35,262	171.682
1,300	38,376	174.096
1,400	41,530	176.51
1,500	44,738	178.665
1,600	47,990	180.82
1,700	51,305	182.772
1,800	54,618	184.724
1,900	57,990	186.5105
2,000	61,400	188.297

TABLE 14.7
Enthalpy and Gibbs Free Energy of Formation and Absolute Entropy for Several Substances (298.15 K, 1 atm)

	$h_f°$ kJ/kmol	$g_f°$ kJ/kmol	$s_f°$ kJ/kmol-K
CH_4	−74,850	−50,790	186.16
H_2O (g)	−241,820	−228,590	188.72
H_2O (l)	−285,830	−237,180	69.95
CO_2	−393,520	−394,380	213.69
CO	−110,530	−137,150	197.54

where T is temperature in K, R_{unv} = 8.314 kJ/kmol-K, and the coefficients of the polynomial are as follows:

$$\alpha = 3.826, \beta = -3.979E - 03, \gamma = 24.558E - 06, \delta = -22.733E - 09, \varepsilon = 6.963E - 12$$

The temperature-dependent part of the ideal gas entropy is given by

$$\bar{s}°(T, T_{ref}) = \int_{T_{ref}}^{T} \frac{\bar{c}_p}{T} dT, \qquad (14.2)$$

which, substituting the polynomial function for the specific heat, becomes

$$\frac{\overline{s}^{\circ}(T,T_{ref})}{R_{unv}} = \int_{T_{ref}}^{T} \frac{\alpha + \beta T + \gamma T^2 + \delta T^3 + \epsilon T^4}{T} dT, \tag{14.3}$$

$$\frac{\overline{s}^{\circ}(T,T_{ref})}{R_{unv}} = \int_{T_{ref}}^{T} \left(\frac{\alpha}{T} + \beta + \gamma T + \delta T^2 + \epsilon T^3\right) dT, \tag{14.4}$$

so that, with $T_{ref} = 0$ K and $\overline{s}^{\circ} = 0$ at T_{ref}, we end up with

$$\frac{\overline{s}^{\circ}(T)}{R_{unv}} = \alpha \ln T + \beta T + \gamma \frac{T^2}{2} + \delta \frac{T^3}{3} + \epsilon \frac{T^4}{4}. \tag{14.5}$$

Finally, for the ideal gas enthalpy, skipping the intermediate steps, we have

$$\overline{h}(T,T_{ref}) = \int_{T_{ref}}^{T} \overline{c}_p dT, \tag{14.6}$$

$$\frac{\overline{h}(T)}{R_{unv}} = \alpha T + \beta \frac{T^2}{2} + \gamma \frac{T^3}{3} + \delta \frac{T^4}{4} + \epsilon \frac{T^5}{5}. \tag{14.7}$$

Examples of specific heat and ideal gas entropy calculations are provided in Tables 14.8 and 14.9.

The data herein is provided only for convenience; that is, to facilitate the example calculations in the book. Much more extensive data can be found in the appendices of Moran and Shapiro (Ref. [1] of Chapter 2), Bejan et al. (Ref. [4] in Chapter 2), and

---

TABLE 14.8
Calculation of CH_4 Specific Heat for $T = 499.8$ K

	Polynomial Coefficients	Temperature Contribution	Polynomial Terms
a	3.826		3.826
β	−3.979E-03	499.8	−1.989
γ	2.4558E-05	2.50E+05	6.135
δ	−2.2733E-08	1.25E+08	−2.838
ϵ	6.963E-12	6.24E+10	0.434
			5.568
$\dfrac{\overline{c}_p}{R_{unv}}$			
			46.29
$\overline{c}_p$ kJ/kmol-K			

TABLE 14.9
Calculation of CH_4 Ideal Gas Entropy for T = 499.8 K

	Polynomial Coefficients	Temperature Contribution	Polynomial Terms
a	3.826	6.214	23.776
β	−3.979E-03	499.8	−1.989
γ	2.4558E-05	1.25E+05	3.067
δ	−2.2733E-08	4.16E+07	−0.946
ε	6.963E-12	1.56E+10	0.109
			24.017
$\dfrac{\overline{s}^{\,\circ}}{R_{unv}}$			
$\overline{s}^{\,\circ}$, kJ/kmol-K			199.67

Kotas (Ref. [1] in Chapter 1). These days, pretty much all sorts of property information can be found on the internet just by googling. In any event, most practical calculations require commercial software tools with their own extensive property package (especially, chemical process flowsheet simulation software). However, such software products have extremely hefty license fees, which render them inaccessible to most individuals, who can have access to them through their organizations (in some cases).

14.1.2 SIMPLE AND USEFUL FORMULAE

For quick estimates, one clearly needs an easy-to-use transfer function to calculate exhaust gas enthalpy and exergy as a function of exhaust gas temperature. Since entropy is needed for exergy calculation, with a transfer function for exergy, a separate transfer function for entropy is not necessary. For combustion with natural gas (mostly methane), composition effect can be ignored with no appreciable loss in fidelity. The transfer functions presented subsequently were originally developed using the US Customary System. In order to prevent united errors and typos, they are provided in their original form. Conversion to the US System can easily be done by the reader, particularly the following:

- 1 Btu/lb is 2.326 kJ/kg
- 1 Btu/lb-R is 4.1868 kJ/kg-K
- Conversion from °C to °F is via 1.8 × °C + 32

Note that gas turbine combustion products typically contain about 11–12% H_2O vapor by volume (natural gas fuel with requisite excess air – see the chapter on combustion). At 59°F (25°C) and 14.7 psia (1 atm) – that is, standard ISO reference conditions – part of the H_2O condenses. This is so because the partial pressure of 11–12%(v) H_2O in the gas mixture is about 1.5 psi. At that pressure, saturation temperature of H_2O is about 115°F (46°C), whereas 59°F (25°C) corresponds to a saturation pressure of 0.25 psi (3.626 bar). Therefore, cooling the gas below 115°F until it reaches 59°F will knock out H_2O from the mixture via condensation until its content reduces to 0.25/14.7 ~ 1.7% by volume. Accounting for the presence of liquid H_2O (or not) makes a difference in exergy calculation.

Since the objective in evaluating exergy of gas turbine exhaust gas is to determine the theoretically possible maximum work output from a bottoming cycle making use of the said exhaust gas, we typically ignore the contribution of latent heat of condensation of water vapor. This is the same principle at work when choosing LHV of gas turbine fuel over HHV; it is not cost-effective to make use of that extra heat content. The difference in calculated exergy is about 10% (i.e., same as HHV − LHV difference for natural gas).

With that in mind, the exergy of gas turbine exhaust gas burning natural gas fuel, to a good approximation, is given by

$$a_{exh}[Btu/lb] = 0.001628 \cdot T_{exh}^{1.60877}\ [in\ °F], \qquad (14.8)$$

with a 0 Btu/lb enthalpy reference of 59°F and H_2O in the mixture in gaseous form. Zero-entropy reference is defined as 59°F and 14.7 psia (since ideal gas entropy is a function of temperature *and* pressure). Identical results can be obtained with the following simple linear correlation:

$$a_{exh} = 0.1858 \cdot T_{exh}\ [in\ °F] - 76.951\ Btu/lb \qquad (14.9)$$

Similarly, the enthalpy of the exhaust gas is given by

$$h_{exh} = 0.3003 \cdot T_{exh}\ [in\ °F] - 55.576\ Btu/lb \qquad (14.10)$$

with a 0 Btu/lb enthalpy reference of 59°F and H_2O in the mixture in gaseous form. With a 0 Btu/lb enthalpy reference of 77°F (25°C), the transfer function becomes

$$h_{exh} = 0.3003 \cdot T_{exh}\ [in\ °F] - 59.897\ Btu/lb \qquad (14.11)$$

The difference in enthalpy calculation between the two reference point assumptions is about 4.3 Btu/lb (10 kJ/kg).

Equations 14.8–14.11 can be used for the temperature range 900–1,200°F (480–650°C) with reasonable accuracy. The error resulting from using these simple equations in lieu of bona fide property calculations should not be more than ±1–2%. The information implicit in Equations 14.10 and 14.11 is the specific heat of the gas turbine exhaust gas; that is, c_p = 0.3003 Btu/lb-R (1.2573 kJ/kg-K), which is approximately, but not exactly, constant in the range of its applicability.

For combined cycle calculations, we also need to estimate the enthalpy of the HRSG stack gas, which can be done by using

$$h_{stack} = 0.2443 \cdot T_{stack} \text{ [in } ^\circ F] - 13.571 \text{ Btu / lb} \tag{14.12}$$

$$h_{stack} = 0.2443 \cdot T_{stack} \text{ [in } ^\circ F] - 17.892 \text{ Btu / lb,} \tag{14.13}$$

with a 0 Btu/lb enthalpy reference of 59°F (in Equation 14.12; 77°F (25°C) in Equation 14.13) and H_2O in the mixture in gaseous form. Equations 14.12 and 14.13 are reasonably accurate for the temperature range 160–225°F (70–110°C). The information implicit in Equations 14.12 and 14.13 is the specific heat of the HRSG stack gas; that is, $c_p = 0.2443$ Btu/lb-R (1.0228 kJ/kg-K), which is approximately, but not exactly, constant in the range of its applicability. For the exergy of the HRSG stack gas, use

$$a_{stck} = 0.0519 \cdot T_{stck} \text{ [in } ^\circ F] - 6.1539 \text{ Btu / lb} \tag{14.14}$$

$$a_{stck} = 0.045 \cdot T_{stck} \text{ [in } ^\circ F] - 5.7748 \text{ Btu / lb,} \tag{14.15}$$

with a 0 Btu/lb enthalpy reference of 59°F (in Equation 14.14; 77°F (25°C) in Equation 14.15) and H_2O in the mixture in gaseous form.

Performance fuel gas heating is a feature of all modern gas turbines for power generation (e.g., in a combined cycle configuration with hot boiler feed water bled from the HRSG). For heated fuel delivered to the gas turbine combustor, the total heat input into the gas turbine control volume is the total of LHV (at 77°F) and the sensible heat evaluated by using

$$h_{fuel} = 0.5832 \cdot T_{fuel} \text{ [in } ^\circ F] - 44.9 \text{ Btu / lb,} \tag{14.16}$$

which is for 100% methane at zero-enthalpy reference temperature of 77°F (25°C). Similar to the gas enthalpies discussed earlier, Equation 14.16 assumes a constant fuel gas specific heat of 0.5832 Btu/lb-R (2.4417 kJ/kg-K). For quick estimates using 0.6 Btu/lb-R (2.5 kJ/kg-K) for fuel gas specific heat is sufficient.

14.2 ENTHALPY

Many experienced engineers use enthalpy in compressor or turbine/expander calculations automatically without giving a lot of thought to it. For most practical purposes, this is just fine. Nevertheless, a deeper understanding of enthalpy might come in handy in certain cases; for example, when discussing compressor or turbine stage aerothermodynamics.

Enthalpy is simply flow energy; that is, the combination of thermal and mechanical energies of a fluid. It can be found via combining the two *Tds* equations (Equations 2.1 and 2.2 in Chapter 2) and integrating as

$$h = u + Pv \tag{14.17}$$

on *specific* (per unit mass, i.e., Btu/lb or kJ/kg) basis or as

$$H = U + P{\cdot}V \tag{14.18}$$

on *total* basis (i.e., Btu or kJ). The internal energy, U, is the macroscopic manifesta-tion of the kinetic energy of molecules constituting the substance in question. It can be derived rigorously from the statistical mechanics.

The second term, $P{\cdot}V$, can be translated into the following form by substituting F/A for pressure, P, and $A{\cdot}x$ for volume, V (note that F is force and A is area, where x is an arbitrary distance perpendicular to area A)

$$P{\cdot}V = F{\cdot}x. \tag{14.19}$$

Force times distance is total work done by the said force over the said distance. When you consider a cylindrical mass of fluid flowing in a pipe with velocity C (see Figure 14.2), U is the internal energy of the fluid mass, $m = \rho V = Ax/v$, and $pV = Fx$ is the energy repre-senting the work done by the fluid to open up for itself a space of $V = Ax$ by exerting its pressure on its surroundings across the flow cross-sectional area of A. While U represents the thermal energy of the fluid mass m, pV represents its mechanical energy, which is *different* from the kinetic energy of the same as given by $\frac{1}{2}mC^2$.

When we write the generic first law equation for an adiabatic compressor or expander (turbine) with constant mass flow rate as

$$\dot{W}_{c/t} = \dot{m}\left(h_1 - h_2\right), \tag{14.20}$$

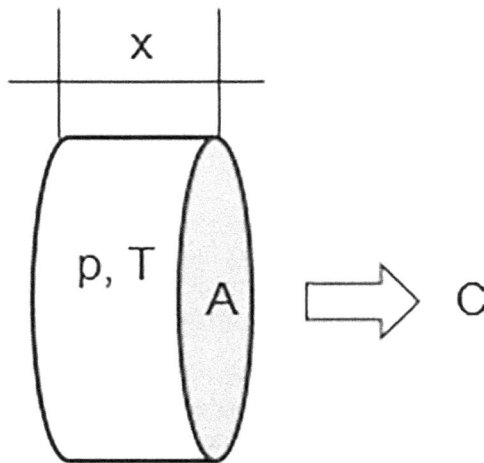

FIGURE 14.2   Cylindrical body of fluid (mass of m) in flow.

(with the implicit assumption that kinetic energy change is negligible) what we really have is

$$\dot{W}_{c/t} = \dot{m}\left(u_1 - u_2\right) + \dot{m}\left(P_1 v_1 - P_2 v_2\right), \text{ or} \tag{14.21}$$

$$w_{c/t} = \left(u_1 - u_2\right) + \left(P_1 v_1 - P_2 v_2\right). \tag{14.22}$$

In other words, work done in the machine, *on* the fluid or *by* the fluid, has two parts: *thermal* (the first term on the right-hand side) and *mechanical* (the second term on the right-hand side). Relative contribution of the mechanical term is illustrated via two examples:

1. Typical high-pressure section of a steam turbine.
2. Flow of hot combustion gas across the first stage stator of a gas turbine.

Calculations are shown in Table 14.10.

As it was stressed in the main body of the book quite frequently, it is the *change* in enthalpy, which has a real meaning because the absolute value is dependent on the – somewhat arbitrary – choice of a zero-enthalpy state. With that in mind, for the two examples in Table 14.10, change in $P \cdot v$ is about 20% of the enthalpy change with the remainder coming from the internal energy, u.

Also note the negligible contribution of kinetic energy at the steam turbine inlet and exit in Table 14.10. Thus, there is no distinction between *total* (also referred to as *stagnation* in the literature) and *static* values of pressure and temperature. The same can be said for the hot gas at the stator inlet with little loss in accuracy. However, acceleration of the hot gas while flowing through the stator nozzle vanes results at a high velocity just before entering the rotor. The change in kinetic energy is of the same order as the change in enthalpy (of which nearly 20% is due to $P \cdot v$). The change in *total* flow energy, $h + C^2/2$, is about 5% loss mainly due to friction, heat transfer

TABLE 14.10
Examples Illustrating $h = u + Pv$

		Steam			Gas		
		HP Throttle	Cold Reheat	Δ	Stator Inlet	Rotor Inlet	Δ
P_{tot}	psia	2,500	485		205	199	
T_{tot}	F	1,112	672		2,475	2,359	
T_{dyn}	F	negligible	negligible		16	427	
P_{stat}	psia	~ same	~ same		~ same	101.6	
h	Btu/lb	1,531	1,343	188	689	524	166
$p \cdot v$	Btu/lb	158.5	117.2	41	199.9	163.9	36
C	ft/s	200	200		500	2,591	
KE	Btu/lb	0.8	0.8		5.0	134.1	−129

to and mixing with cooling air. Since no work is done by the gas flowing across the stator, if the nozzle vanes were uncooled and the flow were frictionless, the total flow energy change would be zero. The change in enthalpy (reduction) would be exactly equal to the change in kinetic energy (increase). Without this translation from thermal to kinetic energy in a turbine and *vice versa* in a compressor, one would not have a turbomachine.

Yet unless we investigate the stage-by-stage performance of a compressor or a turbine, we tend to ignore this fundamental aspect of turbomachines. As clearly illustrated by the steam turbine example in Table 14.10, steam velocity is totally inconsequential in most calculations (unless, of course, they are for stage-by-stage turbine design). Since this ignoring convention is done so frequently and so matter-of-factly, many engineers doing heat balance simulations on a daily basis cannot even cite the typical steam velocity in, say, the main steam pipe of a combined cycle steam turbine. (In passing, 200 ft/s used in Table 14.10 is on the high side.)

One side effect of this practical simplification of the underlying physics is the terminology mess prevalent in the technical literature. Therefore, the reader should keep the following facts in mind (even if he or she keeps working with existing – sometimes downright incorrect – technical shorthand):

1. *Total* temperature is a useful concept facilitated by the simplifying assumption of constant specific heat.
2. Strictly speaking, there is only the *static* temperature, which, unfortunately, *cannot* be measured directly in a flow system (whereas the total temperature *can* be measured).
3. Using an equation of state in rigorous calculation results in total flow energy comprising the following:
   a.   Enthalpy
   b.   Kinetic energy
4. Some texts use the term *total enthalpy* in lieu of *total flow energy*, which is incorrect because of the following:
   a.   Enthalpy is a property of the fluid in question.
   b.   Enthalpy can be calculated from a known equation of state as a function of either of the following:
       i.   *Static* temperature only (ideal gas)
       ii. *Static* temperature *and static* pressure (real fluid)
   c.   Thus, by definition, enthalpy is a *static property*.
5. Whenever you are using a property package, such as ASME steam or JANAF gas tables, to calculate enthalpy, entropy, or another thermodynamic property of a fluid in flow, the following happens:
   a.   The temperature and the pressure of the fluid you obtain from respective transducers are *total* (stagnation) values.
   b.   Therefore, when plugging temperature and pressure values into the property subroutines, either of the following happens:

i. You either make an implicit assumption that the velocity of the fluid in question is low enough to assume that the static and total values are close enough.

ii. You already evaluated the contribution of the fluid kinetic energy and *extracted* the *static* values of temperature and/or pressure.

iii. All calculated properties are, by definition, *static* values.

6. Only changes in enthalpy between two *equilibrium* states are meaningful quantities.

7. Absolute values of enthalpy can be used to represent the *thermal energy* of a fluid; for example, main steam in a steam turbine power plant, gas turbine exhaust gas at the inlet to the HRSG, and so on.

8. In such cases, you *must* be aware of and/or clearly state the applicable *zero-enthalpy reference state*.

9. The most logical choice for the zero-enthalpy reference is the ambient conditions; that is, 59°F and 14.7 psia (15°C and 1 atm) per ISO definition, which is also the dead state for exergy calculations. Another useful zero-enthalpy reference for heat balance calculations is 77°F (25°C) because the fuel heating values are typically quoted at that reference temperature. The difference between the two reference temperatures is 4.3 Btu/lb (10 kJ/kg) in enthalpy.

14.3 CANONICAL ENSEMBLE

The canonical ensemble is essentially a supersystem comprising M macroscopic systems, as described in Figure 14.3. Of those M *identical* systems, only one is the system of interest with the remaining $M - 1$ mental copies forming a heat bath around it. The idea behind the ensemble method, invented by Gibbs, is to replace the *time average* of the system of interest by a *time-independent average* over a large number of systems representative of it. Furthermore, the following are more details:

• Each macroscopic system is characterized by N (number of particles, e.g., molecules) and V (volume).

• Possible energy states for such a system are E_1, E_2, E_3, . . . , E_j (energy *eigenvalues*).

• Each E_j is a function of N and V.

• It is a *quantum-mechanical* system; that is, $E_j = (j - 1)\,\varepsilon$, where ε is the energy difference between two consecutive states.

Note that the ensemble model of Gibbs is analogous to the Boltzmann based on molecules in an isolated container. You can just consider each system as a supermolecule containing trillions of molecules. The supermolecule in question is in a superstate or macrostate compared to the microstate of a molecule.

- Macroscopic system with N molecules and volume V
- There are M such systems in the "heat bath" (the "ensemble")
- For any given system, the remaining M – 1 systems in the ensemble serve as the heat bath
- Thus, each system and the ensemble is at temperature, T
- The ensemble has a total volume of M x V
- The ensemble has M x N molecules in it
- The ensemble has total energy, E_t

FIGURE 14.3   Canonical ensemble.

Each system in the canonical ensemble has the same N and V. Thus, all the systems have the same set of energy states, $E_1, E_2, E_3, \ldots, E_j$. Let us assume that there are the following:

- n_1 systems in energy state E_1
- n_2 systems in energy state E_2
- n_j systems in energy state E_j

The set of numbers $n_1, n_2, \ldots, n_j$ (also referred to as *occupation* numbers, as in "n_j systems *occupy* the energy state E_j") constitute a *distribution* subject to the two constraints:

$$\sum_j n_j = M, \text{ and}$$

(14.23)

$$\sum_j n_j E_j = E_t.$$

(14.24)

The number of energy states of the supersystem (i.e., the canonical ensemble) consistent with a given distribution is given by the combinatorial formula

$$\Omega_n = \frac{(n_1 + n_2 + \ldots)!}{n_1! n_2! \ldots} = \frac{M!}{\prod_j n_j!}.$$

(14.25)

For a given distribution, the probability of observing a quantum state, E_j, in a system picked from a canonical example is the same as the fraction of systems in the ensemble in the state E_j:

$$P_j = \frac{n_j}{M}.$$

(14.26)

But there are many possible distributions so that what is really needed is a *weighted* average of P_j over these distributions:

$$P_j = \frac{\bar{n}_j}{M}. \tag{14.27}$$

Each energy state is assumed to be equally likely. It then follows that each distribution has a weight proportional to Ω_n assigned to it in calculating the average P_j:

$$w_n = \frac{\Omega_n}{\sum_n \Omega_n}, \tag{14.28}$$

$$P_j = \frac{\bar{n}_j}{M} = \frac{\sum_n w_n n_{j,n}}{M}, \tag{14.29}$$

where $n_{j,n}$ denotes the value of n_j in the distribution n, and, by definition,

$$\sum_j P_j = 1.$$

The ensemble averages of energy and pressure can be calculated as

$$\bar{E} = \sum_j P_j E_j, \tag{14.30}$$

$$\bar{p} = \sum_j P_j p_j, \tag{14.31}$$

where p_j is the pressure in energy state E_j, which is defined as

$$p_j = -\left(\frac{\partial E_j}{\partial V}\right)_N, \tag{14.32}$$

so that the work that has to be done on the system to increase the system volume by dV is given by the first law of thermodynamics as

$$dE_j = -dW = -p_j dV. \tag{14.33}$$

In the limit of $M \to \infty$, the most probable distribution, n^*, with the largest Ω_n dominates the computation of the average in Equation 14.29 such that

$$P_j = \frac{\bar{n}_j}{M} = \frac{n_j^*}{M}, \tag{14.34}$$

in other words, $\bar{n}_j = n_j^*$ where n_j^* is the value of n_j in the most probable distribution n^* (i.e., the distribution with the largest Ω_n). The problem is reduced to finding the largest Ω_n, which is done by using the method of undetermined multipliers subject to the constraints given by Equations 14.23 and 14.24. The first step is to take the logarithm of both sides of Equation 14.25 and change the index from j to i. In the limit of $M \to \infty$ (which also means that $n_i \to \infty$), using Stirling's approximation, the result is

$$\ln \Omega_n = \left(\sum_i n_i \right) \ln \left(\sum_i n_i \right) - \sum_i n_i \ln(n_i).$$

(14.35)

The method of undetermined multipliers leads to the derivative

$$\frac{\partial}{\partial n_j} \left(\ln \Omega_n - \alpha \sum_i n_i - \beta \sum_i n_i E_i \right) = 0$$

(14.36)

for $j = 1, 2, \ldots$, which is then solved to find the most probable distribution as

$$n_j^* = M e^{-\alpha} e^{-\beta E_j}.$$

(14.37)

Combining Equations 14.23, 14.24, and 14.37 and substituting into Equation 14.34 results in

$$P_j = \frac{\bar{n}_j}{M} = \frac{e^{-\beta E_j(N,V)}}{\sum_i e^{-\beta E_i(N,V)}},$$

(14.38)

which tells us that the probability of observing a quantum state in a canonical ensemble decreases exponentially with the energy of the quantum state. Equation 14.30 in differential form (holding N constant because the system is closed) becomes

$$d\bar{E} = \sum_j E_j \, dP_j + \sum_j P_j \, dE_j,$$

(14.39)

and making use of Equation 14.38, we end up with

$$d\bar{E} = -\frac{1}{\beta} \sum_j \left(\ln P_j + \ln \sum_j e^{-\beta E_j} \right) dP_j + \sum_j P_j \left(\frac{\partial E_j}{\partial V} \right)_N dV.$$

(14.40)

After some algebra (and noting that $\sum_j P_j = 1$ so that $\sum_j dP_j = 0$), Equation 14.40 can be written as

$$-\frac{1}{\beta} d \left(\sum_j P_j \ln P_j \right) = d\overline{E} + \overline{p}\, dV.$$

(14.41)

Now, we can make the *association* between mechanics and thermodynamics by comparing Equation 14.41 with the Gibbs equation, $dE = TdS - pdV$, which tells us that

$$-\frac{1}{\beta} d \left(\sum_j P_j \ln P_j \right) \leftrightarrow TdS, \text{ or}$$

(14.42)

$$dS \leftrightarrow -\frac{1}{\beta T} d \left(\sum_j P_j \ln P_j \right).$$

(14.43)

Note that when the Gibbs equation was first introduced in Chapter 2 (i.e., Equation 2.1), the conventional symbol for the internal energy, U, was used; that is, $dU = Tds - pdV$ (extensive) or $dU = Tds - pdv$, in per unit mass (intensive) terms. In making the association expressed by Equations 14.42 and 14.43, the underlying assumption is that the quantum energy states are simply the translational kinetic energy states; that is, $\varepsilon = \frac{1}{2}mv^2$. This assumption is strictly valid for monatomic gases. However, as stated earlier in Chapter 4, it would be applicable to diatomic or polyatomic molecules as well (with rotational energy states) because the number density of the quantum states is dominated by the density of translational states; that is, they are more closely spaced (i.e., $\Delta \varepsilon \ll kT$).

For the arguments leading to the determination that the term $1/\beta T$ is a universal constant, k, the reader is referred to Schrödinger (pp. 12–13 in Ref. [2] in Chapter 4). Thus, Equation 14.43 leads us to the canonical ensemble expression for the entropy,

$$S = -k \sum_j P_j \ln P_j .$$

(14.44)

In a closed, isothermal system, the probability that the system has an energy departing significantly from $\overline{E}$ is pretty much nil. Furthermore, $\overline{E}$ is essentially equal to the most probable energy state denoted by E^*. The number of energy states E^* can be found from the combinatorial formula given by Equation 14.25, say, Ω^*. Each P_j in Equation 14.44 can thus be replaced by $1/\Omega^*$ so that Equation 14.44 becomes

$$S = -k\Omega^* \left(\frac{1}{\Omega^*} \ln \frac{1}{\Omega^*} \right) = k \ln \Omega^* .$$

(14.45)

A *microcanonical* ensemble is a degenerate canonical ensemble, in which all systems have the same energy, E. In other words, it is a *restricted* canonical ensemble; that is, a closed, isothermal system that is not allowed to have values of E other than $\overline{E}$.

The only quantum states accessible to the system are those with energy, E. In that case, E is the same for all quantum states, Ω of them, with $P_j = 1/\Omega$ each so that Equation 14.45 reduces to the famous Boltzmann equation:

$$S = k \ln \Omega. \tag{14.46}$$

Note that the denominator of Equation 14.38 defines the canonical ensemble partition function, Q:

$$Q = \sum_j e^{-\beta E_j} = \sum_j e^{-\frac{E_j}{kT}}, \tag{14.47}$$

so that Equation 14.38 can be rewritten as

$$P_j = \frac{e^{-\frac{E_j}{kT}}}{Q}, \tag{14.48}$$

where the numerator is known as the Boltzmann factor. (Equation 14.38 is Equation 4.45 in Chapter 4.) The relationship between Q and W can be established by using the microcanonical ensemble:

$$Q = \sum_j e^{-\frac{E_j}{kT}} = \Omega e^{-\frac{E}{kT}}, \text{ or} \tag{14.49}$$

$$\ln Q = \ln \Omega - \frac{E}{kT}, \tag{14.50}$$

where $E = E^*$. (Equation 14.50 can be shown to lead to Equation 4.44 in Chapter 4.)

14.4   EXERGY – A WAR OF LETTERS TO THE EDITOR

In Chapter 6, it was shown that the main source of irreversibility in heat recovery from the exhaust gas of a gas turbine was the constant-pressure–temperature evaporation characteristic of water. The reason for that was explained graphically using the conceptual heat release diagram for a *single-pressure, no-reheat* (1PNR) HRSG is shown in Figure 6.7. Constant-pressure–temperature evaporation of water in the boiler leads to a significant mismatch between the heat release lines of the hot fluid stream (i.e., gas turbine (GT) exhaust gas) and cold fluid stream (i.e., feed water and steam). Quantification of this irreversibility was also undertaken in Section 5.5 using the heat exchange schematic and heat release diagram in Figure 5.9, which is reproduced herein for the evaporator section of the HRSG (Figure 14.4).

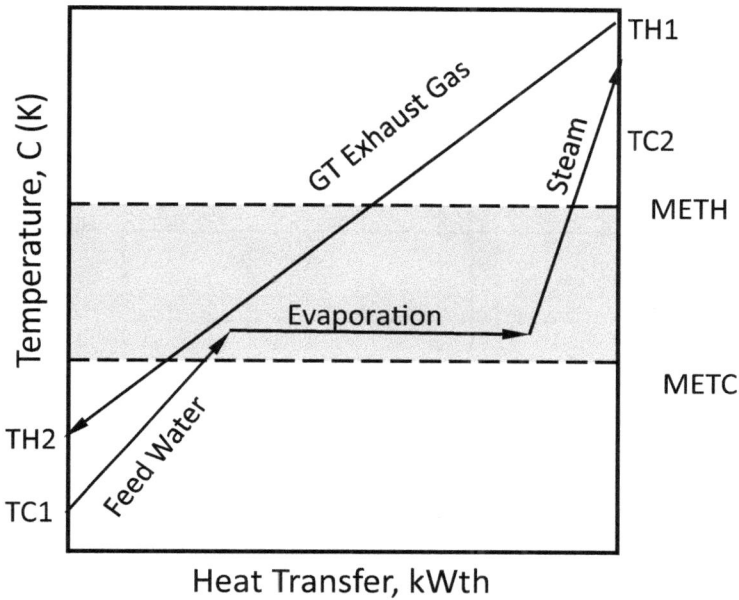

FIGURE 14.4   HRSG evaporator (generic) heat release diagram.

Graphically, the heat transfer irreversibility in the evaporator is quantified by the shaded rectangular area (i.e., refer to Equation 5.55). Even with an aggressive design employing tight approach and pinch temperature differences, the underlying physics prevents a significant reduction of this irreversibility.

A solution to this problem was proposed by Kalina based on switching the bottoming cycle working fluid from water-steam (H_2O) to a mixture of *ammonia* (NH_3) and water (approximately 70/30 by mass, respectively) [1–2]. The evaporator is of a once-through boiler configuration. The vapor is ammonia-rich so that as evaporation proceeds the boiling temperature of the remaining liquid rises, resulting in a better heat transfer match to the GT exhaust gas as shown in Figure 14.5.

There is no doubt that, when a comparison is made on a single evaporator/boiler basis, the ammonia-water mixture as working fluid is going to result in better exergetic efficiency (everything else being the same). In a paper by Stecco and Desideri, the advantage was quantified as 76% versus 71.3% exergetic efficiency (single pressure boiler with ~800 K hot gas inlet) for ammonia-water versus pure water, respectively [3]. In percentage of hot gas exergy terms, the boiler irreversibility was 24% and 28.5% for ammonia-water and pure water, respectively. In other words, everything else being the same, the exergetic efficiency of the (single pressure, no reheat) bottoming cycle can be improved by 4.5 percentage points.

Another possibility to achieve the qualitative effect of ammonia-water mixture working fluid depicted in Figure 14.4 is to change the high-pressure (HP) evaporator design in a three-pressure, reheat (3PRH) HRSG from a *subcritical* to *supercritical*

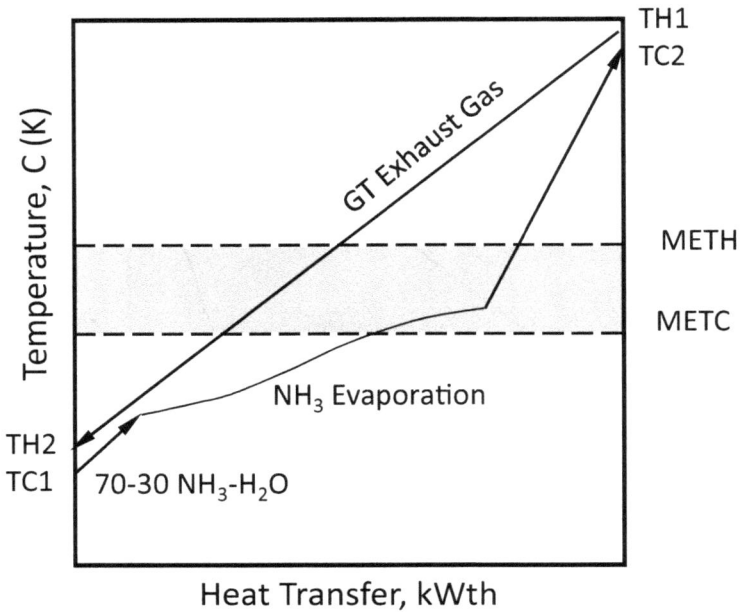

FIGURE 14.5   Kalina evaporator (generic) heat release diagram.

pressure (i.e., above 220.6 bar). This has been investigated in-depth by Gülen using the exergy analysis in an earlier paper [4]. For a conventional (subcritical) 3PRH bottoming cycle (BC), maximum improvement potential afforded by a supercritical HP section was shown to be 2.2 percentage points in BC efficiency, which translated into 0.55 percentage points in combined cycle (CC) efficiency. From a technological and economic feasibility perspective, the maximum practical gain from a supercritical HP section is unlikely to be as high as this entitlement value.

Furthermore, in the same paper, it was also shown that a much more significant impact is easily achievable by an advanced steam cycle (still subcritical) and larger HRSG, along with lower condenser pressure (as permitted by the site conditions); for example, 2.9 and 0.73 BC and CC percentage points of efficiency, respectively – that is, *larger than the entitlement for the supercritical HP section*. (Of that increase, nearly half or 0.35 points in CC efficiency was attributable to the condenser pressure.)

Almost two decades before the cited paper by Gülen (ibid.), in an exchange of letters between Drs. Maher Elmasri (president of Thermoflow Inc.) and Alexander Kalina (the inventor of the Kalina cycle) in the *Gas Turbine World*, the exact same observations were discussed. These letters, published consequently in three issues of the magazine in 1995 and 1996 are reproduced in Figures 14.6 through 14.10. the analysis presented by Dr. Elmasri is a prime example of the correct use of exergy analysis. The readers are highly encouraged to study these letters.

NO WAY AROUND BASIC LAWS OF THERMODYNAMICS

I was surprised to find the incredible and unreasonable claims made in the Kalina Cycle article published in the July-August 1995 issue of Gas Turbine World.

The claims made are unrealistic and defy fundamental thermodynamic logic. To begin with, a 3-pressure reheat steam cycle utilizing exhaust gases from an F-class gas turbine has an exergy efficiency of over 70%. That means that the power output from the steam turbine generator terminals is over 70% of the maximum possible power output that can be accomplished by a bottoming cycle consisting of perfectly reversible machinery, where all turbines, generators and motors are 100% efficient, all bearings frictionless, and temperature differences in all heat exchangers at zero.

To estimate the possible potential for improvement using novel bottoming cycles, we need to determine how much room for improvement exists and where it may be found. To this end, I have performed an exergy analysis of a 3-pressure, reheat, bottoming steam cycle with steam conditions of 1800 psig/1025°F, IP and reheat at 400 psig/1025°F, and LP admission at 55 psig/530°F with condenser at 0.5 psia.

Steam is generated in a heat recovery boiler using the exhaust of a GE 7231 FA gas turbine at ISO conditions. Industry-standard assumptions about equipment design and various losses are made. The bottoming cycle converts 33.65% of the exhaust energy to electricity, which corresponds to converting 73.13% of the exhaust exergy. The remaining exergy is lost by four principal mechanisms as shown in the pie-chart.

The biggest exergy loss is the 10% in the heat recovery boiler. This can be reduced in the steam cycle by reducing pinch temperature differences, assumed at 20°F, but the cost would be prohibitive. The Kalina cycle offers an innovative and brilliant way to reduce it by reducing the average temperature difference between the gas and working fluid in the boiler while maintaining reasonable pinch differences. However, that average temperature difference cannot go to zero or else the required surface area of the boiler would need to become infinite. Furthermore, the 10% boiler loss includes working fluid pressure drops that would remain finite. I can reasonably estimate, without proof or calculation, that this 10% loss can be cut to half at the most, providing a 5% reduction in exergy losses.

The next largest exergy loss, about 9%, is in the steam turbine generator. It includes internal fluid friction and exhaust losses, as well as mechanical and electrical losses. I doubt that this can be reduced by an ammonia-water mixture turbine, and suspect, in fact, that it may even increase due to the smaller blading.

Next in the loss breakdown is the condenser. At 0.5 psia the steam cycle is rejecting heat at

only 20°F above ambient. I doubt that this can be reduced with an ammonia/water mixture. Further, I suspect that it may increase, since the mixture does not condense at constant temperature like pure steam.

Last of the main loss mechanisms comes the stack. At only 196°F in the steam cycle, it only costs 3% of the exergy. I cannot see much room for improvement there either.

Overall, my numbers indicate that the Kalina method may improve the exergy efficiency of a bottoming cycle from about 73% for three-pressure reheat steam to about 78%. This would correspond to less than 7% increase in bottoming cycle output, which translates to about 2.5% increase in plant output and about 1.25 percentage points in net plant efficiency. The claim of 25% increased power output is about three times my estimate of the potential improvement.

Dr. Maher Elmasri President
Thermoflow Inc.
Wellesley, Mass.

Energy Analysis of Bottoming Cycle Three-Pressure Reheat

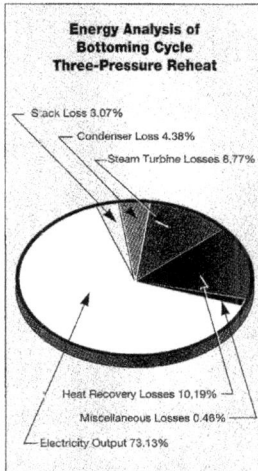

- Stack Loss 3.07%
- Condenser Loss 4.38%
- Steam Turbine Losses 8.77%
- Heat Recovery Losses 10.19%
- Miscellaneous Losses 0.48%
- Electricity Output 73.13%

VESTED INTEREST IN KALINA COMBINED CYCLES

GE signed in 1993 an exclusive worldwide licensing agreement with Exergy, Inc. to pursue the commercialization of the Kalina Cycle technology for combined cycle applications and is presently in the process of identifying a site for

a commercial-scale plant to demonstrate the technology.

The demonstration plant is expected to be in the 40 to 140 MW (combined cycle) size range, such as would be powered by an LM6000, 6B or 7EA gas turbine, and is planned to confirm the life-cycle cost, performance and operational characteristics of the Kalina Cycle technology, and to provide a further foundation for commercializing the technology. Our objective is to have this plant in operation by 1998.

Since GE has a vested interest in the combined cycle application of the Kalina Cycle, we believe it is necessary to point out the following inaccuracies in your July-August 1995 article on this subject:

■ The emphasis in the article is on the 7FA model. We indeed expect a substantial net combined cycle efficiency improvement — approximately 2.5 percentage "points" — however, the corresponding combined cycle power output increase is approximately 5 percent and not the 23 percent cited in the article. The actual values expected are 233 MW (STAG 107FA steam cycle) and 263 MW (STAG 107FA with Kalina Cycle).

■ The bottoming cycle output is expected to increase by approximately 13 percent with the 7FA. The 23 percent value cited would be expected with an LM6000 gas turbine — the greater percentage increase being due to its lower gas turbine exhaust temperature.

■ The 57.3% efficiency cited for a 7FA Kalina Cycle system is inaccurate. A more realistic estimate would be 55.8% (compared with 53% for a 6FA combined cycle with a conventional steam system). This decrease in cycle efficiency in comparison to the efficiency of a 7FA system is due to the smaller gas/vapor turbine size and gear losses.

■ The premium of 10% in the total combined cycle plant capital cost cited for the Kalina Cycle vs. a conventional steam cycle may be a reasonable objective for the technology after it has matured. Since the power output is 5% greater, this translates into a premium of 5% on a $/kW basis, not the reduction cited (which assumes a 25% increase in combined cycle power output capability).

Finally, GE disclaims any discussion in the article of specific terms of its license from Exergy. GE would be pleased to discuss any potential opportunities for application of the Kalina Cycle to combined cycle systems with interested parties.

GE continues to work closely with Exergy to pursue the commercialization of the Kalina Cycle for combined cycle applications and believes that it is a promising technology.

Robert Bjorge
Manager-Gas Turbine
Combined Cycle Programs
GE Power Systems
Schenectady, New York

FIGURE 14.6   *Gas Turbine World* (*GTW*) November–December 1995 (Part 1).

Kalina cycle criticism fundamentally flawed

Exergetical analysis of a thermodynamic cycle is a powerful tool for process improvement. But it must be carried out correctly. The letter from Dr. Maher El-masri critical of the Kalina Cycle (Nov-Dec 1995 GTW) is an example of how this tool may be misused.

Dr. Elmasri drew up a pie chart to show the exergetical losses in a triple pressure Rankine bottoming cycle and stated that, according to his assessments, the exergetical advantage of the Kalina bottoming cycle over the triple pressure Rankine bottoming cycle with a 7FA gas turbine is only half the value quoted by General Electric.

Dr. Elmasri did not make an exergetical analysis of the Kalina Cycle. In his own words, his assessment of the Kalina Cycle is just an estimate "without proof or calculation". Thereafter, in at least two instances he states, "I doubt that this (loss) can be reduced and suspect that it may even increase."

It is absolutely clear that doubts and suspicions are not the same as an objective scientific analysis. In fact, the exergetical analysis of the 3-pressure reheat bottoming Rankine Cycle presented by Dr. Elmasri is flawed and inaccurate.

First of all, it is absolutely wrong to consider output across the terminals as the power output of a cycle. Part of the power output measured at the terminals has to be returned back to the cycle in the form of pump work. The work performed by the circulating pumps increases exergy of the working fluid and, thus, presents a stream of exergy which is recirculated back to the cycle.

Another part of the terminal's output, which must be spent to operate a cooling tower, also increases overall available exergy. It too represents a stream of recirculating exergy. In his evalua-

tion, Dr. Elmasri completely disregards any exergy losses for auxiliaries.

Data presented by Dr. Elmasri need correcting as well. It is instructive to compare his data with the exergetical analysis of the 3-pressure Rankine Cycle performed by GE, and to compare his guesses with the KCS6D4 Kalina Bottoming Cycle analysis made by Exergy.

Dr. Elmasri states that exergy losses in the boiler of the Rankine Cycle, including stack losses, present 13.26% of the total exergy input. According to GE's calculations, this figure is 15.80%. For the Kalina Cycle, total loss of exergy in the HRSG and intercooler

Comparisons of Rankine vs. Kalina Cycle Exergy Losses

Summary of exergy losses (as percentage of exergy input) for 3-pressure Rankine Cycle quoted by Dr. Elmasri and calculated by GE compared with losses calculated by Exergy for a KCS6D4 Kalina Cycle

Design Area	Rankine Dr. Elmasri	Rankine Gen Electric	Kalina Exergy
HRSG, Stack, Intercooler	13.26 %	15.80 %	6.40 %
Turbines	8.77%	7.70%	4.30%
Condenser & Distillation	4.38%	5.70%	4.80%
Balance of Plant	0.46%	4.80%	4.80%
Total of Losses	26.87 %	34.00 %	20.30 %
Exergetical Efficiency	73.13 %	66.00 %	79.70 %

(including stack losses) is only 6.40% of the total exergy input.

With respect to exergy losses in the turbine, Dr. Elmasri assesses those losses at 8.77%; GE's computations show this loss as 7.70%. The main error of Dr. Elmasri is that he is confusing the exergetical efficiency of a vapor turbine with its isentropic efficiency. The exergetical efficiency of a vapor turbine is significantly higher than its isentropic efficiency.

Exergetical efficiency shows the real exergy destruction in the turbine, while isentropic efficiency shows both exergy

destruction and the failure to convert the exergy of the working fluid into mechanical work. Unconverted (but not destroyed) exergy increases the exergy of the turbine exhaust stream.

If this exergy is not utilized or recuperated, then it will be lost and the exergetical efficiency of a steam turbine will be equal to its isentropic efficiency. But that is not the case here. In the 3-pressure reheat Rankine Cycle, undestroyed exergy after the HP turbine increases the exergy of this turbine exhaust which results in less heat consumption in the following reheater.

Undestroyed exergy after the IP turbine increases exergy of the inlet at the LP turbine — although there is some loss of exergy when low pressure steam is admitted and mixed with the LP turbine exhaust. Finally, all the exergy of the LP turbine exhaust stream is fully destroyed in the condenser.

In the Kalina Cycle, in contrast, the exergy of the HP turbine exhaust is recuperated in the reheater (identically to the Rankine Cycle); the exergy of the IP turbine exhaust is recuperated in the intercooler and, what is most important, the exergy of the exhaust of the LP turbine is not lost in the condenser as it is in the Rankine Cycle.

Instead, this exergy is utilized and recuperated in the distillation and condensation subsystem. As a result, the total losses of exergy in turbines of the Kalina Cycle equal 4.30% of the total exergy input.

It is also wrong to expect, as Dr. Elmasri does, that exergy losses in water-ammonia turbines will be any different from those in the steam turbine. Water and ammonia have practically equal molecular weights and therefore turbine blading is identical for steam and water-ammonia turbines.

He also makes an error in assessing the exergy losses in the condenser of the Rankine Cycle. His assumption that the pressure of condensation is equal to 0.5 psia is unrealistically low and, as a result, his assessment of the loss of ex-

FIGURE 14.7   *GTW* March–April 1996 (Part 2, page 1).

Readers Feedback

ergy in the condenser as 4.38% is underestimated. According to GE's calculations, this loss is equal to 5.70%.

In assessing the exergy loss in the condenser of the Kalina Cycle, Dr. Elmasri again commits a grave error. Condensation in the Kalina Cycle occurs at variable temperatures, provides a better match between condensing steam and cooling water temperatures than does a Rankine Cycle. As a result, the final temperature of condensation in the Rankine Cycle is defined by the final, i.e. highest temperature of cooling water, while for the Kalina Cycle this temperature is defined by the initial, i.e. lowest temperature of cooling water.

Dr. Elmasri did not perform exergetical analysis on the Kalina Cycle, despite the fact that data on this cycle has been widely published. Exergy has performed such an analysis by itself and jointly with GE (see table).

As shown, the Kalina Cycle has an efficiency exceeding the efficiency of the triple pressure Rankine Cycle by 13.70% with a correspondingly higher power output. An accurate assessment of Kalina Cycle performance, based on General Electric's 7FA gas turbine, is presented in the letter by Dr. Robert Bjorge of GE Power Systems published in the same issue of the magazine (Nov-Dec GTW).

This is not the first time that Dr. Elmasri has suffered from premature articulation in applying erroneous exergetical analyses to the bottoming cycles of combined cycle systems. In an article in the Journal of Engineering for Gas Turbine and Power (Vol. 109 April 1987) he stated, "Further reductions of heat transfer irreversibilities are possible through increasing the number of evaporator pressures or using binary fluid mixtures, but the improvements over an optimized two-pressure system at typical exhaust temperatures are unlikely to exceed 3-4 percent of the exhaust gas availability, which corresponds to about one percentage point gain in the first-law efficiency."

Actual development of bottoming cycles has demonstrated the contrary. Dr. Elmasri was wrong then, comparing the 2-pressure Rankine Cycle to a 3- pressure Rankine Cycle, and he is wrong now comparing a triple-pressure Rankine Cycle to the Kalina Cycle.

The errors made by Dr. Elmasri

would not have occurred if, instead of estimates, suspicions, and doubts, he had taken the time to make the calculations before publishing his criticism — or had at least studied the published analyses.

The size of this letter precludes us from submitting a detailed presentation of exergetical analysis of the Kalina and Rankine cycles. However, in the near future, we do plan to publish such a detailed analysis, to be prepared jointly by Exergy and GE.

Dr. Alexander Kalina
President and CEO, Energy Inc.

(Rebuttal by Dr. Elmasri)

Claims do not stand up to close scrutiny

First, Dr. Kalina is upset about my use of estimates ".... without proof or calculation" as well as by my doubts that certain exergy losses can be reduced below certain reasonable limits. He calls that ".. not the same as objective scientific analysis".

This posture is ignorant or misleading, or both; since I would have expected Dr. Kalina to understand that the issues being addressed are ones of engineering practicality and tradeoffs, and not of scientific possibility.

The science of thermodynamics tells us that a cycle is within the realm of possibility as long as its exergy efficiency does not exceed 100%. However, since fluids are viscous and since heat transfer requires a finite temperature difference, 100% exergy efficiency is unattainable in practice.

So, is 95% attainable? Is 90% possible? Is 85% reasonable? The answer is that it depends on engineering practicality and on cost tradeoffs, not on the science of thermodynamics.

If we confine our discussion to vapor-turbine bottoming cycles that produce power from gas turbine exhaust heat, the major engineering limitations are turbo-generator efficiency and minimum stack temperature to ensure buoyancy of flue gases; the major capital-cost limitations are flow circuit complexity and size of heat exchangers and pipes to minimize temperature differences and pressure losses.

The engineering limitations, in and of

themselves, allow exergy efficiency up to the mid 80's if cost is not a consideration. The economic tradeoffs take it down from there, to 66% for a conservative 3-pressure reheat steam cycle or to 73% for a more efficient, more costly but reasonable 3-pressure reheat steam cycle.

There is nothing in the science of thermodynamics or in the body of engineering know-how that prevents us from making a steam bottoming cycle with an exergy efficiency well above 80% by using multiple pressures and reheats. What prevents us from doing that is not science, but the fact that the complexity and cost do not justify the gain in efficiency.

Dr. Kalina's invention holds the promise of raising efficiency, without the prohibitive cost of multiple steam pressures and reheats, by using the superior variable-temperature profile of an ammonia-water mixture at constant pressure.

Brilliant as that is, it is important to understand that any efficiency gain it promises can also be obtained by pure steam in a complex, expensive, circuit. Its potential advantage over pure steam is in the possibility of a superior cost/efficiency combination, not in fundamental thermodynamics.

In his letter, Dr. Kalina claims that variable temperature condensation provides a fundamental thermodynamic advantage. On the contrary, whereas the variable temperature boiling of a mixture is suited to the variable temperature heat source of turbine exhaust gases, the constant temperature condensation of pure steam is better matched to the constant temperature environment, to which heat rejection must occur.

If there is any intrinsic thermodynamic advantage, it lies with pure steam. For pure steam to benefit from this intrinsic advantage, however, requires large cooling water flow with a small temperature range.

Thus, any advantage of variable temperature condensation is purely economic, since it allows better use of a smaller flow rate of cooling water with a larger temperature range. This can indeed yield a satisfactory combination of thermodynamic merit and capital cost and is commonly employed in pure steam plants with multiple LPT paths.

If the Kalina Cycle is to prove superi-

FIGURE 14.8 *GTW* March–April 1996 (Part 2, page 2).

or to pure steam, that superiority lies in the promise of raising efficiency with less cost and complexity, and not in fundamental science. If Dr. Kalina does not already realize this, it is time he did. If he does indeed realize it, then he should stop trying to mislead those who have not studied thermodynamics in depth by mumbling all manner of obfuscating mumbo-jumbo about exergy.

After all, exergy analysis does not change the result of any cycle calculation, it is simply a means to perform a postmortem of the results. Dr. Kalina's comments about how you should "correctly" handle exergetical analysis of a pump or a turbine are simply intended to confuse.

There is no such thing as correct or incorrect exergetical analysis, since the various exergy efficiency definitions are not universally agreed upon and are immaterial to the bottom line result of how a cycle performs, which engineers who have never uttered the word "exergy" are fully capable of calculating.

The true benefit of exergy analysis is to provide insight into why the efficiency is what it is and where improvements should be sought. The power of this tool is that it allows us to make reasonable estimates of how much we can improve upon a cycle without the need to compute the improved cycle in detail, which is precisely how I used it to produce the estimates that annoy Dr. Kalina. Since he prefers to use exergy as a means of obfuscation rather than of elucidation, I must propose that Dr. Kalina rename his company "Anergy" rather than presuming to call it "Exergy".

Dr. Kalina tries to raise doubts about

the analysis outlined in my previous letter by comparing it with a GE analysis showing 66% exergy efficiency for a triple pressure pure steam bottoming cycle. He forgets to state the assumptions used by GE to provide the 66% figure, so I repeated my analysis using typical

Comparison of Conservative and Advanced Steam Bottoming Cycle Assumptions

The assumptions made below for the high efficiency cycle add to the capital cost of plant design but are all well within the boundaries of extensive, known steam plant experience and practice.

Parameter	Conservative 3P Reheat Cycle	High Efficiency 3P Reheat Cycle
High pressure	1465 psia/1000 F	1815 psia/1025 F
Reheat	365 psia/1000 F	415 psia/1025 F
Low pressure	65 psia/518 F	75 psia/535 F
Pinch points	30 F	20 F
Approach subcooling	10 F	5 F
Continuous blowdown	1%	None
Boiler heat loss	1%	0.25%
Superheater pressure loss	4%	3%
Reheat circuit pressure loss	10%	6%
Turbine efficiency	SCC*	SCC+2%*
Condenser	0.75 psia	0.5 psia
Energy efficiency	30.5%	33.6%
Exergy efficiency	66.4%	73.1%

* SCC is the published 1974 procedure for calculation of turbine efficiency of Spencer, Cotton & Cannon. The 2% improvement reflects current state-of-the-art.

conservative assumptions for which GE is well known and obtained the same results for a conservative cycle (66% exergy efficient) and a high-efficiency cycle that formed the basis of my original letter (73% exergy efficient).

Referring to the letter of GE's Dr. Robert Bjorge published alongside mine in Nov-Dec 1995 GTW, I note that he expects an approximately 13% improvement from the Kalina Cycle over a 3-pressure 7FA steam bottoming cycle. If I assume that he is comparing to the 66% exergy-efficient steam cycle that Dr. Kalina ascribes to GE, then he is projecting 75% exergy-efficiency for the Kalina Cycle.

The analysis in my letter projected

78% exergy-efficiency for the Kalina Cycle (1.07 times 73%). It seems that, Dr. Kalina should have been grateful for my analysis instead of unleashing his torrent of vitriol!

That ignorant and mindless torrent extends to a statement about "premature articulation applying erroneous exergetical analyses" in a 1987 paper wherein I state that increasing the number of pressures beyond two would result in only 3-4 percent of additional exergy recovery. It continues "Dr. Elmasri was wrong then, comparing the 2-pressure Rankine Cycle to a 3-pressure Rankine Cycle and he is wrong now...".

Well, perhaps Dr. Kalina should read GE paper GER-3574E published in August 1994, where Table 12 shows that the 2-pressure combined cycle has only 1.1% less output and efficiency than the 3-pressure. This translates to a difference of about 3% between the steam cycles, in conformity with my "premature" analysis done ten years ago !

In conclusion, I note that Dr. Kalina's letter no longer repeats his exaggerated claim of a 25% advantage for his cycle over pure steam for a GE 7FA, which had prompted my initial letter as well as that of Dr. Robert Bjorge. I hope that he will henceforth promote his excellent invention by using reasonable claims that can stand up to intelligent scrutiny.

I further hope that he can henceforth refrain from trying to bully and intimidate the engineering community by hurling crude insults upon anyone who dares to provide an objective analysis of his system.

Dr. Maher A. Elmasri
President, ThermoĮlow Inc.
■

FIGURE 14.9   *GTW* March–April 1996 (Part 2, page 3).

Readers Feedback

Rebuttal lacking in both reasons and arguments

In the last issue of your magazine (Mar-Apr 1996 GTW), you published a rebuttal by Dr. Maher Elmasri to my letter in that same issue responding to his attack (Nov-Dec 1995 GTW) on the validity of the Kalina Cycle. I have no intention of entering into a prolonged series of exchanges but must speak out.

In his letter, Dr. Elmasri states that it is wrong (of me) to expect him to carry out exergetical analyses and calculations to support his comments because the efficiencies of power plants depend on economics.

But in making any comparison between a three-pressure Rankine Cycle and the Kalina Cycle, it is self-evident that such a comparison is valid only if the economics of the two systems are identical, i.e. only if both have the same $ per kW installed cost.

We designed the Kalina Bottoming Cycle with an exergetical efficiency of 79.7% net, to have the same installed cost per kW as General Electric's 3P Rankine Cycle which has an exergetical efficiency of 66% net. This position is so self-evident that I did not even deem it necessary to mention in my letter. Therefore, Dr. Elmasri's position that it is impossible and unnecessary to rely on calculations rather than on estimates is indefensible.

In my comments, I also pointed out that in a proper exergetical analysis, the work of the circulating pump has to be subtracted from the output of the vapor turbine and then reverted as energy input into the system.

Dr. Elmasri's rejoinder is that "there is no such thing as correct or incorrect exergetical analysis." If one were to accept this statement, then exergetical analysis makes no sense at all. Therefore, it is impossible to agree with this position taken by Dr. Elmasri.

Another point I made is that he confuses the exergetical efficiency of the vapor turbine with its isentropic efficiency. In his rebuttal, Dr. Elmasri does not address this subject at all. He also does not address the fact that it is wrong to expect energy losses in the water/ammonia turbine to be different from those in the steam turbine. Thus, by default, he concedes that I am correct.

With respect to energy losses in the condenser, Dr. Elmasri has practically agreed with my position that it is advantageous for a power system to have variable temperatures of condensation.

In his rebuttal, Dr. Elmasri states that I projected the exergetical efficiency of the Kalina Cycle to be 75% whereas the tabular data clearly shows

Readers Feedback

Gas Turbine World welcomes comments from our readers. The editors reserve the right to accept or reject any letter, and to edit them for style, clarity and length. In the interest of clarifying technical disagreement, or to resolve differences in opinion, the editors may also solicit letters of rebuttal for publication in the same issue.

it as 79.7% which is exactly a 13.7% increase over the 66% efficiency for the 3P Rankine Cycle. It is important to note that my reference is net efficiency whereas Dr. Elmasri operates with gross efficiency.

Further, his point that the efficiency of a 3P Rankine Cycle can be increased at the expense of higher costs is self-evident and trivial. The same is true for the Kalina Cycle. But again, if the Kalina Cycle is to have the same cost per installed kW as the Rankine Cycle with an exergetical efficiency of 66%, then the net exergetical efficiency of the Kalina Cycle would be 79.7%.

To sum up, Dr. Elmasri did not rebut

even a single point in my letter. He agreed with me on the loss of efficiency in the condenser. And he did not disprove or even address my comments on 1) the disregarding of auxiliary losses, 2) erroneous estimates in vapor turbine performance, and 3) assessment of energy losses in the heat recovery boiler.

I certainly do appreciate Dr. Elmasri's characterization of the Kalina Cycle as "brilliant." I am, however, disappointed that Dr. Elmasri, in his rebuttal, lacking reasons and arguments to challenge my statements, had to resort to a personal attack on me.

Dr. Alexander Kalina
President and CEO, Exergy, Inc.

GTW owes a full and thorough rectification

Your article on "expanding global EPC business" in the March-April issue was of great interest.

Let me point out, however, that the complete gas turbine upgrading and conversion for the Mahmoudia and Damanhour power plant projects in Egypt (page 48) were undertaken by NEM on a turnkey basis, and not by Parsons!

Egypt's state utility, EEA, operates both plants. NEM was given responsibility for converting them from open cycle to combined cycle, and for upgrading the existing Frame 5 gas turbines to an output of 25 MW each.

The conversion to combined cycle resulted in increasing the total (module) output from 92 to 150 MW. The net result was to increase total output from 275 MW to 430 MW.

The Damanhour plant is powered by a single combined cycle module consisting of four gas turbines, four HRSGs, and one steam turbine. Two combined cycle modules of the same type are installed at Mahmoudia.

Ing. A. J. Zweers
MBA NEM bv

FIGURE 14.10 *GTW* May–June 1996 (Part 3).

REFERENCES

1. Kalina, A.I., Combined Cycle System with Novel Bottoming Cycle, *ASME Journal of Engineering Gas Turbines Power*, 106, pp. 737–742, 1984.
2. Marston, C.H., and Hyre, M., Gas Turbine Bottoming Cycles: Triple-Pressure Steam Versus Kalina, *ASME Journal of Engineering Gas Turbines Power*, 117, pp. 10–15, 1995.
3. Stecco, S.S., and Desideri, U., Considerations on the Design Principles for a Binary Mixture Heat Recovery Boiler, *ASME Journal of Engineering Gas Turbines Power*, 114, pp. 701–706, 1992.
4. Gülen, S.C., Performance Entitlement of Supercritical Steam Bottoming Cycle, *ASME Journal of Engineering Gas Turbines Power*, 135, p. #124501, 2013.

Index

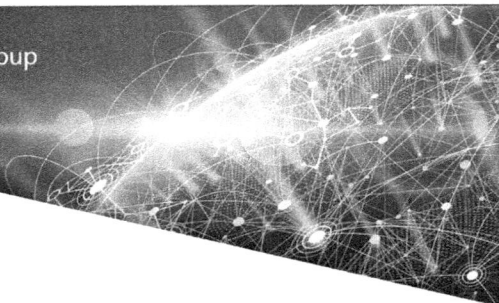

For Product Safety Concerns and Information please contact our EU
representative GPSR@taylorandfrancis.com
Taylor & Francis Verlag GmbH, Kaufingerstraße 24, 80331 München, Germany